T0140489

Lecture Notes
in Control and Information Sciences 335

Editors: M. Thoma · M. Morari

Krzysztof Kozłowski (Ed.)

Robot Motion and Control

Recent Developments

With 200 Figures

 Springer

Series Advisory Board

F. Allgöwer · P. Fleming · P. Kokotovic · A.B. Kurzhanski ·
H. Kwakernaak · A. Rantzer · J.N. Tsitsiklis

Editor

Professor Dr.-Ing. habil. Krzysztof Kozłowski
Poznan University of Technology
Institute of Control and Systems Engineering
ul. Piotrowo 3a
60-965 Poznań
Poland
Krzysztof.Kozlowski@put.poznan.pl

British Library Cataloguing in Publication Data
International Workshop on Robot Motion and Control (4th : Puszczykowo, Poland : 2004)
    Robot motion and control : recent developments. -
    (Lecture notes in control and information sciences ; 335)
    1.Robots - Control systems - Congresses 2.Robots - Motion - Congresses
    I.Title II.Kozłowski, Krzysztof
    629.8'92
ISBN-13 9781846284045
ISBN-10 184628404X

Library of Congress Control Number: 2006923560

Lecture Notes in Control and Information Sciences ISSN 0170-8643
ISBN-10: 1-84628-404-X   e-ISBN: 1-84628-405-8   Printed on acid-free paper
ISBN-13: 978-1-84628-404-5

© Springer-Verlag London Limited 2006

MATLAB® is the registered trademark of The MathWorks, Inc., 3 Apple Hill Drive, Natick, MA 01760-2098, USA. http://www.mathworks.com

Apart from any fair dealing for the purposes of research or private study, or criticism or review, as permitted under the Copyright, Designs and Patents Act 1988, this publication may only be reproduced, stored or transmitted, in any form or by any means, with the prior permission in writing of the publishers, or in the case of reprographic reproduction in accordance with the terms of licences issued by the Copyright Licensing Agency. Enquiries concerning reproduction outside those terms should be sent to the publishers.

The use of registered names, trademarks, etc. in this publication does not imply, even in the absence of a specific statement, that such names are exempt from the relevant laws and regulations and therefore free for general use.

The publisher makes no representation, express or implied, with regard to the accuracy of the information contained in this book and cannot accept any legal responsibility or liability for any errors or omissions that may be made.

Typesetting: Data conversion by editor.
Final processing by PTP-Berlin Protago-TeX-Production GmbH, Germany (www.ptp-berlin.com)
Cover-Design: design & production GmbH, Heidelberg

Printed in Germany

9 8 7 6 5 4 3 2 1

Springer Science+Business Media
springer.com

Contents

Part II Control and Mechanical Systems

6 Novel Adaptive Control of Partially Modeled Dynamic Systems

József K. Tar, Imre J. Rudas, Ágnes Szeghegyi, Krzysztof Kozłowski. . . . 99

7 Example Applications of Fuzzy Reasoning and Neural Networks in Robot Control

Waldemar Wróblewski 113

8 Adaptive Control of Kinematically Redundant Manipulator along a Prescribed Geometric Path

Mirosław Galicki 129

Part III Climbing and Walking Robots

Part IV Multi-agent Systems and Localization Methods

Part I

Control and Trajectory Planning
of Nonholonomic Systems

1

Trajectory Tracking for Nonholonomic Vehicles

Pascal Morin and Claude Samson

INRIA, 2004 Route des Lucioles, 06902 Sophia-Antipolis Cedex, France
Pascal.Morin@inria.fr, Claude.Samson@inria.fr

1.1 Introduction

For many years, the control of nonholonomic vehicles has been a very active research field. At least two reasons account for this fact. On one hand, wheeled-vehicles constitute a major and ever more ubiquitous transportation system. Previously restricted to research laboratories and factories, automated wheeled-vehicles are now envisioned in everyday life (e.g. through car-platooning applications or urban transportation services), not to mention the military domain. These novel applications, which require coordination between multiple agents, give rise to new control problems. On the other hand, the kinematic equations of nonholonomic systems are highly nonlinear, and thus of particular interest for the development of nonlinear control theory and practice. Furthermore, some of the control methods initially developed for nonholonomic systems have proven to be applicable to other physical systems (e.g. underactuated mechanical systems), as well as to more general classes of nonlinear systems.

The present paper addresses feedback motion control of nonholonomic vehicles, and more specifically *trajectory tracking*, by which we mean the problem of stabilizing the state, or an output function of the state, to a desired reference value, possibly time-varying. The trajectory tracking problem so defined incorporates most of the problems addressed in the control literature: output feedback regulation, asymptotic stabilization of a fixed-point and, more generally, of admissible non-stationary trajectories, practical stabilization of general trajectories. A notable exception is the *path following* problem, which will not be considered here because it is slightly different in nature. This problem is nonetheless important for applications, and we refer the reader to e.g. [7, 29] for related control design results.

The methods reviewed in the paper cover a large range of applications: position control of vehicles (e.g. car-platooning, "cruising mode" control), position and orientation control (e.g. parking, stabilization of pre-planned reference trajectories, tracking of moving targets). For controllable linear

K. Kozłowski (Ed.): Robot Motion and Control, LNCIS 335, pp. 3–23, 2006.
© Springer-Verlag London Limited 2006

systems, linear state feedbacks provide simple, efficient, and robust control solutions. By contrast, for nonholonomic systems, different types of feedback laws have been proposed, each one carrying its specific advantages and limitations. This diversity is partly justified by several theoretical results, recalled further in the paper, which account for the difficulty/impossibility of deriving feedback laws endowed with all the good properties of linear feedbacks for linear control systems. As a consequence, the choice of a control approach for a given application is a matter of compromise, depending on the system characteristics and the performance requirements. At this point, simulations can provide useful complementary guidelines for the choice of the control law. Due to space limitations, we are not able to include simulation results in this paper, but we refer the interested reader to [23] where a detailed simulation study for a car like-vehicle, based on the control laws here proposed, is given.

While the present paper reviews most of the classical trajectory tracking problems for nonholonomic vehicles, it is by no means a complete survey of existing control methods. Besides paper size considerations, those here discussed are primarily based on our own experience, and reflect our preferences. For survey-like expositions, we refer the reader to e.g. [7,12]. The paper's scope is also limited to "classical" nonholonomic vehicles, for which the "hard" nonlinearities arise exclusively from the kinematics. More general nonholonomic mechanical systems, in the sense of e.g. [4], are not considered here.

The paper is organized as follows. Several models are introduced in Section 1.2, with some of their properties being recalled. Section 1.3 is the core of the paper: the main trajectory tracking problems are reviewed from both the application and control design viewpoints. In particular, advantages and limitations inherent to specific types of feedback controllers are discussed. Finally, some concluding remarks are provided.

1.2 Modeling of Vehicles' Kinematics

In this section, some aspects of the modeling of nonholonomic vehicles are recalled and illustrated in the case of unicycle and car-like vehicles. The properties reviewed in this section apply (or extend) to most wheeled vehicles used in real-life applications.

1.2.1 Kinematics w.r.t. an Inertial Frame

Wheeled mechanical systems are characterized by non-completely integrable velocity constraints $\langle \alpha_j(q), \dot{q} \rangle = 0$, $q \in Q$, with Q the mechanical configuration space (manifold) and the α_j's denoting smooth mappings. These constraints are derived from the usual wheel's rolling-without-slipping assumption. Under very mild conditions (satisfied for most systems of practical interest), these constraints are equivalent to

$$\dot{q} = \sum_{i=1}^{m} u_i X_i(q), \tag{1.1}$$

where the u_i's denote "free" variables, the X_i's are smooth vector fields (v.f.) orthogonal to the α_j's, and $m < \dim(Q)$. A state space reduction to a submanifold M of the mechanical configuration space Q (see e.g. [6] for more details) yields a control model in the same form, with the following properties which will be assumed to hold from now on.

Properties:
P.1 $m < n := \dim(M)$,
P.2 the X_i's are linearly independent at any $q \in M$,
P.3 the X_i's satisfy the Lie Algebra Rank Condition on M, i.e. for any $q \in M$,

$$\mathrm{span}\{X_i(q), [X_i, X_j](q), [X_i, [X_j, X_k]](q), \ldots\} = \mathbb{R}^n.$$

Recall that Property **P.3** ensures that System (1.1) is locally controllable at any point, and globally controllable if M is connected (see e.g. [25, Prop. 3.15]).

A basic example is the unicycle-like robot of Fig. 1.1, whose kinematic model with respect to the inertial frame $\mathcal{F}_0 = (0, i_0, j_0)$ is given by:

$$\dot{q} = u_1 \begin{pmatrix} \cos\theta \\ \sin\theta \\ 0 \end{pmatrix} + u_2 \begin{pmatrix} 0 \\ 0 \\ 1 \end{pmatrix} \tag{1.2}$$

with $q = (x, y, \theta)'$, u_1 the signed longitudinal velocity of the vehicle's body, and u_2 its angular velocity.

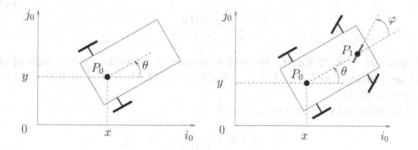

Fig. 1.1. The unicycle (l) and car (r)-like vehicles

A second example is the car-like vehicle of Fig. 1.1. A kinematic model for this system is

$$\dot{q} = u_1 \begin{pmatrix} \cos\theta \\ \sin\theta \\ \dfrac{\tan\varphi}{\ell} \\ 0 \end{pmatrix} + u_2 \begin{pmatrix} 0 \\ 0 \\ 0 \\ 1 \end{pmatrix} \tag{1.3}$$

with $q = (x, y, \theta, \varphi)'$, φ the steering wheel angle, and ℓ the distance between P_0 and P_1. An equivalent, but slightly simpler, model is given by

$$\dot{q} = u_1 \begin{pmatrix} \cos\theta \\ \sin\theta \\ \zeta \\ 0 \end{pmatrix} + u_2 \begin{pmatrix} 0 \\ 0 \\ 0 \\ 1 \end{pmatrix} \tag{1.4}$$

with $q = (x, y, \theta, \zeta)'$ and $\zeta := (\tan\varphi)/\ell$. Note that the control variable u_2 in (1.4) differs from the one in (1.3) by a factor $(1 + \tan^2\varphi)/\ell$.

1.2.2 Kinematics w.r.t. a Moving Frame

A generic property of vehicles is the invariance (or symmetry) with respect to the Lie group of rigid motions in the plane. More precisely, following [4], the state space M can usually be decomposed as a product $M = G \times S$, where $G = \mathbb{R}^2 \times \mathbb{S}^1 \approx SE(2)$ is associated with the vehicle's body configuration (position and orientation) in the plane, and S is associated with "internal" state variables of the vehicle. With this decomposition, one has

$$q = \begin{pmatrix} g \\ s \end{pmatrix}, \qquad g = (x, y, \theta)' \in G, s \in S \tag{1.5}$$

and System (1.1) can be written as

$$\begin{aligned} \dot{g} &= \sum_{i=1}^{m} u_i X_i^g(g, s), \\ \dot{s} &= \sum_{i=1}^{m} u_i X_i^s(s). \end{aligned} \tag{1.6}$$

For example, $S = \varnothing$ for the unicycle-like vehicle whereas, in the case of the car-like vehicle, $S = \mathbb{S}^1$ with $s = \varphi$ for System (1.3), and $S = \mathbb{R}$ with $s = \zeta$ for System (1.4).

G is endowed with the group product

$$g_1 g_2 := \begin{pmatrix} \begin{pmatrix} x_1 \\ y_1 \end{pmatrix} + R(\theta_1) \begin{pmatrix} x_2 \\ y_2 \end{pmatrix} \\ \theta_1 + \theta_2 \end{pmatrix} \tag{1.7}$$

with $g_i = (x_i, y_i, \theta_i)' \in G$ $(i = 1, 2)$, and $R(\theta)$ the rotation matrix of angle θ. For any fixed s the "vector fields" $X_i^g(g, s)$ are left-invariant w.r.t. this group product, i.e. for any fixed $g_0 \in G$ and any solution $t \longmapsto (g(t), s(t))$ to (1.6), $t \longmapsto (g_0 g(t), s(t))$ is also a solution to (1.6), associated with the same control input. By using this invariance property, it is simple to show that the kinematics w.r.t. a moving frame $\mathcal{F}_r = (0_r, i_r, j_r)$ (see Fig. 1.2) is given by

$$\dot{g}_e = \sum_{i=1}^{m} u_i X_i^g(g_e, s) - \bar{R}(g_e, g_r)\dot{g}_r,$$

$$\dot{s} = \sum_{i=1}^{m} u_i X_i^s(s) \tag{1.8}$$

with

$$g_e := g_r^{-1} g = \left(R(-\theta_r) \begin{pmatrix} x - x_r \\ y - y_r \end{pmatrix} \\ \theta - \theta_r \right) \text{ and } \bar{R}(g_e, g_r) := \begin{pmatrix} R(-\theta_r) \begin{pmatrix} -y_e \\ x_e \end{pmatrix} \\ 0 \quad\quad 1 \end{pmatrix}.$$

$$\tag{1.9}$$

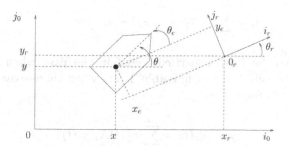

Fig. 1.2. Kinematics w.r.t. a moving frame

Note that System (1.8) can also be written as

$$\dot{q}_e = \sum_{i=1}^{m} u_i X_i(q_e) + P(q_e, t) \tag{1.10}$$

with $q_e = (g_e, s)$ and $P(q_e, t) = (-\bar{R}(g_e, g_r)\dot{g}_r, 0)$.

System (1.8) is a generalization of System (1.6), and corresponds to the kinematic model w.r.t. the moving frame $\mathcal{F}_r$. Let us mention a few important properties of this system. First of all, it is defined for any configuration of the vehicle. Then, as a consequence of the invariance property evoked above, the control v.f. of this system are the same as those of System (1.6).

1.2.3 Tracking Error Models

A trajectory tracking problem typically involves a reference trajectory $q_r :$ $t \longmapsto q_r(t) = (g_r, s_r)(t)$, with $t \in \mathbb{R}_+$. Then, one has to define a suitable representation for the tracking error. This step is all the more important that an adequate choice significantly facilitates the control design. As pointed out earlier, g_e given by (1.9) is a natural choice for the tracking error associated with g. Usually the set S is a product such as $\mathbb{S}^1 \times \mathbb{S}^1 \times \cdots$ or $\mathbb{R} \times \mathbb{R} \times \cdots$, so that it is also endowed with a "natural" (abelian) group structure. A simple choice for the tracking error associated with s is $s_e := s - s_r$. Note, however, that depending on the control v.f.'s structure, this is not always the best choice. Now, with the tracking error defined by

$$q_e := \begin{pmatrix} g_e \\ s_e \end{pmatrix} := \begin{pmatrix} g_r^{-1}g \\ s - s_r \end{pmatrix}$$

we obtain the following *tracking error model*, deduced from (1.8),

$$\begin{aligned}
\dot{g}_e &= \sum_{i=1}^{m} u_i X_i^g(g_e, s_e + s_r) - \bar{R}(g_e, g_r)\dot{g}_r, \\
\dot{s}_e &= \sum_{i=1}^{m} u_i X_i^s(s_e + s_r) - \dot{s}_r.
\end{aligned} \tag{1.11}$$

This model can be further particularized in the case when the reference trajectory is "feasible" (or "admissible"), i.e. when there exist smooth time functions u_i^r such that

$$\forall t, \quad \dot{q}_r(t) = \sum_{i=1}^{m} u_i^r(t) X_i(q_r(t)).$$

Then, (1.11) becomes

$$\begin{aligned}
\dot{g}_e &= \sum_{i=1}^{m} u_i^e X_i^g(g_e, s_e + s_r) + \sum_{i=1}^{m} u_i^r \left(X_i^g(g_e, s_e + s_r) - \tilde{A}(g_e) X_i^g(0, s_r) \right), \\
\dot{s}_e &= \sum_{i=1}^{m} u_i^e X_i^s(s_e + s_r) + \sum_{i=1}^{m} u_i^r \left(X_i^s(s_e + s_r) - X_i^s(s_r) \right)
\end{aligned} \tag{1.12}$$

with $u_i^e := u_i - u_i^r$ and

$$\tilde{A}(g_e) := \begin{pmatrix} I_2 & \begin{pmatrix} -y_e \\ x_e \end{pmatrix} \\ 0 & 1 \end{pmatrix}.$$

The choice $s_e = s - s_r$ is natural when the v.f. X_i are affine in s. This property is satisfied for several vehicles' models, under an appropriate choice of coordinates, and in particular by the car's model (1.4). In this latter case, Eq. (1.12) is given by

$$\begin{aligned}
\dot{g}_e &= u_1^e \begin{pmatrix} \cos\theta_e \\ \sin\theta_e \\ \zeta_e + \zeta_r \end{pmatrix} + u_1^r \begin{pmatrix} \cos\theta_e - 1 + y_e\zeta_r \\ \sin\theta_e - x_e\zeta_r \\ \zeta_e \end{pmatrix}, \\
\dot{\zeta}_e &= u_2^e.
\end{aligned} \tag{1.13}$$

1.2.4 Linearized Systems

A classical way to address the control of a nonlinear error model like (1.12) is
to consider its linearization at the equilibrium $(q_e, s_e, u^e) = 0$. It is given by

$$\dot{q}_e = \sum_{i=1}^{m} u_i^r(t) A_i(s_r(t)) q_e + B(s_r(t)) u^e \qquad (1.14)$$

with $B(s_r) = (X_1(0, s_r) \cdots X_m(0, s_r))$ and the A_i's some matrices easily
determined from (1.12). A first observation is that System (1.14) is
independent of g_r. This is again a consequence of the invariance property
recalled in Section 1.2.2. Another well known property is that this linearized
system is neither controllable nor stabilizable at fixed points (i.e. $u^r = 0$), since
in this case the system reduces to $\dot{q}_e = B u^e$ with B a $n \times m$ constant matrix,
and from Property **P.1**, $m < n$. However, along non-stationary reference
trajectories, the linearized system can be controllable. Consider for instance
the car's error model (1.13). In this case, the matrices A_1, A_2, and B in (1.14)
are given by

$$A_1(s_r) = \begin{pmatrix} 0 & \zeta_r & 0 & 0 \\ -\zeta_r & 0 & 1 & 0 \\ 0 & 0 & 0 & 1 \\ 0 & 0 & 0 & 0 \end{pmatrix}, \quad B(s_r) = \begin{pmatrix} 1 & 0 \\ 0 & 0 \\ \zeta_r & 0 \\ 0 & 1 \end{pmatrix}$$

and $A_2(s_r) = 0$. By applying classical results from linear control theory (see
e.g. [8, Sec. 5.3]), one can show for example that if, on a time-interval $[t_0, t_1]$,
ζ_r and u_1^r are smooth functions of time and u_1^r is not identically zero, then
the linearized system is controllable on $[t_0, t_1]$. In other words, this system is
controllable for "almost all" reference inputs u^r. As shown in [31], this is a
generic property for analytic Systems (1.1) satisfying Property **P.3**.

1.2.5 Transformations into Chained Systems

Let us close this section with a few remarks on local models. It is well known,
from [24,32], that the kinematic models (1.1) of several nonholonomic vehicles,
like unicycle and car-like vehicles, can be locally transformed into *chained
systems* defined by

$$\begin{cases} \dot{x}_1 = v_1, \\ \dot{x}_2 = v_2, \\ \dot{x}_k = v_1 x_{k-1} \quad (k = 3, \ldots, n). \end{cases} \qquad (1.15)$$

For example, the car's model (1.4) can be transformed into System (1.15) with
$n = 4$, with the coordinates x_i $(i = 1, \ldots, 4)$ and inputs v_1, v_2 defined by

$$\begin{aligned} (x_1, x_2, x_3, x_4) &= (x, \zeta/(\cos^3 \theta), \tan \theta, y), \\ (v_1, v_2) &= (u_1 \cos \theta, (u_2 + 3u_1 \zeta^2 \tan \theta)/(\cos^3 \theta)). \end{aligned} \qquad (1.16)$$

Beside its (apparent) simplicity, an important property of System (1.15) is that its v.f. are left-invariant, in the sense of Section 1.2.2, w.r.t to the group operation on $\mathbb{R}^n$ defined by

$$xy = \begin{pmatrix} x_1 + y_1 \\ x_2 + y_2 \\ x_k + y_k + \sum_{j=2}^{k-1} \frac{y_1^{k-j} x_j}{(k-j)!} \quad (k = 3, \ldots, n) \end{pmatrix}$$

with $x, y \in \mathbb{R}^n$. In these new coordinates, the group invariance suggests to define the tracking error vector as $x_e := x_r^{-1} x$ (compare with (1.9)). A difficulty, however, comes from that the change of coordinates is only locally defined, on a domain related to the inertial frame $\mathcal{F}_0$. While this is not a strong limitation for fixed-point stabilization, it becomes a major issue when the reference trajectory is not compelled to stay within the domain of definition of the change of coordinates. A way to handle this difficulty consists in considering a local transformation associated with the tracking error model (1.8). More precisely, whenever System (1.1) can be transformed into a chained system, then it follows from the formulation (1.10) of System (1.8) that the latter can also be transformed into a chained system with an added perturbation term $P(x_e, t)$. This new error model is then well defined whenever the vehicle's configuration g is in a (semi-global) neighborhood of the configuration g_r associated with the reference frame.

1.3 An Overview of Trajectory Tracking Problems

Consider a linear control system

$$\dot{x} = Ax + Bu \tag{1.17}$$

with the pair (A, B) controllable, and a matrix K such that $A + BK$ is Hurwitz-stable. Consider also an admissible reference trajectory $t \mapsto x_r(t)$, with $\dot{x}_r = Ax_r + Bu^r$. Then, the feedback law

$$u(x, x_r, u^r) := u^r + K(x - x^r) \tag{1.18}$$

applied to System (1.17) yields $\dot{x}_e = (A + BK)x_e$, with $x_e := x - x_r$. Since $A + BK$ is Hurwitz-stable, the feedback law (1.18) asymptotically stabilizes any admissible reference trajectory. Does there exist similar "universal" continuous feedback controls $u(x, x_r, u^r)$ for System (1.1)? It has been known for a long time (Brockett [5]) that the answer to this question is negative, because fixed-points (for which $u^r = 0$) cannot be asymptotically stabilized by continuous pure-state feedbacks. However, they can always be asymptotically stabilized by (periodic) *time-varying* continuous state feedbacks [9]. Then, we may ask whether there exist universal time-varying continuous feedback laws $u(x, x_r, u^r, t)$ that make *any* admissible reference trajectory asymptotically stable. A recent result shows that the answer to this question is again negative.

Theorem 1. *[13] Consider System* (1.1) *with* $m = 2$, *and assume Properties* **P.1-P.3**. *Then, given a continuous feedback law* $u(x, x_r, u^r, t)$, *with* $\partial u / \partial t$ *and* $\partial^2 u / (\partial u^r \partial t)$ *well defined everywhere and bounded on* $\{(x, x_r, u^r, t) : x = x_r, u^r = 0\}$, *there exist admissible reference trajectories which are not asymptotically stabilized by this control*[1].

Since universal asymptotic stabilization of admissible reference trajectories is not possible, what else can be done? Three major possibilities have been explored in the dedicated control literature.

1. Output feedback control. This consists in stabilizing only a part of the system's state. A typical application example is the car-platooning problem in the cruising mode for which controlling the vehicle's orientation directly is not compulsory (more details in the next section).
2. Asymptotic stabilization of specific trajectories. By restricting the set of admissible trajectories, via the imposition of adequate extra conditions, the problem of asymptotic stabilization considered earlier becomes amenable. Two types of trajectories have been more specifically addressed: trajectories reduced to fixed points ($u^r = 0$), corresponding to parking-like applications, and trajectories for which u^r does not converge to zero.
3. Practical stabilization. The idea is to relax the asymptotic stabilization objective. For many applications, practical stabilization yielding ultimately bounded and small tracking errors is sufficient. Not only the theoretical obstruction revealed by Theorem 1 no longer holds in this case, but also any trajectory, not necessarily admissible, can be stabilized in this way.

We now review in more details these different trajectory tracking problems and strategies.

1.3.1 Output Feedback Control

As pointed out in Section 1.2.4, the linearization of a control system (1.1) satisfying Property **P.1**, at any equilibrium $(q, u) = (q_0, 0)$, is neither controllable nor asymptotically stabilizable. This accounts for the difficulty of controlling such a system. In some applications, however, it is not necessary to control the full state q, but only a vector function $h := (h_1, \ldots, h_p)'$ of q and, possibly, t. In particular, if $p \leqslant m$, then the mapping

$$u \longmapsto \frac{\partial h}{\partial q} \dot{q} = \frac{\partial h}{\partial q} \sum_{i=1}^{m} u_i X_i(q) = \frac{\partial h}{\partial q} X(q) u \tag{1.19}$$

with $X(q) = (X_1(q) \cdots X_m(q))$, may be onto, as a result of the full rankedness of the matrix $\frac{\partial h}{\partial q} X(q)$. If this is the case, h can be easily controlled via its time-derivative. To illustrate this fact, consider the car-platooning problem,

---

[1] A few (weak) uniformity assumptions w.r.t. x_r are also required in the definition of asymptotic stability (see [13] for more details).

for which a vehicle is controlled so as to follow another vehicle which moves with a positive longitudinal velocity. This problem can be solved via the asymptotic stabilization of a point P attached to the controlled vehicle to a point P_r attached behind the leading reference vehicle (see Fig. 1.3). With the notation of this figure, the function h above can then be defined as $h(q,t) = (x_P(q) - x_r(t), y_P(q) - y_r(t))'$ or, if only relative measurements are available, by

$$h(q,t) = R(-\theta_r)\begin{pmatrix} x_P(q) - x_r(t) \\ y_P(q) - y_t(t) \end{pmatrix}. \tag{1.20}$$

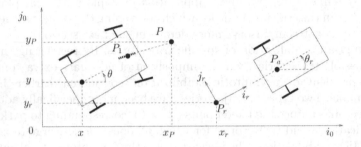

Fig. 1.3. Car-platooning

The control objective is to asymptotically stabilize h to zero. With the output function h defined by (1.20) and the control model (1.3), a direct calculation shows that the determinant of the square matrix $\frac{\partial h}{\partial q}X(q)$ in (1.19) is equal to $d/(\cos\varphi)$ with d the distance between P_1 and P. Since

$$\dot{h} = \frac{\partial h}{\partial q}X(q)u + \frac{\partial h}{\partial t},$$

the feedback law

$$u = \left(\frac{\partial h}{\partial q}X(q)\right)^{-1}\left(Kh - \frac{\partial h}{\partial t}\right)$$

with K any Hurwitz stable matrix, is well defined when $d > 0$ and $|\varphi| < \pi/2$, and yields $\dot{h} = Kh$, from which the asymptotic and exponential stability of $h = 0$ follows. Such a control strategy, which basically consists in virtually "hooking" the controlled vehicle to the leader by tying the points P and P_r together, is simple and works well as long as the longitudinal velocity of the reference vehicle remains positive. However, it is also intuitively clear that it does not work when the reference vehicle moves backward (with a negative longitudinal velocity), because the orientation angle difference $\theta - \theta_r$ and the steering wheel angle φ then tend to grow away from zero, so as to enter new stability regions near π. This is similar to the jack-knife effect for a tractor trailer. To avoid this effect, an active orientation control is needed.

1.3.2 Stabilization of Specific Trajectories

Stabilization of both the position and the orientation implies controlling the full state q. Asymptotic stabilization of trajectories has been successfully addressed in two major cases:

1. The reference trajectory is reduced to a fixed point. This corresponds to a parking-like problem.
2. The reference trajectory is admissible (feasible) and the longitudinal velocity along this trajectory does not tend to zero.

Fixed-point Stabilization

There is a rich control literature on the fixed point stabilization problem for nonholonomic vehicles. Brockett's theorem [5], (and its extensions [27, 34]...) has first revealed the non-existence of smooth pure-state feedback asymptotic stabilizers.

Theorem 2. *[5, 27] Consider a system* (1.1) *satisfying the properties* **P.1–P.2**, *and an equilibrium point q_0 of this system. Then, there exists no continuous feedback $u(q)$ that makes q_0 asymptotically stable.*

Following [28], in which time-varying continuous feedback was used to asymptotically stabilize a unicycle-like robot, the genericity of this type of control law has been established.

Theorem 3. *[9] Consider a system* (1.1) *satisfying the property* **P.3**, *and an equilibrium point q_0 of this system. Then, there exist smooth time-varying feedbacks $u(x, t)$, periodic w.r.t. t, that make q_0 asymptotically stable.*

Since then, many studies have been devoted to the design of feedback laws endowed with similar stabilizing properties. Despite a decade of research effort in this direction, it seems that the following dilemma cannot be avoided:

- Smooth (i.e. differentiable or at least Lipschitz-continuous) asymptotic stabilizers can be endowed with good robustness[2] properties and low noise sensitivity, but they yield slow (not exponential) convergence to the considered equilibrium point (see e.g. [16]).
- Stability and uniform exponential convergence can be obtained with feedback laws which are only continuous, but such controllers suffer from their lack of robustness, and their high sensitivity to noise.

---

[2] The type of robustness we are more specifically considering here is the property of preserving the closed-loop stability of the desired equilibrium against small structured modeling errors, control delays, sampling of the control law, fluctuations of the sampling period, etc...

To illustrate this dilemma, we provide and discuss below different types of asymptotic stabilizers which have been proposed in the literature. Explicit control laws are given for the car-like vehicle. For simplicity, the desired equilibrium is chosen as the origin of the frame $\mathcal{F}_0$, and the coordinates (1.16) associated with the chained system (1.15) are used. Let us start with smooth time-varying feedbacks, as proposed e.g. in [26, 28, 33]...

Proposition 1. *[29] Consider some constants $k_i > 0$ ($i = 1, \ldots, 4$) such that the polynomial $p(s) := s^3 + k_2 s^2 + k_3 s + k_4$ is Hurwitz. Then, for any Lipschitz-continuous mapping $g : \mathbb{R}^3 \longrightarrow \mathbb{R}$ such that $g(0) = 0$ and $g(y) > 0$ for $y \neq 0$, the Lipschitz-continuous feedback law*

$$\begin{cases} v_1(x, t) = -k_1 x_1 + g(x_2, x_3, x_4) \sin t, \\ v_2(x, t) = -|v_1| k_2 x_2 - v_1 k_3 x_3 - |v_1| k_4 x_4 \end{cases} \tag{1.21}$$

makes $x = 0$ asymptotically stable for the chained system (1.15) with $n = 4$.

While Lipschitz-continuous feedbacks are reasonably insensitive to measurement noise, and may be robust w.r.t. modeling errors, their main limitation is slow convergence rate. Indeed, most trajectories converge to zero only like $t^{-1/\alpha}$ ($\alpha \geqslant 1$). This has been the main motivation for investigating feedbacks which are only continuous. In this context, homogeneous (polynomial) v.f. have played an important role, especially those of degree zero which yield exponential convergence once asymptotic stability is ensured. It is beyond the scope of this paper to provide an introduction to homogeneous systems, and we refer the reader to [11] for more details. Let us only mention that the control theory for homogeneous systems is an important and useful extension of the classical theory for linear systems. The association of homogeneity properties with a time-varying feedback, for the asymptotic stabilization of nonholonomic vehicles, was first proposed in [15]. This approach is quite general since it applies to any system (1.1) with the property **P.3** [17]. An example of such a controller is proposed next (see e.g. [16] for complementary results).

Proposition 2. *[19] Consider some constants $k_i > 0$ ($i = 1, \ldots, 5$) such that the polynomial $p(s) := s^3 + k_2 s^2 + k_3 s + k_4$ is Hurwitz. For any $p, d \in \mathbb{N}^*$, denote by $\rho_{p,d}$ the function defined on $\mathbb{R}^3$ by*

$$\rho_{p,q}(\bar{x}_2) := \left(|x_2|^{p/r_2(d)} + |x_3|^{p/r_3(d)} + |x_4|^{p/r_4(d)} \right)^{1/p}$$

with $\bar{x}_2 := (x_2, x_3, x_4)$, $r(d) := (1, d, d+1, d+2)'$. Then, there exists $d_0 > 0$ such that, for any $d \geqslant d_0$ and $p > d + 2$, the continuous feedback law

$$\begin{cases} v_1(x, t) = -\left(k_1(x_1 \sin t + |x_1|) + k_5 \rho_{p,d}(\bar{x}_2)\right) \sin t, \\ v_2(x, t) = -\dfrac{|v_1| k_2 x_2}{\rho_{p,d}(\bar{x}_2)} - \dfrac{v_1 k_3 x_3}{\rho_{p,d}^2(\bar{x}_2)} - \dfrac{|v_1| k_4 x_4}{\rho_{p,d}^3(\bar{x}_2)} \end{cases} \tag{1.22}$$

makes $x = 0$ $\mathcal{K}$-exponentially stable for the chained system (1.15) with $n = 4$.

The property of $\mathcal{K}$-exponential stability evoked in the above proposition means that along any solution to the controlled system, $|x(t)| \leqslant k(|x_0|)e^{-\gamma t}$, with $\gamma > 0$ and k some continuous, positive, strictly increasing function from $\mathbb{R}_+$ to $\mathbb{R}_+$, with $k(0) = 0$. While this property ensures uniform exponential convergence of the trajectories to the origin, it is not equivalent to the classical definition of uniform exponential stability because k is not necessarily smooth. It is also interesting to compare the feedbacks (1.21) and (1.22). In particular, the second component v_2 of (1.22) is very similar to the component v_2 of (1.21), except that the constant gains k_i are replaced by the state dependent control gains $k_i/\rho_{p,d}^{i-1}(\bar{x}_2)$. The fact that these "gains" tend to infinity when x tends to zero (even though the overall control is well defined by continuity at $x = 0$), accounts for the lack of robustness and high noise sensitivity of this type of feedback. A more specific statement concerning the robustness issue is as follows.

Proposition 3. *[14] For any $\varepsilon > 0$, there exist v.f. $Y_1^\varepsilon, Y_2^\varepsilon$ on $\mathbb{R}^4$, with $\|Y_i^\varepsilon(x)\| \leqslant \varepsilon$ ($i = 1, 2; x \in \mathbb{R}^4$), such that the origin of the system*

$$\dot{x} = v_1(x,t)(X_1 + Y_1^\varepsilon) + v_2(x,t)(X_2 + Y_2^\varepsilon) \tag{1.23}$$

with X_1 and X_2 the v.f. of the chained system of dimension 4, and v_1, v_2 given by (1.22), is not stable.

In other words, whereas the origin of System (1.23) is asymptotically stable when $Y_1^\varepsilon = Y_2^\varepsilon = 0$, the slightest perturbation on the control v.f. may invalidate this result. Note that a small inaccuracy about the robot's geometry may very well account for such a perturbation in the modeling of the system. The practical consequence is that, instead of converging to the origin, the state will typically converge to a limit cycle contained in a neighborhood of the origin. It is possible to give an order of magnitude for the size of such a limit cycle, as a function of ε (assumed small). In terms of original coordinates (x, y, θ, φ) of the car's model (1.3), we obtain for the control law (1.22)

$$\text{size}(x) \approx \varepsilon^{1/(d+1)}, \text{size}(y) \approx \varepsilon^{(d+2)/(d+1)}, \text{size}(\theta) \approx \varepsilon, \text{size}(\varphi) \approx \varepsilon^{d/(d+1)}$$
$$\tag{1.24}$$

with $\text{size}(z)$ denoting the order of magnitude of the limit cycle in the z-direction. Relation (1.24) shows how differently the state components are affected, and also how the modeling errors effects are amplified for the components x and φ. Note, however, that (1.24) corresponds to the worst case so that there also exist structured uncertainties which preserve the asymptotic stability of the origin. An analysis of the sensitivity to state measurement noise would yield similar results, with ε interpreted as the maximum amplitude of the noise.

In [3], a control approach based on the use of hybrid feedbacks has been proposed to ensure exponential convergence to the origin, with the stability of the origin being preserved against small perturbations of the control v.f., like those considered in (1.23). The hybrid continuous/discrete feedbacks there considered are related to time-varying continuous feedback $v(x,t)$, except that the dependence on the state x is updated only periodically. The result given in [3], devoted to the class of chained systems, has been extended in [18] to any analytic system (1.1) satisfying Property **P.3**. For example, the following result is shown in the latter reference.

Proposition 4. *[18] Let $T, k_1, \ldots, k_4$ be some constants such that $T > 0$ and $|k_i| < 1 \ \forall i$. Then the hybrid-feedback $v(x(kT), t)$, $k \in \mathbb{N} \cap (t/T - 1, t/T]$ with v defined, in the coordinates x_i of the chained system of dimension 4, by*

$$\begin{cases} v_1(x,t) = ((k_1 - 1)x_1 + 2\pi\rho(x)\sin(2\pi t/T))/T, \\ v_2(x,t) = \left((k_2 - 1)x_2 + 2(k_3 - 1)\frac{x_3}{\rho(x)}\cos(2\pi t/T) \right. \\ \qquad\qquad \left. + 8(k_4 - 1)\frac{x_4}{\rho^2(x)}\cos(4\pi t/T)\right)/T \end{cases} \qquad (1.25)$$

and $\rho(x) = a_3|x_3|^{1/3} + a_4|x_4|^{1/4}$ ($a_3, a_4 > 0$) is a $\mathcal{K}(T)$-exponential stabilizer for the car, robust w.r.t. unmodeled dynamics.

The property of $\mathcal{K}(T)$-exponential stability means that, for some constants $K, \eta > 0$ and $\gamma < 1$, each solution $x(t, 0, x_0)$ of the controlled system with initial condition x_0 at $t = 0$ satisfies, for any $k \in \mathbb{N}$ and $s \in [0, T)$, $|x((k+1)T, 0, x_0)| \leqslant \gamma|x(kT, 0, x_0)|$ and $|x(kT + s, 0, x_0)| \leqslant K|x(kT, 0, x_0)|^\eta$. While it implies the exponential convergence of the solutions to the origin, it is neither equivalent to the classical uniform exponential stability, nor to the $\mathcal{K}$-exponential stability.

Unfortunately, the robustness to unmodeled dynamics evoked in Proposition 4 relies on the perfect timing of the control implementation. The slightest control delay, or fluctuation of the sampling period, destroys this property with the same effects as those discussed previously for continuous homogeneous feedbacks like (1.22).

To summarize, asymptotic fixed-point stabilization is theoretically possible but, in practice, no control solution designed to this goal is entirely satisfactory due to the difficulty, inherent to this type of system, of ensuring robust stability and fast convergence simultaneously.

Non-stationary Admissible Trajectories

The difficulty of stabilizing fixed-points comes from the fact that the linearized system at such equilibria is not controllable (and not stabilizable either). However, as shown in Section 1.2.4, the system obtained by linearizing the tracking error equations around other reference trajectories may be controllable. For a car-like vehicle, this is the case for example when the longitudinal velocity u_1^r associated with the reference trajectory is different from zero. Under some extra conditions on u_1^r (e.g. if $|u_1^r|$ remains larger than a positive constant), such trajectories can then be *locally* asymptotically stabilized by using feedbacks calculated for the linearized error system. The following proposition provides a nonlinear version of such a feedback for the car-like vehicle, with a proven large domain of stability. This result can be found in [21] (modulo a slight extension of the stability domain), following a technique already used in [30].

Proposition 5. *[21] Consider the tracking error model* (1.13) *associated with the car model* (1.4), *and assume that along the reference trajectory* ζ_r *is bounded. Then, the feedback law*

$$
\begin{cases}
u_1^e = -k_1|u_1^r|\left(x_e \cos\theta_e + y_e \sin\theta_e + \frac{\zeta}{k_2}\frac{\sin}{\cos^3}(\frac{\theta_e}{2})\right), \\[2mm]
u_2^e = -k_3 u_1^r \frac{\sin}{\cos^3}(\frac{\theta_e}{2}) - k_4|u_1^r|z + k_2 F_x \cos^2\frac{\theta_e}{2}\sin\theta_e - 2k_2 F_y \cos^4\frac{\theta_e}{2} \\[2mm]
\qquad - k_2 F_\theta \left(x_e(\frac{3}{4}\sin^2\theta_e - \cos^4\frac{\theta_e}{2}) - 2y_e \sin\theta_e \cos^2\frac{\theta_e}{2}\right)
\end{cases}
$$

$$(1.26)$$

with $k_1,\ldots,k_4 > 0$, $z := \zeta_e + 2k_2(-x_e \sin\frac{\theta_e}{2} + y_e \cos\frac{\theta_e}{2})\cos^3\frac{\theta_e}{2}$, *and*

$$
F_x = u_1^e \cos\theta_e + u_1^r(\cos\theta_e - 1 + y_e\zeta_r), \quad F_y = u_1^e \sin\theta_e + u_1^r(\sin\theta_e - x_e\zeta_r),
$$
$$
F_\theta = u_1^e \zeta + u_1^r \zeta_e
$$

makes the origin of System (1.13) *stable. Furthermore, if* u^r *is differentiable with* u^r *and* $\dot{u}^r$ *bounded, and* u_1^r *does not tend to zero as* t *tend to infinity, then the origin is also globally asymptotically stable on the set* $\mathbb{R}^2 \times (-\pi,\pi) \times \mathbb{R}$.

Note that the sign of u_1^r is not required to be constant, so that reference trajectories involving both forward and backward motions can be stabilized (compare with Section 1.3.1).

A simpler controller can be obtained either by working on a linearized error system, or by linearizing the expression (1.26) w.r.t. the state variables. This yields the feedback control

$$
\begin{cases}
u_1^e = -k_1|u_1^r|(x_e + \zeta_r\theta_e/(2k_2)), \\
u_2^e = 2k_2 u_1^r \zeta_r x_e - 2k_2 k_4|u_1^r|y_e - u_1^r(2k_2 + k_3/2)\theta_e - k_4|u_1^r|\zeta_e
\end{cases}
\tag{1.27}
$$

and one can show that it is a *local* asymptotic stabilizer for System (1.13) if, for example, $|u_1^r(t)| > \delta > 0\ \forall t$.

Finally, let us mention that, for both controllers (1.26) and (1.27), if u_1^r is a constant different from zero, then the convergence of the tracking error to zero is exponential with a rate proportional to $|u_1^r|$.

1.3.3 Practical Stabilization

Asymptotic stability is certainly a desirable property for a controlled system, but several results recalled in the previous sections point out that it cannot always be obtained, or robustly ensured, in the case of nonholonomic systems.

i) Theorem 1 basically tells us that no continuous feedback can asymptotically stabilize all admissible trajectories. This leads to the difficult question of choosing a controller when the reference trajectory is not known *a priori*.

ii) The difficult compromise between stability robustness and fast convergence, arising when trying to asymptotically stabilize a desired fixed configuration, was pointed out in Section 1.3.2.

iii) Asymptotic stabilization of non-admissible trajectories is not possible, by definition, although it may be useful, for some applications, to achieve some type of tracking of such trajectories. Consider for example a two-car platooning situation with the leading car engaged in a sequence of maneuvers. A way of addressing this problem consists in trying to stabilize a virtual frame attached to the leading car, at a certain distance behind it. This corresponds to the situation of Fig. 1.3 with (P_r, i_r, j_r) representing the virtual frame. The reference trajectory is then $g_r = (x_r, y_r, \theta_r)$ and $s_r = 0$, with (x_r, y_r) the coordinates of the point P_r. It is simple to verify that, except for pure longitudinal displacements of the leading car, the resulting trajectory of the virtual frame is not admissible.

We present in this section a control approach that we have been developing for a few years, and which allows to address the trajectory tracking problem in a novel way. This method is potentially applicable to all nonholonomic vehicles, and it has already been tested experimentally on a unicycle-like robot [1]. We illustrate it below for the problem of tracking another vehicle with a car, and refer to [22] for the general setting. As indicated in Point *iii)* above, this problem can be addressed by defining a reference trajectory (g_r, s_r), with $s_r = 0$ and g_r the configuration of a frame located behind the reference vehicle (Fig. 1.3). No assumption is made on g_r so that the associated reference trajectory may, or may not, be admissible. It may also be reduced to a fixed-point.

The control approach is based on the concept of transverse function [20].

Definition 1. *Let $p \in \mathbb{N}$ and $\mathbb{T} := \mathbb{R}/2\pi\mathbb{Z}$. A smooth function $f : \mathbb{T}^p \longrightarrow M$ is a* transverse function *for System* (1.1) *if,*

$$\forall \alpha \in \mathbb{T}^p, \quad \mathrm{rank} H(\alpha) = n \quad (= \dim(M)) \tag{1.28}$$

with

$$H(\alpha) := \left(X_1(f(\alpha)) \cdots X_m(f(\alpha)) \, \frac{\partial f}{\partial \alpha_1}(\alpha) \cdots \frac{\partial f}{\partial \alpha_p}(\alpha) \right). \tag{1.29}$$

Transverse functions allow to use $\dot{\alpha}_1, \ldots, \dot{\alpha}_p$ as additional (virtual) control inputs. This is related to the idea of controlled oscillator in [10]. For the car model (1.4), the usefulness of these complementary inputs is explicited in the following lemma.

Lemma 1. *Consider the tracking error model (1.10) associated with the car model (1.4) and the reference trajectory $(g_r, s_r = 0)$. Let $f : \mathbb{T}^p \longrightarrow \mathbb{R}^2 \times \mathbb{S}^1 \times \mathbb{R}$ denote a smooth function. Define the "neighbor" state z by*

$$
z := \begin{pmatrix} \begin{pmatrix} x_e \\ y_e \end{pmatrix} - R(\theta_e - f_3) \begin{pmatrix} f_1 \\ f_2 \end{pmatrix} \\ \theta_e - f_3 \\ \zeta - f_4 \end{pmatrix} \tag{1.30}
$$

and the augmented control vector $\bar{u} := (u_1, u_2, -\dot{\alpha}_1, \ldots \ldots, -\dot{\alpha}_p)'$. Then,

$$
\dot{z} = A(z, f) \left(H(\alpha)\bar{u} + B(z)P(q_e, t) + u_1 C(z) \right) \tag{1.31}
$$

with $H(\alpha)$ the matrix specified in relation (1.29),

$$
A(z, f) := \begin{pmatrix} R(z_3) & \left(R(z_3) \begin{pmatrix} f_2 \\ -f_1 \end{pmatrix} \ 0 \right) \\ 0 & I_2 \end{pmatrix}, \quad B(z) := \begin{pmatrix} R(-z_3) & 0 \\ 0 & I_2 \end{pmatrix}
$$

and $C(z) := (0, 0, z_4, 0)'$.

If f in (1.30) is a transverse function, then $A(z, f)H(\alpha)$ is a full-rank matrix and it is simple to use (1.31) in order to derive a control which asymptotically stabilizes $z = 0$. Such a control law is pointed out by the following proposition.

Proposition 6. *With the notations of Lemma 1, assume that f is a transverse function for the car model (1.4), and consider the dynamic feedback law*

$$
\bar{u} := H^{\dagger}(\alpha) \left(-B(z)P(q_e, t) - (A(z, f))^{-1}Z(z) \right) \tag{1.32}
$$

with $H^{\dagger}(\alpha)$ a right-inverse of $H(\alpha)$ and

$$
Z(z) := (k_1 z_1, k_2 z_2, 2k_3 \tan(z_3/2), k_4 x_4)' \quad (k_i > 0).
$$

Then, for any reference trajectory g_r such that $\dot{g}_r$ is bounded,

1. *$z = 0$ is exponentially stable for the controlled system (1.31),*
2. *any trajectory q_e of the controlled tracking error model (1.10) converges to $f(\mathbb{T}^p)$ and, with an adequate choice of $\alpha(0)$, this set is exponentially stable.*

This result shows that it is possible to stabilize the tracking error q_e to the set $f(\mathbb{T}^p)$ for *any* reference trajectory g_r. Since $\mathbb{T}^p$ is compact and f is smooth, $f(\mathbb{T}^p)$ is bounded. In particular, if $f(\mathbb{T}^p)$ is contained in a small neighborhood of the origin, then the tracking error q_e is ultimately small, *whatever* the trajectory g_r. This leaves us with the problem of determining transverse functions. In [22], a general formula is proposed for driftless systems. A family of transverse functions, computed in part from this formula, is pointed out below. These functions are defined on $\mathbb{T}^2$, i.e. they depend on two variables α_1, α_2. It is clear that for the car model, two is the smallest number of variables for which (1.28) can be satisfied. However, using more variables can also be of interest in practice (see [2] for complementary results in this direction), and some research effort could be devoted to explore this issue further.

Lemma 2. *For any* $\varepsilon > 0$, *and* $\eta_1, \eta_2, \eta_3 > 0$ *such that* $6\eta_2\eta_3 > 8\eta_3 + \eta_1\eta_2$, *the function* f *defined by*

$$f(\alpha) = \left(\bar{f}_1(\alpha), \bar{f}_4(\alpha), \arctan(\bar{f}_3(\alpha)), \bar{f}_2(\alpha)\cos^3 f_3(\alpha)\right)' \qquad (1.33)$$

with $\bar{f} : \mathbb{T}^2 \longrightarrow \mathbb{R}^4$ *defined by*

$$\bar{f}_1(\alpha) = \varepsilon(\sin\alpha_1 + \eta_2\sin\alpha_2),$$
$$\bar{f}_2(\alpha) = \varepsilon\eta_1\cos\alpha_1,$$
$$\bar{f}_3(\alpha) = \varepsilon^2\left(\frac{\eta_1\sin 2\alpha_1}{4} - \eta_3\cos\alpha_2\right),$$
$$\bar{f}_4(\alpha) = \varepsilon^3\left(\eta_1\frac{\sin^2\alpha_1\cos\alpha_1}{6} - \frac{\eta_2\eta_3\sin 2\alpha_2}{4} - \eta_3\sin\alpha_1\cos\alpha_2\right)$$

is a transverse function for the car model (1.4).

1.4 Conclusion

Trajectory stabilization for nonholonomic systems is a multi-faceted problem. In the first place, the classical objective of *asymptotic stabilization* (combining stability and convergence to the desired trajectory) can be considered only when the reference trajectory is known to be *admissible*. The difficulties, in this favorable case, are nonetheless numerous and epitomized by the non-existence of universal stabilizers. For some trajectories, endowed with the right properties (related to motion persistency, and fortunately often met in practice) the problem can be solved by applying classical control techniques. This typically yields linear controllers derived from a linear approximation of the trajectory error system, or slightly more involved nonlinear versions in order to expand the domain of operation for which asymptotic stabilization can be proven analytically. For other trajectories, such as fixed configurations, less classical control schemes (time-varying ones, for instance) have to be used, with mitigated practical success though, due to the impossibility of complying with the performance/robustness compromise as well as in the linear case. Basically, one has to choose between fast convergence, accompanied with high sensitivity to modeling errors and measurement noise, and slow convergence, with possibly more robustness. In the end, when the reference trajectory and its properties are not known in advance (except for its admissibility), so that it may not belong to the two categories evoked above, the practitioner has no other choice than trying to guess which control strategy will apply best to its application, or work out some empirical switching strategy, with no absolute guarantee of success in all situations. This is not very satisfactory, all the more so because there are applications for which the control objective is more naturally expressed in terms of tracking a *non-admissible* trajectory (we gave the example of tracking a maneuvering vehicle). In the latter case, none of the control techniques extensively studied during last fifteen years towards the goal of asymptotic stabilization is suitable. These considerations led us to propose another point of view according to which the relaxation of this stringent, and sometimes unworkable, goal into a more pragmatic one of *practical stabilization*, with ultimate boundedness of the tracking errors instead of convergence to zero, can be beneficial to enlarge the set of control possibilities and address the trajectory tracking problem in a re-unified way. The *transverse function* approach described in Section 3.3 has been developed with this point of view. It allows to derive smooth feedback controllers which uniformly ensure ultimate boundedness of the tracking errors, with arbitrary pre-specified tracking precision, *whatever the reference trajectory* (admissibility is no longer a prerequisite).

References

1. G. Artus, P. Morin, C. Samson. Tracking of an omnidirectional target with a nonholonomic mobile robot. In: *Proc. IEEE Conf. on Advanced Robotics (ICAR)*, 2003, pp. 1468–1473.
2. G. Artus, P. Morin, C. Samson. Control of a maneuvering mobile robot by transverse functions. In: *Proc. Symp. on Advances in Robot Kinematics (ARK)*, 2004.
3. M.K. Bennani, P. Rouchon. Robust stabilization of flat and chained systems. In: *Proc. European Control Conference*, 1995, pp. 2642–2646.
4. A.M. Bloch, P.S. Krishnaprasad, J.E. Marsden, R.M. Murray. Nonholonomic mechanical systems with symmetry. *Archive for Rational Mechanics and Analysis*, vol. 136, 1996, pp. 21–99.
5. R.W. Brockett. Asymptotic stability and feedback stabilization. In: R.S. Millman, R.W. Brockett and H.J. Sussmann (Eds.), *Differential Geometric Control Theory*. Birkauser, 1983.
6. G. Campion, B. d'Andrea Novel, G. Bastin. Modelling and state feedback control of nonholonomic constraints. In: *Proc. IEEE Conf. on Decision and Control*, 1991, pp. 1184–1189.
7. C. Canudas de Wit, B. Siciliano, G. Bastin (Eds.). *Theory of robot control.* Springer Verlag, 1996.
8. C.-T. Chen. *Linear system theory and design.* Oxford University Press, 1984.
9. J.-M. Coron. Global asymptotic stabilization for controllable systems without drift. *Mathematics of Control, Signals, and Systems*, vol. 5, 1992, pp. 295–312.
10. W.E. Dixon, D.M. Dawson, E. Zergeroglu, F. Zhang. Robust tracking and regulation control for mobile robots. *Int. Journal of Robust and Nonlinear Control*, vol. 10, 2000, pp. 199–216.
11. H. Hermes. Nilpotent and high-order approximations of vector field systems. *SIAM Review*, vol. 33, 1991, pp. 238–264.
12. I. Kolmanovsky, N.H. McClamroch. Developments in nonholonomic control problems. *IEEE Control Systems*, 1995, pp. 20–36.
13. D.A. Lizárraga. Obstructions to the existence of universal stabilizers for smooth control systems. *Mathematics of Control, Signals, and Systems*, vol. 16, 2004, pp. 255–277.
14. D.A. Lizárraga, P. Morin, C. Samson. Non-robustness of continuous homogeneous stabilizers for affine control systems. In: *Proc. IEEE Conf. on Decision and Control*, 1999, pp. 855–860.

15. R.T. M'Closkey, R.M. Murray. Nonholonomic systems and exponential convergence: some analysis tools. In: *Proc. IEEE Conf. on Decision and Control*, 1993, pp. 943–948.
16. R.T. M'Closkey, R.M. Murray. Exponential stabilization of driftless nonlinear control systems using homogeneous feedback. *IEEE Trans. on Automatic Control*, vol. 42, 1997, pp. 614–628.
17. P. Morin, J.-B. Pomet, C. Samson. Design of homogeneous time-varying stabilizing control laws for driftless controllable systems via oscillatory approximation of Lie brackets in closed-loop. *SIAM Journal on Control and Optimization*, vol. 38, 1999, pp. 22–49.
18. P. Morin, C. Samson. Exponential stabilization of nonlinear driftless systems with robustness to unmodeled dynamics. *Control, Optimization & Calculus of Variations*, vol. 4, 1999, pp. 1–36.
19. P. Morin, C. Samson. Control of non-linear chained systems. From the Routh-Hurwitz stability criterion to time-varying exponential stabilizers. *IEEE Trans. on Automatic Control*, vol. 45, 2000, pp. 141–146.
20. P. Morin, C. Samson. A characterization of the Lie algebra rank condition by transverse periodic functions. *SIAM Journal on Control and Optimization*, vol. 40, no. 4, 2001, pp. 1227–1249.
21. P. Morin, C. Samson. Commande. In: J.-P. Laumond (Editor), *La robotique mobile*. Hermes, 2001 (in French).
22. P. Morin, C. Samson. Practical stabilization of driftless systems on Lie groups: the transverse function approach. *IEEE Trans. on Automatic Control*, vol. 48, 2003, pp. 1496–1508.
23. P. Morin, C. Samson. Trajectory tracking for non-holonomic vehicles: overview and case study. In: K. Kozlowski (Editor), *Proc. 4th Int. Workshop on Robot Motion and Control (RoMoCo)*, 2004, pp. 139–153.
24. R.M. Murray, S.S. Sastry. Steering nonholonomic systems in chained form. In: *Proc. IEEE Conf. on Decision and Control*, 1991, pp. 1121–1126.
25. H. Nijmeijer, A.J. Van der Schaft. *Nonlinear Dynamical Control Systems*. Springer Verlag, 1991.
26. J.-B. Pomet. Explicit design of time-varying stabilizing control laws for a class of controllable systems without drift. *Systems & Control Letters*, vol. 18, 1992, pp. 467–473.
27. E.P. Ryan. On Brockett's condition for smooth stabilizability and its necessity in a context of nonsmooth feedback. *SIAM Journal on Control and Optimization*, vol. 32, 1994, pp. 1597–1604.
28. C. Samson. Velocity and torque feedback control of a nonholonomic cart. *Int. Workshop in Adaptative and Nonlinear Control: Issues in Robotics*, 1990. Also in LNCIS, vol. 162, Springer Verlag, 1991.
29. C. Samson. Control of chained systems. Application to path following and time-varying point-stabilization. *IEEE Trans. on Automatic Control*, vol. 40, 1995, pp. 64–77.
30. C. Samson, K. Ait-Abderrahim. Feedback control of a nonholonomic wheeled cart in cartesian space. In: *Proc. IEEE Conf. on Robotics and Automation*, 1991, pp. 1136–1141.
31. E.D. Sontag. Universal nonsingular controls. In: *Systems & Control Letters*, vol. 19, 1992, pp. 221–224.
32. O.J. Sørdalen. Conversion of the kinematics of a car with n trailers into a chained form. In: *Proc. IEEE Conf. on Robotics and Automation*, 1993, pp. 382–387.

33. A.R. Teel, R.M. Murray, G. Walsh. Nonholonomic control systems: from steering to stabilization with sinusoids. In: *Proc. IEEE Conf. on Decision and Control*, 1992, pp. 1603–1609.
34. J. Zabczyk. Some comments on stabilizability. *Applied Mathematics & Optimization*, vol. 19, 1989, pp. 1–9.

2

Posture Stabilization of a Unicycle Mobile Robot – Two Control Approaches

Krzysztof Kozłowski, Jarosław Majchrzak, Maciej Michałek, and
Dariusz Pazderski

Chair of Control and Systems Engineering, Poznań University of Technology
ul. Piotrowo 3a, 60-965 Poznań, Poland `name.surname@put.poznan.pl`

2.1 Introduction

State feedback control issues for nonholonomic systems are still very challenging for control researchers. Among the group of nonholonomic systems one can number wheeled mobile vehicles, manipulators with nonholonomic gears, free-floating robots, underwater vessels, nonholonomic manipulator's grippers, dynamically balanced hopping robots and others [4, 10, 15]. Difficulties in designing effective stabilizers arise from nonintegrable kinematic constraints imposed on system evolution. These constraints impose restriction on admissible velocities of controlled dynamic systems, preserving however their controllability. Moreover, lower dimensionality of the control space $U \subset \mathbb{R}^m$ in comparison to the configuration space $Q \subset \mathbb{R}^n$ $(n > m)$ causes difficulties in control design, especially for stabilization task problems [5]. Despite the problems mentioned, many different feedback control strategies for nonholonomic kinematics in automatics and robotics literature have been proposed – see for example [7], [18] or [8]. Still, some important problems, like robustness to control signals limitations existance, intuitive contoller parameter tunning and good control quality during transient stage seem to remain open issues and involve further research. In this paper two different stabilization approaches to the mentioned problems, with alternative solution in comparison to existing strategies, are described. The presented approaches are applied to derive two stabilizers to solve the stabilization task for a unicycle mobile robot, taking into account control limitations. The first controller, which may be classified as a Time-Varying Oscillatory (TVO) stabilizer, is based on an idea of stabilization initially proposed by Dixon *et al.* and next generalized by Morin and Samson [17]. The second approach results from a novel concept called the Vector Field Orientation (VFO) approach introduced for the first time in [12]. For both stabilizers, control signal limitations will be considered at the kinematic level as restrictions imposed on the maximal velocity of the robot wheels. The effectiveness of the proposed control strategies will be compared and illustrated by simulation results.

K. Kozłowski (Ed.): Robot Motion and Control, LNCIS 335, pp. 25–54, 2006.
© Springer-Verlag London Limited 2006

2.2 Kinematics

This work considers a unicycle mobile robot, which can be treated as a reduced two-wheeled differentially driven mobile vehicle with angular, Ω, and linear, V, velocities chosen as the input controls. Assuming nonexistence of lateral slippage of the robot wheels, the kinematic model of the unicycle mobile robot can be described in the following manner:

$$
\begin{bmatrix} \dot{\varphi} \\ \dot{x} \\ \dot{y} \end{bmatrix} = \begin{bmatrix} 1 \\ 0 \\ 0 \end{bmatrix} u_1 + \begin{bmatrix} 0 \\ \cos\varphi \\ \sin\varphi \end{bmatrix} u_2, \tag{2.1}
$$

where $q \triangleq [\varphi\; x\; y]^T \in \mathcal{Q} \subset \mathbb{R}^3$ denotes the state vector and $u_1 = \Omega, u_2 = V \in \mathbb{R}$ describe the control inputs (see Fig. 2.1). Model (2.1) belongs to a class of underactuated nonholonomic driftless systems with two inputs described in the following general form:

$$
\dot{q} = g_1 u_1 + g_2(q) u_2, \tag{2.2}
$$

where $g_1, g_2(q)$ are basic vector fields – generators. From the last two equations of (2.1) one can derive the assumed nonitegrable velocity constraint:

$$
A(q)\dot{q} = 0, \qquad A(q) = [0\; \sin\varphi\; -\cos\varphi], \tag{2.3}
$$

where $\dot{q} \triangleq [\dot{\varphi}\; \dot{x}\; \dot{y}]^T \in \mathbb{R}^3$ is the generalized state velocity and A is called the constraint matrix. It is well known that the system (2.1) is fully controllable in $\mathcal{Q}$, however all the generalized state velocities $\dot{q}$ accessible during time evolution exclude some set of instantaneously inadmissible movement directions. This constraint together with less number of inputs in relation to the state vector dimension cause difficulties in solving control tasks, especially for the point stabilization task [5].

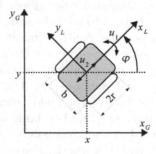

Fig. 2.1. Mobile robot in the global frame $\{G\}$

2.3 Posture Stabilization – Two Control Approaches

In this paper posture stabilization in Cartesian space of system (2.1) is considered. In order to determine the posture error the following vector $e(\tau) \in \mathbb{R}^3$ is defined:

$$e(\tau) \triangleq \begin{bmatrix} e_1(\tau) \\ e_2(\tau) \\ e_3(\tau) \end{bmatrix} \triangleq q_t - q(\tau) = \begin{bmatrix} \varphi_t - \varphi(\tau) \\ x_t - x(\tau) \\ y_t - y(\tau) \end{bmatrix}, \qquad (2.4)$$

where q_t and $q(\tau)$ determine the reference (desired) and actual state vectors, respectively, and the time variable is denoted by τ.

In the sequel two alternative control approaches, which allow to solve practical or asymptotic posture stabilization for nonholonomic kinematics (2.1) are described.

The control task considered in this paper, called hereafter a *stabilization task*, can be defined as follows.

Definition 1 (Stabilization task) *Find bounded controls $u_1(\tau), u_2(\tau)$ for kinematics (2.1), such that for initial condition $e(0) \in \mathcal{E} \subset \mathbb{R}^3$ the Euclidean norm of the error $e(\tau)$ tends to some constant $\varepsilon > 0$ as $\tau \to \infty$:*

$$\lim_{\tau \to \infty} \|e(\tau)\| = \varepsilon, \qquad (2.5)$$

where ε is an assumed error envelope, which can be made arbitrarily small. For $\varepsilon > 0$ the above statement defines the practical stabilization task, and for $\varepsilon = 0$ – the asymptotic convergence task.

2.3.1 Oscillatory-based Time-varying Control Law

The first control method (TVO) is based on a tuned oscillator idea introduced by Dixon *et al.* [9]. It should be noted that this approach can be seen as a particular case of a more general theory developed by Morin and Samson [17], taking advantage of so-called transverse functions. The main feature concerning these control schemes lies in virtual periodic signals tracked by the state vector of the system.

Model Transformation

In control theory area much work has been devoted to the stabilization problem of abstract mathematical objects which are not directly related to physical systems. Such approach allows to generalize control solutions applicable to some class of systems which are, in general, much simpler than original ones. In particular, so-called chained canonical form were investigated during the last fifteen years. As pointed out by many authors (see for example [14, 17, 22]) mathematical models used to describe different physical systems which include, for example, kinematics of most terrain vehicles, may be transformed to the chained form using properly chosen transformation.

In this section we present a control law taking advantage of mathematical properties of the system known as Brockett's nonholonomic integrator [6]. This is a driftless nilpotent system for which controllability in a short time is ensured by the first order Lie bracket. The nonholonomic integrator can be written in the following form (cf. [9]):

$$\dot{x}^* = v, \tag{2.6}$$

$$\dot{x}_3 = x^{*T} J v, \tag{2.7}$$

where $x = \begin{bmatrix} x_1 & x_2 & x_3 \end{bmatrix}^T = \begin{bmatrix} x^{*T} & x_3 \end{bmatrix}^T \in \mathbb{R}^3$ is a state vector, $v = \begin{bmatrix} v_1 & v_2 \end{bmatrix}^T \in \mathbb{R}^2$ denotes an input and $J = \begin{bmatrix} 0 & -1 \\ 1 & 0 \end{bmatrix}$ is a skew-symmetric matrix.

In order to transform kinematic equation (2.4) to the form of the nonholonomic integrator the following nonlinear transformation considered by Dixon *et al.* [9] may be used (for derivation of the transformation see [20]):

$$x \triangleq -P(q, e) e, \tag{2.8}$$

where P defines a global diffeomorphism with respect to the origin

$$P\left(\theta, \tilde{\theta}\right) \triangleq \begin{bmatrix} 1 & 0 & 0 \\ 0 & c\varphi & s\varphi \\ 0 & e_1 c\varphi + 2s\varphi & e_1 s\varphi - 2c\varphi \end{bmatrix} \in \mathbb{R}^{3 \times 3}, \tag{2.9}$$

where $c\varphi \equiv \cos\varphi$ and $s\varphi \equiv \sin\varphi$.

Next, taking the time derivative of x^* and using (2.8) one can obtain a relation between the original, u, and auxiliary, v, input signals as follows:

$$v = T(q, e) u, \tag{2.10}$$

where $T(q, e) \in \mathbb{R}^{2 \times 2}$ is the control transformation matrix defined as

$$T(q, e) = \begin{bmatrix} 1 & 0 \\ L(q, e) & 1 \end{bmatrix}, \tag{2.11}$$

with $L(q, e) = -e_2 \sin\varphi + e_3 \cos\varphi$, such that $|L(q, e)|$ means a projection of the distance between the actual and desired positions of the robot calculated in the direction perpendicular to the robot's heading.

Since T is globally invertible it is straightforward to get the following inverse formula, which can be directly used for control design:

$$u = T^{-1}(q, e) v. \tag{2.12}$$

Summarizing, as a result of transformations (2.8) and (2.10) it is possible to solve the posture stabilization task for kinematics (2.1) by developing the control law which stabilizes the system given by (2.7). This approach is presented in the next subsection.

Control Law Development

The main concept of TVO stabilizer, as well as other controllers using transverse functions, consists of decreasing the regulation error, x, indirectly by tracking additional virtual signals x_d.

Hence, taking into account system (2.7) the following signal $z \in \mathbb{R}^3$ may be introduced:

$$z \triangleq \begin{bmatrix} z^* \\ z_3 \end{bmatrix} = x - \begin{bmatrix} x_d \\ -x_d^T J x^* \end{bmatrix}, \tag{2.13}$$

where $z^* = [z_1 \ z_2]^T \in \mathbb{R}^2$ and $x_d = [x_{d1} \ x_{d2}]^T \in \mathbb{R}^2$ denote auxiliary time-varying bounded signals which will be defined later.

It is very important to note that relation (2.13) can be explained by means of differential geometry, since it determines a left-invariant group operation for the control system (2.7) (see [16, 17]).

The auxiliary task of control consists of asymptotic (exponential) stabilization of (2.13) defined as

$$\forall_{\tau > 0} \ \|z\| \leqslant \gamma \|z(0)\| \exp(-\beta\tau), \tag{2.14}$$

where γ and $\beta > 0$ are some positive constants.

According to (2.14) and Definition (2.13) one can conclude that

$$\lim_{\tau \to \infty} z^* = 0 \Rightarrow \lim_{\tau \to \infty} x^* = x_d \tag{2.15}$$

and

$$\lim_{\tau \to \infty} z_3 = 0 \Rightarrow \lim_{\tau \to \infty} x_3 = 0. \tag{2.16}$$

These relations show that the accuracy of regulation in the steady state is determined by signal x_d. Moreover, x_d significantly influences transient behavior during the regulation process since it is tracked by x^* according to (2.15). This problem will be discussed later.

In order to develop the control law asymptotically stabilizing z the following Lyapunov function candidate is proposed:

$$V \triangleq \frac{1}{2} z^T z. \tag{2.17}$$

Next, taking the time derivative of V one can get

$$\dot{V} = z^T \dot{z} = z^{*T} \dot{z}^* + z_3 \dot{z}_3. \tag{2.18}$$

Calculating the time derivative of z^* from (2.13) the following relation can be obtained:

$$\dot{z}^* = v - \dot{x}_d. \tag{2.19}$$

It should be noted that the regulation task for the subsystem $z^* = x^* - x_d$ is relatively easy and can be solved using the following control signal:

$$v = -k_1 z^* + \dot{x}_d, \tag{2.20}$$

where $k_1 > 0$ is a constant parameter. Using (2.20) and (2.13) in (2.18) yields

$$\dot{V} = -k_1 z^{*T} z^* + z_3 \dot{z}_3. \tag{2.21}$$

Next, term $\dot{z}_3$ can be calculated according to (2.13) as follows (see details in Appendix):

$$\dot{z}_3 = x_d^T J \dot{x}_d + 2k_1 x^{*T} J x_d. \tag{2.22}$$

Here signal $\dot{x}_d$ can be interpreted as an additional control signal and can be used for asymptotic stabilization of coordinate z_3. In order to calculate $\dot{x}_d$ we assume that x_d is originated by a tunable oscillator [9] according to the following equation:

$$x_d = \boldsymbol{\Psi}\boldsymbol{\xi}, \tag{2.23}$$

where

$$\boldsymbol{\Psi} = \begin{bmatrix} \psi_1 & 0 \\ 0 & \psi_2 \end{bmatrix} \tag{2.24}$$

is a gain matrix with scalar functions $\psi_1(\tau)$ and $\psi_2(\tau) > 0$, which may be changed during regulation process, and $\boldsymbol{\xi}$ is a solution of the following differential equation:

$$\dot{\boldsymbol{\xi}} = u_\omega J \boldsymbol{\xi} \tag{2.25}$$

with u_ω determining an instantaneous frequency of $\boldsymbol{\xi}$ and the initial condition

$$\boldsymbol{\xi}(0)^T \boldsymbol{\xi}(0) = 1. \tag{2.26}$$

As can be seen, (2.25) describes an undamped linear oscillator with constant amplitude of signal $\boldsymbol{\xi}$, such that $\forall_{\tau > 0} \boldsymbol{\xi}^T(\tau)\boldsymbol{\xi}(\tau) = 1$. Similarly to the control law given by Morin and Samson [17] frequency u_ω can be regarded as the third control signal (apart from v_1 and v_2) that makes the system to be virtually fully actuated.

An analytical formula describing u_ω can be obtained using the time derivative of (2.23), namely

$$\dot{x}_d = \dot{\boldsymbol{\Psi}}\boldsymbol{\xi} + \boldsymbol{\Psi}\dot{\boldsymbol{\xi}} \tag{2.27}$$

and relations (2.22) and (2.20). A a result the following equation can be written

$$\dot{z}_3 = \boldsymbol{\xi}^T \boldsymbol{\Psi}^T J \dot{\boldsymbol{\Psi}}\boldsymbol{\xi} + 2k_1 x^{*T} J x_d - \psi_1 \psi_2 u_\omega. \tag{2.28}$$

Now, it is straightforward to show that applying u_ω written as

$$u_\omega = \frac{-w + \boldsymbol{\xi}^T \boldsymbol{\Psi}^T J \dot{\boldsymbol{\Psi}}\boldsymbol{\xi} + 2k_1 x^{*T} J x_d}{\psi_1 \psi_2} \tag{2.29}$$

in (2.28) leads to a decoupled subsystem, namely

$$\dot{z}_3 = w, \tag{2.30}$$

where w is a scalar function which is a new input. In order to ensure exponential stabilization of z_3 we propose to set

$$w = -k_2 z_3, \tag{2.31}$$

where $k_2 > 0$ is a constant controller parameter. Consequently, taking into account (2.31) and (2.30) allows to rewrite the time derivative of V as

$$\forall_{\tau>0,\|z\|\neq0} \; \dot{V} = -k_1 z^{*T} z^* - k_2 z_3^2 < 0. \tag{2.32}$$

Then, assuming that $\beta = \min\{k_1, k_2\}$, the following upper bound of $\dot{V}$ can be obtained

$$\dot{V} \leqslant -\beta z^T z = -2\beta V. \tag{2.33}$$

As a consequence one can prove that V tends to zero exponentially, namely

$$\forall_{\tau>0} \; V(\tau) = V(0) \exp(-2\beta\tau). \tag{2.34}$$

Proposition 1 *Assuming that k_1, k_2, ψ_1, and $\psi_2 > 0$, ψ_1, ψ_2, $\dot{\psi}_1$, and $\dot{\psi}_2 \in \mathcal{L}_\infty$, the controller given by (2.20), (2.29), (2.31), (2.23), (2.25), and (2.26) stabilizes system (2.13) exponentially in terms of (2.14).*

Proof. The result given by Proposition 1 is a direct consequence of relation (2.34). However, to complete the proof it is necessary to show that all signals in the controlled system are bounded. First, we examine frequency u_ω taking into account relation (2.29) and $\|\boldsymbol{\xi}\| = 1$. Then using the following relations:

$$\left| \boldsymbol{\xi}^T \boldsymbol{\Psi}^T \boldsymbol{J} \dot{\boldsymbol{\Psi}} \boldsymbol{\xi} \right| \leqslant \rho_1 \triangleq \left| \psi_1 \dot{\psi}_2 - \dot{\psi}_1 \psi_2 \right| \in \mathcal{L}_\infty \tag{2.35}$$

and

$$\left| z^{*T} \boldsymbol{J} \boldsymbol{x}_d \right| \leqslant \rho_2 \|z^*\|, \tag{2.36}$$

with

$$\rho_2 \triangleq \max\{\psi_1, \psi_2\} \in \mathcal{L}_\infty. \tag{2.37}$$

one can easily show that

$$|u_\omega| \leqslant \frac{k_2|z_3| + 2k_1\rho_2\|z^*\| + \rho_1}{\psi_1\psi_2} \leqslant \rho_3 \frac{\|z\| + \frac{\rho_1}{\rho_3}}{\psi_1\psi_2} \leqslant \rho_3 \frac{\|z(0)\|\exp(-\beta\tau) + \frac{\rho_1}{\rho_3}}{\psi_1\psi_2}, \tag{2.38}$$

where $\rho_3 = \sqrt{2}\max\{k_1, 2k_2\rho_2\}$. As a consequence, $u_\omega \in \mathcal{L}_\infty$ if only ψ_1 and ψ_2 satisfy the assumptions given by Proposition 1. Next, considering $\dot{\boldsymbol{x}}_d$ it is easy to show (see (2.27)) that $u_\omega \in \mathcal{L}_\infty$ implies $\left\|\dot{\boldsymbol{\xi}}\right\| \in \mathcal{L}_\infty$ and consequently $\|\dot{\boldsymbol{x}}_d\| \in \mathcal{L}_\infty$. Finally, it is straightforward to conclude that $\|v\| \in \mathcal{L}_\infty$. □

Considering (2.29) one can see that frequency of oscillation u_ω is strictly related to functions ψ_1 and ψ_2, which determine the amplitude of auxiliary signal x_d. Therefore, the selection of these functions taking into account the evolution of error z highly affects system behavior during transient states. In order to limit overshoots it is reasonable to choose $\psi_1 \psi_2$ to be high enough to imply non-oscillatory behavior of the closed-loop system. On the other hand, according to (2.15) amplitude of x_d determines the accuracy of regulation in the steady-state. As a result, high precision of regulation needs to assume $\lim_{\tau \to \infty} \psi_i(\tau) = \varepsilon_i$, where ε_i denotes a positive small constant (small enough for the desired accuracy).

In order to do that, similarly to [9], the following gain functions $\psi_1(\tau)$ and $\psi_2(\tau)$ explicitly depending on time are proposed:

$$\psi_i(\tau) = \psi_{i0} \exp(-\alpha_i \tau) + \varepsilon_i, \text{ for } i = 1, 2, \tag{2.39}$$

where $\psi_{i0} > 0$, $\alpha_i > 0$, and $\varepsilon_i > 0$ are scalar coefficients determining the initial and limit values of the function ψ_i and its convergence rate, respectively. It is clear that, since functions (2.39) satisfy the assumptions of Proposition 1, the proof of it is still valid.

Considering the tuning methods introduced by functions (2.39) one can prove according to (2.15) and (2.16) that

$$\lim_{\tau \to \infty} |x_1(\tau)| \leqslant \varepsilon_1, \quad \lim_{\tau \to \infty} |x_2(\tau)| \leqslant \varepsilon_2, \quad \lim_{\tau \to \infty} x_3(\tau) = 0. \tag{2.40}$$

Finally, we return to the stabilization problem of a nonholonomic robot (2.1) in Cartesian space. According to the error definition given by (2.4) and transformation (2.8) one can conclude that $\lim_{\tau \to \infty} x = 0$ implies that $\lim_{\tau \to \infty} e = 0$. The above considerations result in the following proposition:

Proposition 2 *Assuming that ψ_1 and ψ_2 are defined by (2.39), the controller given by (2.20), (2.29), (2.23), (2.25), and (2.26) with transformations (2.8) and (2.10) ensures boundness of errors e in the sense that*

$$\lim_{\tau \to \infty} |e_1(\tau)| \leqslant \varepsilon_1, \quad \lim_{\tau \to \infty} \|e^*(\tau)\| \leqslant \varepsilon_2 \sqrt{\frac{\varepsilon_1^2}{4} + 1}, \tag{2.41}$$

where $e^* \triangleq [e_2 \ e_3]^T$.

Remark 1. The convergence rate determined by the α_i coefficient may be choosen arbitrary, at least from a theoretical point of view. However, taking into account (2.38) and considering the term written as

$$\frac{\|z(0)\| \exp(-\beta \tau)}{\psi_1 \psi_2} \tag{2.42}$$

one can easily conclude that it decreases for all times if

$$\alpha_1 + \alpha_2 < \beta. \tag{2.43}$$

Hence, using this condition for selection of scaling functions ψ_1 and ψ_2 allows to prevent from increasing $|u_\omega|$ during the regulation process.

Remark 2. Exponential convergence of an auxiliary error z (see (2.14)) implies exponential convergence of the transformed state vector x to a neighborhood of the desired point with the radius determined by ε_i (cf. also [9]). Consequently, considering inverse transformation P^{-1} one can deduce a similar conclusion governing convergence of error e. Moreover, the convergence rate can be determined easily, since it is directly related to the selection of k_1, k_2, as well as α_i parameters (compare (2.34) and (2.39)).

Remark 3. Assuming that $\varepsilon_i \equiv 0$ and $\dot{\psi}_1\psi_2 = \psi_1\dot{\psi}_2$ which yields to $\rho_1 = 0$, then it is possible to obtain an asymptotic convergence result in Cartesian space. It can be easily proved using (2.38) that $|u_\omega|$ remains bounded for all times if only condition (2.43) is satisfied. However, this result is based on the fact that for the initial condition $\|z(0)\| > 0$ the closed-loop system will never reach the equilibrium point in finite time. Hence, the controller considered here does not asymptotically stabilize the system at the origin. It means that if $\|z(0)\| > 0$ it is necessary to select $\psi_1\psi_2 > 0$ in order to avoid singularity. Therefore, in the case when $\varepsilon_i \equiv 0$ the control law considered here is time-differentiable everywhere except the origin.

Simplified Version of the Controller

As one can see for the TVO controller, the terms concerning the time-derivatives of ψ_1 and ψ_2 are used explicitly. However, it would be simpler to neglect these terms, which allows to change ψ_i during regulation without calculating their time-derivatives. Moreover, it has been observed during extensive simulations that neglecting $\dot{\psi}_i$ in the control law may result in less oscillatory behavior. This phenomena can be explained taking into account the numerator of expression (2.29). Since the following inequality holds:

$$\left| -w + 2k_1 x^{*T} J x_d \right| + \left| \xi^T \Psi^T J \dot{\Psi} \xi \right| \geqslant \left| -w + 2k_1 x^{*T} J x_d \right| \tag{2.44}$$

removing the term $\xi^T \Psi^T J \dot{\Psi} \xi$ should result in a lower upper bound of $|u_\omega|$.

This statement is illustrated in Fig. 2.2 for exponential scaling of ψ_i. The controller parameters were selected as $k_1 = k_2 = 10$, $\xi(0) = \frac{\sqrt{2}}{2}[1\ 1]^T$, $\psi_1(0) = \psi_2(0) = 0.5$, $\alpha_1 = \alpha_2 = 5$ and $\varepsilon_1 = \varepsilon_2 = 0.05$. As can be seen, convergence rate in both cases is quite similar, but if matrix $\dot{\Psi}$ is used explicitly, the transient behavior appears to be more oscillatory than in the case when $\dot{\Psi}$ is dropped. Note that the steady-state errors in both cases are limited to the nonzero values and are in the same assumed neighborhood of the origin if only $\lim_{\tau \to \infty} \dot{\Psi}(\tau) = 0 \in \mathbb{R}^{2 \times 2}$.

As a consequence based on the presented observation the following proposition can be justified:

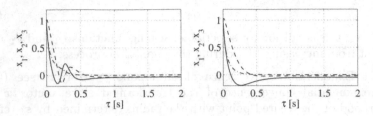

Fig. 2.2. Stabilization errors for the TVO controller obtained for the nonholonomic integrator: $x_1(-)$, $x_2(--)$ and $x_3(-.-)$. Left: the original controller, right: the simplified controller (without using $\dot{\boldsymbol{\Psi}}$ in the control law)

Proposition 3 *Assuming that gain coefficients* $k_1, k_2 > 0.5$, $\psi_i > 0$, *and* $\dot{\psi}_i$ *can be lower bounded by an exponentially time-varying function such that* $\lim_{\tau \to \infty} \dot{\psi}_i(\tau) = 0$, *the modified control law written as*

$$v = -k_1 z^* + \boldsymbol{\Psi}\dot{\boldsymbol{\xi}}, \tag{2.45}$$

$$u_\omega = \frac{k_2 z_3 + 2k_1 z^{*T} \boldsymbol{J} \boldsymbol{x}_d}{\psi_1 \psi_2} \tag{2.46}$$

ensures exponential convergence of $\boldsymbol{z}$.

Proof. In order to show stability of system (2.13) we use the positive scalar function defined by (2.17). Then calculating its time derivative and using (2.45)-(2.46) one can get

$$\dot{V} = -k_1 z^{*T} z^* - k_2 z_3^2 - z^{*T} \dot{\boldsymbol{\Psi}}\boldsymbol{\xi} + z_3 \boldsymbol{\xi}^T \boldsymbol{\Psi}^T \boldsymbol{J} \dot{\boldsymbol{\Psi}} \boldsymbol{\xi}. \tag{2.47}$$

One can easily conclude that the signs of terms $z^{*T} \dot{\boldsymbol{\Psi}}\boldsymbol{\xi}$ and $\boldsymbol{\xi}^T \boldsymbol{\Psi}^T \boldsymbol{J} \dot{\boldsymbol{\Psi}} \boldsymbol{\xi}$ are not determined. Hence, we may find the upper bound of them using relation (2.36) and

$$z^{*T} \dot{\boldsymbol{\Psi}}\boldsymbol{\xi} \leqslant \rho_4 \|z^*\|, \tag{2.48}$$

where

$$\rho_4 \overset{\Delta}{=} \max\left\{ \left|\dot{\psi}_1\right|, \left|\dot{\psi}_2\right| \right\}. \tag{2.49}$$

Consequently $\dot{V}$ satisfies

$$\dot{V} \leqslant -k_1 \|z^*\|^2 - k_2 z_3^2 + \rho_4 \|z^*\| + \rho_1 |z_3| \tag{2.50}$$

with ρ_1 defined by (2.35).

Introducing the upper bound of the last two terms as

$$\rho_4 \|z^*\| \leqslant \tfrac{1}{2}\rho_4^2 + \tfrac{1}{2}\|z^*\|^2, \quad \rho_1 |z_3| \leqslant \tfrac{1}{2}\rho_1^2 + \tfrac{1}{2}z_3^2 \tag{2.51}$$

implies that

$$\dot{V} \leqslant (-k_1 + 0.5)\|z^*\|^2 + (-k_2 + 0.5)z_3^2 + 0.5\left(\rho_4^2 + \rho_1^2\right). \tag{2.52}$$

Next, redefining β as

$$\beta \triangleq \min\{k_1 - 0.5, \; k_2 - 0.5\} \tag{2.53}$$

and defining

$$\rho_5^2(\tau) \triangleq 0.5\left(\rho_4^2 + \rho_1^2\right) \leqslant \rho_5^2(0)\exp\left(-2\alpha_{\min}\tau\right), \tag{2.54}$$

where $\alpha_{\min} \triangleq \min\{\alpha_1, \alpha_2\}$ yields

$$\dot{V} \leqslant -\beta\left\|z\right\|^2 + 0.5\rho_5^2(0)\exp\left(-2\alpha_{\min}\tau\right). \tag{2.55}$$

Then time-derivative of V becomes

$$\dot{V} \leqslant -2\beta V + \rho_5^2(0)\exp\left(-2\alpha_{\min}\tau\right). \tag{2.56}$$

Next calculating the solution of inequality (2.56) (see Appendix) one can obtain that

$$V(\tau) \leqslant V(0)\exp\left(-2\beta\tau\right) + \frac{0.5\rho_5^2(0)}{\beta - \alpha_{\min}}\left[\exp\left(-2\alpha_{\min}\tau\right) - \exp\left(-2\beta\tau\right)\right], \tag{2.57}$$

where an assumption that $\beta \neq \alpha_{\min}$ has been used. This result leads to the following upper bound of $\left\|z(\tau)\right\|$:

$$\left\|z(\tau)\right\| \leqslant \left\|z(0)\right\|\exp\left(-\beta\tau\right) + \frac{\sqrt{2}}{2}\rho_5(0)\sqrt{\frac{\exp\left(-2\alpha_{\min}\tau\right) - \exp\left(-2\beta\tau\right)}{\beta - \alpha_{\min}}}. \tag{2.58}$$

From (2.58) one can conclude that tracking error $\left\|z(\tau)\right\|$ tends to zero as time goes to infinity (at the limit disturbance term under the square root disappears) despite the fact that scalar function V is not globally decreasing for all initial conditions $\left\|z(0)\right\|$ (note that $\dot{V}$ may be positive in some time interval).

Next, based on the stability result given by (2.58) and making calculations similar to those presented in the proof of Proposition 1 it is clear to see that $\left\|v\right\|, u_\omega \in \mathcal{L}_\infty$ if $\psi_i, \dot{\psi}_i \in \mathcal{L}_\infty$. □

Remark 4. Assuming that k_1 and k_2 are selected properly in order to satisfy (2.43), while β is defined by (2.53), the upper bound of $|u_\omega|$ decreases during the regulation process. Hence, the maximal value $|u_\omega|$ can be estimated by calculating $|u_\omega(0)|$ as follows (for simplicity it is supposed that $\rho_3 = 2\sqrt{2}k_1\rho_2$ and $\varepsilon_1 = \varepsilon_2 \approx 0$ are negligible with respect to ψ_{10} and ψ_{20}):

$$|u_{\omega max}| \leqslant 2\sqrt{2}k_1\left\|z(0)\right\|\frac{\psi_{max0}}{\psi_{min0}^2}, \tag{2.59}$$

where $\psi_{\min 0} \triangleq \min\{\psi_{10}, \psi_{20}\}$, $\psi_{max0} \triangleq \max\{\psi_{10}, \psi_{20}\}$.

According to this result one can chose ψ_{10} and ψ_{20} high enough in order to limit oscillatory behavior during transient states. On the other hand selection of high values of ψ_{10} or ψ_{20} may lead to high error at the beginning of the regulation process according to assumption (2.13). Moreover, for typical applications, since ψ_1 shapes the orientation error (compare (2.13) and (2.8)), it is reasonable to choose $\psi_{10} \leqslant \pi$.

It should also be mentioned that transient states are dependent on selection of $\boldsymbol{\xi}(0)$, which determines the initial direction of convergence of vector $\boldsymbol{x}^*$ in auxiliary state space. As a result, different $\boldsymbol{\xi}(0)$ leads to different paths of the robot observed in Cartesian space, namely it can move forward or backward, with one turn or more, etc.

Control Scaling

In many control applications saturation of input signals occurs. Therefore, developing a controller without taking into account this limitation may not guarantee good results (see for example [20, 21]).

Here we consider a method to rescale the control signal for a TVO controller used for stabilization of an affine driftless system. The presented algorithm can be used to decrease or increase the value of the control signals (in this case amplitudes as well as frequency) without lack of stability from a theoretical point of view. It may be useful to scale the control signal to be within range of permissible values.

First, let us assume that the original control signal $\boldsymbol{v}$ calculated according to (2.45) is rescaled by some positive scalar function $\mu(\tau) \in \mathcal{L}_\infty$ as

$$\boldsymbol{v}(\tau) = \mu(\tau)\left[-k_1 \boldsymbol{z}^*(\tau) + \boldsymbol{\Psi}(\tau_s)\boldsymbol{J}\boldsymbol{\xi}(\tau_s)u_\omega(\tau)\right] \tag{2.60}$$

with

$$u_\omega(\tau) = \frac{k_2 z_3(\tau) + 2k_1 \boldsymbol{z}^{*T}(\tau)\boldsymbol{J}\boldsymbol{\Psi}(\tau_s)\boldsymbol{\xi}(\tau_s)}{\psi_1(\tau_s)\psi_2(\tau_s)}, \tag{2.61}$$

where

$$\tau_s \overset{\Delta}{=} \tau_s(\tau) \overset{\Delta}{=} \int_0^\tau \mu(\sigma)\,d\sigma \tag{2.62}$$

denotes scaled time. It should be noted that introducing time scaling affects only auxiliary variables depending explicitly on time, namely related with signal $\boldsymbol{x}_d$. Therefore, at instant time τ, $\boldsymbol{x}_d$ and consequently $\boldsymbol{\xi}$ and $\boldsymbol{\Psi}$ are calculated at τ_s. Accordingly, in the definition of $\boldsymbol{z}(\tau)$ (see Eq. (2.13)) $\boldsymbol{x}_d(\tau)$ is replaced by $\boldsymbol{x}_d(\tau_s)$. This proposition is a result of using frequency of $\boldsymbol{\xi}$ as a third control signal. Hence, rescaling physical input $\boldsymbol{v}$ must result in proportional scaling $\dot{\boldsymbol{\xi}}$, that simply leads to introduction the virtual time τ_s.

As can be proved, introducing time-scaling with respect to $\boldsymbol{x}_d$ yields

$$\dot{\boldsymbol{\xi}}(\tau_s(\tau)) = \mu(\tau)\tfrac{d}{d\tau_s}\boldsymbol{\xi}(\tau_s), \quad \dot{\boldsymbol{\Psi}}(\tau_s(\tau)) = \mu(\tau)\tfrac{d}{d\tau_s}\boldsymbol{\Psi}(\tau_s), \tag{2.63}$$

where $\boldsymbol{\xi}(\tau_s)$ denotes a solution of the oscillator equation determined for time τ_s, namely

$$\frac{d}{d\tau_s}\boldsymbol{\xi} = u_\omega(\tau)\, \boldsymbol{J}\boldsymbol{\xi}(\tau_s). \tag{2.64}$$

Substituting Eqs. (2.60)-(2.61) to (2.19) and (2.22) allows to show that the time-derivative of $\boldsymbol{z} = \begin{bmatrix} \boldsymbol{z}^{*T} & z_3 \end{bmatrix}^T$ becomes

$$\dot{\boldsymbol{z}}^*(\tau) = \mu\left[-k_1 \boldsymbol{z}^*(\tau) - \frac{d}{d\tau_s}\boldsymbol{\Psi}(\tau_s)\,\boldsymbol{\xi}(\tau_s)\right] \tag{2.65}$$

and

$$\dot{z}_3 = \mu\left[-k_2 z_3(\tau) + \boldsymbol{\xi}^T(\tau_s)\,\boldsymbol{\Psi}^T(\tau_s)\,\frac{d}{d\tau_s}\boldsymbol{\Psi}(\tau_s)\,\boldsymbol{\xi}(\tau_s)\right]. \tag{2.66}$$

Next, it is straightforward to prove that

$$\dot{V} = \mu\left[-k_1 \left\|\boldsymbol{z}^*(\tau)\right\|^2 - k_2 z_3^2(\tau) - \boldsymbol{z}^{*T}(\tau)\,\frac{d}{d\tau_s}\boldsymbol{\Psi}(\tau_s)\,\boldsymbol{\xi}(\tau_s) + \right. \tag{2.67}$$

$$\left. + z_3(\tau_s)\,\boldsymbol{\xi}^T(\tau_s)\,\boldsymbol{\Psi}^T(\tau_s)\,\boldsymbol{J}\frac{d}{d\tau_s}\boldsymbol{\Psi}(\tau_s)\,\boldsymbol{\xi}(\tau_s)\right].$$

Similarly to calculation made in the proof of Proposition 3 one can prove that

$$\forall_{\tau>0}\; V(\tau) \leqslant V(0)\exp\left(-2\mu\beta\tau\right) + \frac{0.5\rho_5^2(0)}{\beta - \alpha_{\min}}\left[\exp\left(-2\alpha_{\min}\tau_s\right) - \exp\left(-2\beta\tau_s\right)\right]. \tag{2.68}$$

Finally, one can conclude that convergence of function V can be changed proportionally to μ. Moreover, in the same way the time evolution of $\boldsymbol{\xi}$ and $\boldsymbol{\Psi}$ is rescaled (compare (2.63)). Hence, $\boldsymbol{z}$ may evolve faster (for $\mu > 1$) or slower (for $0 < \mu < 1$) with respect to the original solution (i.e. when $\mu = 1$). It is very important to note that the shape of the trajectory describing the evolution of $\boldsymbol{z}$ in the state space is independent on choosing $\mu > 0$.

Remark 5. Signal and time scaling presented here seems to be a very effective solution for controlling a driftless system. Using properly scaling function μ one can easily make regulation to be faster or slower in order to avoid control signal saturation and limit its frequency. Consequently, it is simpler to tune the controller in order to obtain non-oscillatory results theoretically and next rescale the control signal taking into account control limitations which affect the physical object.

2.3.2 Control Law Based on Vector Field Orientation Approach

Now, the problem of asymptotic stabilization (in the sense of Definition 1) will be considered. Without lack of generality we assume, that the reference point is at the origin:

$$q_t \triangleq [0\ 0\ 0]^T. \tag{2.69}$$

In the next section the Vector Field Orientation (VFO) approach will be described. It allows to derive a simple controller, which makes the posture error (2.4) asymptotically converge to zero. The VFO strategy for the unicycle robot was introduced for the first time in [12].

VFO Approach

The VFO concept comes directly from an intuitive geometrical interpretation of the structure of controlled kinematics (2.1) and its possible time evolution in response to specific controls $u_1(\cdot)$ and $u_2(\cdot)$. The main idea involves decomposition of Eq. (2.1) into two subsystems Σ_1 and Σ_2:

$$\Sigma_1: \quad \dot{\varphi} = u_1, \tag{2.70}$$

$$\Sigma_2: \quad \dot{q}^* = g_2^*(q)u_2, \quad \text{where} \quad \dot{q}^* \triangleq \begin{bmatrix} \dot{x} \\ \dot{y} \end{bmatrix}, \ g_2^*(q) \triangleq \begin{bmatrix} \cos\varphi \\ \sin\varphi \end{bmatrix}. \tag{2.71}$$

The first 1-D subsystem is linear. The second one (2-D) is highly nonlinear. One can find that the direction of time evolution of state variables x, y in $\mathbb{R}^2$ depends on the direction of vector field $g_2^*(q)$:

$$Dir\{\dot{q}^*\} = Dir\{g_2^*(q)\}, \tag{2.72}$$

where $\dot{q}^* \triangleq [\dot{x}\ \dot{y}]^T$ and $Dir\{\zeta\}$ denotes the direction of ζ in $\mathbb{R}^N$ (here in $\mathbb{R}^2$). Since both components of $g_2^*(q)$ depend on the first state variable φ, the current direction (and orientation[1]) of $g_2^*(q)$ can be changed by changing the actual value of φ. From (2.70) it results that this change can be accomplished relatively easily with the first input signal u_1. Due to the particular form of the vector field $g_2^*(q)$ in (2.71), all accessible directions in $\mathbb{R}^2$ as functions of φ variable include all possible directions on the plane. Therefore, one can say that $g_2^*(q)$ is fully orientable in $\mathbb{R}^2$. Since φ directly affects the orientation of $g_2^*(q)$, it can be called the *orienting variable*. Since input u_1 directly drives the orienting variable φ, it can be called the *orienting control*. It is easy to find that the second input u_2 drives the sub-state $q^* \triangleq [x\ y]^T$ along the current direction of $g_2^*(q)$. One can say that u_2 *pushes* the sub-state q^* along this vector field. Hence u_2 will be called the *pushing control*. The proposed interpretation and terminology allows to describe the VFO control methodology for the system described by (2.70) and (2.71). First we have to introduce an additional vector field

$$h(e(\tau)) \triangleq \begin{bmatrix} h_1(e(\tau)) \\ h_2(e(\tau)) \\ h_3(e(\tau)) \end{bmatrix} \triangleq \begin{bmatrix} h_1(e(\tau)) \\ h^*(e(\tau)) \end{bmatrix} \in \mathbb{R}^3, \tag{2.73}$$

---

[1] Strictly speaking, the *orientation* of some vector field ζ means its *direction* in $\mathbb{R}^N$ along with its sense.

which will be called the *convergence vector field*. Let us assume that this vector determines an instantaneous convergence direction (orientation) which should be followed by the controlled system to reach the reference goal point q_t. At the moment we assume that h is given. The VFO control strategy can be explained as follows.

Since h defines the convergence direction (and orientation), it is desirable to put the direction (orientation) of the generalized velocity vector field $\dot{q}$ of the controlled system (2.1) onto the direction of h. As will be shown, one can accomplish this task by the first input u_1. Simultaneously, the subsystem (2.71) should be pushed by the second input u_2, along the vector field $\dot{q}^*$ being oriented. Moreover, it seems to be reasonable to push the subsystem (2.71) only proportionally to the current orthogonal projection of $h^* \triangleq [h_2 \ h_3]^T$ onto the instantaneous direction of $g_2^*(q)$ (or of $\dot{q}^*$ due to (2.72)). Next, the whole vector field h should be designed, to guarantee tending of φ to its reference value at the limit as x and y reach their reference values. Mathematically, the VFO strategy explained above can be written in the following form:

$$find\ u_1 : \left\{ \lim_{\tau \to \infty} \dot{q}(\tau) \,||\, h(e(\tau)) \Leftrightarrow \lim_{\tau \to \infty} \dot{q}(\tau)\, k(\tau) = h(e(\tau)) \right\},$$

$$find\ u_2 : \left\{ \| \dot{q}^* \| \propto \| h^* \| \cos \alpha \right\},$$

where $k(\tau) \neq 0$ is a scalar function, $\alpha \triangleq \angle(g_2^*(q), h^*)$, the expression $a\,||\,b$ denotes that both vector fields a and b are parallel, and $\propto$ is a proportionality operator. According to the VFO strategy, the conditions which ensure matching of the directions of the two vector fields h and $\dot{q}$ will be derived. First of above relations can be rewritten as follows (for simplicity some arguments are omitted):

$$find\ u_1 : \left\{ \lim_{\tau \to \infty} \begin{bmatrix} \dot{\varphi}(\tau) \\ \dot{x}(\tau) \\ \dot{y}(\tau) \end{bmatrix} k(\tau) = \begin{bmatrix} h_1 \\ h_2 \\ h_3 \end{bmatrix} \overset{(2.1)}{\Rightarrow} \lim_{\tau \to \infty} \begin{bmatrix} u_1 k(\tau) \\ \cos \varphi(\tau) u_2 k(\tau) \\ \sin \varphi(\tau) u_2 k(\tau) \end{bmatrix} = \begin{bmatrix} h_1 \\ h_2 \\ h_3 \end{bmatrix} \right\}.$$

Combining the last two relations, one obtains two so-called *VFO orienting conditions*:

$$find\ u_1 : \left\{ \begin{array}{c} u_1\, k(\tau) = h_1 \\ \lim_{\tau \to \infty} \varphi(\tau) = \text{Atan2}\left(sgn(k(\tau))h_3,\ sgn(k(\tau))h_2 \right) \end{array} \right\}, \quad (2.74)$$

where Atan2 $(.,.)$ denotes the four-quadrant inverse tangent function and:

$$sgn(a) \triangleq \left\{ \begin{array}{l} 1,\ \text{for } a \geqslant 0, \\ -1,\ \text{for } a < 0. \end{array} \right. \quad (2.75)$$

Conditions (2.74) should be met to ensure placing the direction of $\dot{q}$ onto the direction of h and will be directly used in the sequel for design purposes of the first control signal u_1. Function $k(\tau)$, which appears in the above conditions is not needed to be known explicitly[2], however its sign is helpful to properly shape transient states of the whole control system. We postulate the following equality:

$$sgn(k(\tau)) \triangleq sgn(e_{20}), \tag{2.76}$$

where $e_{20} \equiv e_2(0)$ denotes the initial value of the error $e_2(\tau) \stackrel{(2.4)}{=} x_t - x(\tau)$.

VFO Controller

By an appropriate definition of signal u_1, the first relation in (2.74) can be fulfilled instantaneously. The second one, however, can be generally met only at the limit as $\tau \to \infty$. Hence we introduce an *auxiliary orientation variable*:

$$\varphi_d \triangleq \mathrm{Atan2}\,(sgn(e_{20})h_3,\, sgn(e_{20})h_2) \tag{2.77}$$

and the *auxiliary orientation error*:

$$e_{1d} \triangleq \varphi_d - \varphi. \tag{2.78}$$

Now, to meet the second condition in (2.74) it suffices to show that error e_{1d} tends to zero. Therefore, we propose to define the first component of the convergence vector as follows:

$$h_1(e(\tau)) \triangleq k(\tau)[k_1 e_{1d} + \dot{\varphi}_d], \tag{2.79}$$

where $k_1 > 0$ is a design coefficient and the feedforward term can be computed as follows

$$\dot{\varphi}_d \stackrel{(2.77)}{=} \frac{\dot{h}_3 h_2 - h_3 \dot{h}_2}{h_2^2 + h_3^2}, \qquad h_2^2 + h_3^2 \neq 0, \tag{2.80}$$

where the terms $\dot{h}_2$ and $\dot{h}_3$ are determined in the Appendix. Finally, to meet the first relation in (2.74), it suffices to take:

$$u_1 \triangleq \frac{h_1}{k(\tau)} \stackrel{(2.79)}{=} k_1 e_{1d} + \dot{\varphi}_d. \tag{2.81}$$

The derived control law (2.81) should guarantee that $\lim_{\tau \to \infty} e_{1d} = 0$ (this will be proved in the sequel).

Now the two last components h_2 and h_3 of the convergence vector field h will be defined. Let us introduce the following proposition:

$$h^*(e(\tau)) = \begin{bmatrix} h_2(e(\tau)) \\ h_3(e(\tau)) \end{bmatrix} \triangleq k_p e^* + \dot{q}_{vt}^*, \qquad e^* \triangleq \begin{bmatrix} e_2 \\ e_3 \end{bmatrix}, \tag{2.82}$$

---

[2] As will be shown, function $k(\tau)$ does not appear in the final definition of the control signals.

where $k_p > 0$ is a design coefficient. The last term $\dot{q}_{vt}^*$ is called the *virtual reference velocity* and is defined as follows:

$$\dot{q}_{vt}^* \overset{\Delta}{=} -\eta \, \| e^* \| \, sgn(e_{20}) \, g_{2t}^*, \qquad 0 < \eta < k_p, \qquad (2.83)$$

where

$$g_{2t}^* \overset{\Delta}{=} g_2^*(q_t) \overset{(2.71)}{=} \begin{bmatrix} \cos \varphi_t \\ \sin \varphi_t \end{bmatrix} \overset{(2.69)}{=} \begin{bmatrix} 1 \\ 0 \end{bmatrix}. \qquad (2.84)$$

It will be shown, that the additional coefficient η in (2.83) is helpful in shaping transient states (see Fig. 2.3).

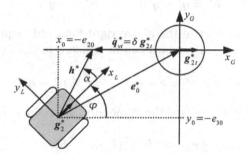

Fig. 2.3. Mobile robot during the VFO control process for $k_p = 1$, $q_t = [0\ 0\ 0]^T$, and $\delta = -\eta \, \| e^* \| \, sgn(e_{20})$

According to the VFO strategy it remains to define the pushing control u_2. Recalling the considerations of Section 2.3.2 we propose to take

$$u_2 \overset{\Delta}{=} \| h^* \| \cos \alpha, \qquad (2.85)$$

where $\alpha \overset{\Delta}{=} \angle(g_2^*, h^*)$ and hence

$$\cos \alpha \overset{\Delta}{=} \frac{g_2^{*T} h^*}{\| g_2^* \| \, \| h^* \|} \qquad \text{for} \quad \| h^* \| \neq 0. \qquad (2.86)$$

Substituting (2.86) into (2.85) allows to obtain a simpler form of control signal u_2:

$$u_2 = \frac{g_2^{*T} h^*}{\| g_2^* \|} \overset{(2.1)}{=} h_2 \cos \varphi + h_3 \sin \varphi. \qquad (2.87)$$

Now, we can formulate the following proposition.

Proposition 4 *Let there be given the reference point (2.69). Assuming that $e^*(0) \in \mathbb{R}^2 \setminus 0$ and $\forall_{\tau < \infty} \| h^*(e(\tau)) \| \neq 0$, the VFO controller (2.81) and (2.87) guarantees asymptotic convergence of the posture error (2.4) to zero as $\tau \to \infty$.*

Proof. First, let us consider the *orienting variable* φ behavior. Substituting (2.81) to (2.1) yields:

$$\dot{e}_{1d} + k_1 e_{1d} = 0. \tag{2.88}$$

One concludes that the orienting variable φ exponentially tends to the auxiliary direction angle φ_d:

$$\lim_{\tau \to \infty} e_{1d}(\tau) = 0. \tag{2.89}$$

Now we will take into account the position error e^*. For the posture stabilization task we have:

$$e^* \stackrel{\Delta}{=} q_t^* - q^* \quad \Rightarrow \quad \dot{e}^* = -\dot{q}^*. \tag{2.90}$$

Using (2.82) one can rewrite the above right-hand-side equation as follows: $\dot{e}^* = -\dot{q}^* + h^* - k_p e^* - \dot{q}_{vt}^*$, which can be ordered in the following way:

$$\dot{e}^* + k_p e^* = r - \dot{q}_{vt}^*, \quad r \stackrel{\Delta}{=} h^* - \dot{q}^*. \tag{2.91}$$

Making simple calculations (see the Appendix) one may derive the following useful relation:

$$\|r\|^2 = \|h^*\|^2 (1 - \cos^2 \alpha), \tag{2.92}$$

where $\cos \alpha$ is defined by (2.86) and (see the Appendix)

$$\lim_{\varphi \to \varphi_d} (1 - \cos^2 \alpha) = 0. \tag{2.93}$$

Now we introduce the following positive-definite Lyapunov function candidate: $V(e^*) \stackrel{\Delta}{=} \frac{1}{2} e^{*T} e^*$. The time derivative of this function can be estimated as follows (to simplify notation, we use $\gamma = \sqrt{1 - \cos^2 \alpha} \in [0, 1]$ and $\delta = -\eta \|e^*\| sgn(e_{20})$):

$$
\begin{aligned}
\dot{V} &= e^{*T} \dot{e}^* \stackrel{(2.91)}{=} e^{*T} (-k_p e^* + r - \dot{q}_{vt}^*) \stackrel{(2.83)}{=} e^{*T} (-k_p e^* + r - \delta g_{2t}^*) = \\
&= -k_p \|e^*\|^2 + e^{*T} r - \delta e^{*T} g_{2t}^* \leqslant -k_p \|e^*\|^2 + \|e^*\| \|r\| + |\delta| \|e^*\| \|g_{2t}^*\| = \\
&\stackrel{(2.84)}{=} -k_p \|e^*\|^2 + \|e^*\| \|r\| + |\delta| \|e^*\| = \\
&\stackrel{(2.92)}{=} -k_p \|e^*\|^2 + \|e^*\| \|h^*\| \gamma + |\delta| \|e^*\| = \\
&\stackrel{(2.82)}{=} -k_p \|e^*\|^2 + \|e^*\| [\|k_p e^* + \dot{q}_{vt}^*\| \gamma + |\delta|] \leqslant \\
&\leqslant -k_p \|e^*\|^2 + \|e^*\| [(k_p \|e^*\| + |\delta|)\gamma + |\delta|] = \\
&= -k_p(1 - \gamma) \|e^*\|^2 + |\delta|(1 + \gamma) \|e^*\| = -[k_p - k_p \gamma - \eta - \eta \gamma] \|e^*\|^2.
\end{aligned}
$$

The above time derivative is negative-definite, if the term in the brackets is positive. It gives the following convergence condition:

$$\gamma < (k_p - \eta)/(k_p + \eta) \quad \Rightarrow \quad \lim_{\tau \to \infty} \|e^*(\tau)\| \to 0. \tag{2.94}$$

Since $0 < \eta < k_p$ and relations (2.93) and (2.89) hold, one concludes:

$$\exists_{\tau_\gamma > 0} \; : \; \forall_{\tau > \tau_\gamma} \quad \gamma < (k_p - \eta)/(k_p + \eta) \tag{2.95}$$

and the norm $\| e^*(\tau) \|$ converges asymptotically (exponentially for $\tau > \tau_\gamma$) to zero as $\tau \to \infty$.

In the rest of the proof we consider the convergence of the robot orientation angle φ (called the orientation variable) to the reference angle $\varphi_t \overset{(2.69)}{=} 0$, at least at the limit $\| e^* \| \to 0$. Due to (2.89), to show the convergence of φ to zero it suffices to show the convergence to zero for φ_d (Eq. (2.77)). According to Definition (2.77) it suffices to show that component h_3 always tends to zero faster than component h_2 [19]. Recalling (2.82) and (2.83) we have for $\varphi_t = 0$:

$$h_2 \overset{\Delta}{=} k_p e_2 - \eta \| e^* \| \, sgn(e_{20}), \qquad h_3 \overset{\Delta}{=} k_p e_3. \tag{2.96}$$

Moreover, it is easy to show that (see the Appendix)

$$\lim_{\varphi \to \varphi_d} \begin{cases} \dot{x} = h_2, \\ \dot{y} = h_3, \end{cases} \overset{(2.90)}{\Rightarrow} \lim_{\varphi \to \varphi_d} \begin{cases} \dot{e}_2 = -h_2, \\ \dot{e}_3 = -h_3. \end{cases} \tag{2.97}$$

Substituting (2.96) into the above right-hand-side relations one obtains:

$$\lim_{\varphi \to \varphi_d} \begin{cases} \dot{e}_2 + k_p e_2 = \eta \| e^* \| \, sgn(e_{20}), \\ \dot{e}_3 + k_p e_3 = 0. \end{cases}$$

It is obvious that e_3 tends to zero faster than e_2. Taking into account (2.96) it is also clear that at the limit $\| e^* \| \to 0$ the component $h_3 \to 0$ always faster than h_2. Finally, one concludes:

$$\lim_{\| e^* \| \to 0} \varphi_d(e^*) \to 0 \overset{(2.89)}{\Rightarrow} \lim_{\tau \to \infty} \varphi(\tau) = 0 \overset{(2.69)}{\Rightarrow} \lim_{\tau \to \infty} e_1(\tau) = 0.$$

If $\forall_{\tau < \infty} \| h^*(\tau) \| \neq 0$, the term $\varphi_d \in \mathcal{L}_\infty$. Now, since h_2, h_3, φ and $\dot{\varphi} \in \mathcal{L}_\infty$, control signals (2.81) and (2.87) are bounded and $\lim_{\tau \to \infty} u_1(\tau), u_2(\tau) = 0$. $\square$

Remark 6. Fulfilling $\| h^* \| \neq 0$ during transient stage depends on the effectiveness of the orienting process – the shorter time interval τ_γ in (2.95), the earlier the convergence of e^* becomes exponential (the norm $\| h^* \|$ will not cross zero in finite time). Hence to guarantee that $\forall_{\tau < \infty} \| h^* \| \neq 0$, the following sequential strategy can be applied: **S1)** use the orienting control (2.81) together with $u_2 \equiv 0$ to fulfill condition (2.94), **S2)** use the complete VFO stabilizer given by Eqs. (2.81) and (2.87) ensuring the exponential convergence of $\| e^* \|$ to zero.

Due to (2.82), Definition (2.77) is not determined when the controlled vehicle is at the reference point q_t^*, which means $h_2 = h_3 = 0 \Rightarrow e_2 = e_3 = 0$.

Hence, point $e = 0$ is not an equilibrium of the closed loop system, and the proposed VFO controller can only be called the *almost stabilizer* (according to the terminology introduced in [3]). It is worth noting that the indeterminacy of the type $\varphi_d = \text{Atan2}(0,0)$ never occurs if the condition (2.95) is met, because, assuming $\| e^*(0) \| \neq 0$, it would only occur at the limit $\tau \to \infty$, which theoretically means never [1, 3]. Although from a practical point of view and in the case when $\| e^*(0) \| = 0$, it is desirable to introduce additional definitions of signals φ_d and $\dot{\varphi}_d$ at the origin. As a consequence, one introduces the following proposition of a discontinuous (piecewise continuous) asymptotic stabilizer, well-defined in the whole error space.

Proposition 5 (VFO stabilizer) *For the given reference point (2.69), the VFO controller (2.81) and (2.87) globally asymptotically stabilizes the point $e = 0$ if*

$$\varphi_d, \dot{\varphi}_d \triangleq \begin{cases} (2.77), (2.80) & \text{for } e^* \neq 0, \\ \varphi_t, \dot{\varphi}_t & \text{for } e^* = 0. \end{cases} \tag{2.98}$$

2.4 Control Limitations

In practice, limitations of control signals always exist. Hence, not all control values are feasible in a real system. Therefore, in this section control signal saturation will be explicitly taken into account and robustness of the considered stabilization controllers to this saturation will be examined. Now we define the limitations imposed on the inputs of the controlled kinematics (2.1).

Although in Eq. (2.1) it is assumed that the inputs to the system are, respectively, the angular, $u_1 = \Omega$, and linear, $u_2 = V$, velocities of the robot platform, in practice and in the case of differentially driven vehicle, one can physically affect only the configuration velocities, which are the left, ω_L, and right, ω_R, robot angular wheel velocities. Since both wheels simultaneously affect signals Ω and V, one cannot independently impose constant limitations Ω_{max} and V_{max} on these inputs (they are related as $\Omega_{max} = f(V_{max})$). Therefore, one has to define the limitations concerning configuration velocities ω_L and ω_R. Let us assume that both wheels and their drives are identical, with r denoting the wheel radius and b denoting the length of the wheel axle (see Fig. 2.1). Parameter $\omega_{max} > 0$ is the maximal feasible angular velocity of each wheel. The well-known linear relation between the inputs and respective configuration velocities is as follows:

$$u = \boldsymbol{\Phi}\omega, \tag{2.99}$$

where

$$u \triangleq \begin{bmatrix} u_1 \\ u_2 \end{bmatrix}, \quad \boldsymbol{\Phi} \triangleq \begin{bmatrix} r/b & -r/b \\ r/2 & r/2 \end{bmatrix}, \quad \omega \triangleq \begin{bmatrix} \omega_R \\ \omega_L \end{bmatrix}.$$

Denoting by $\boldsymbol{u}_c = [u_{1c}\ u_{2c}]^T$ the control vector computed (non-saturated) by one of the presented stabilizers, the computed (non-saturated) configuration velocities follow:

$$\boldsymbol{\omega}_c = \boldsymbol{\Phi}^{-1}\boldsymbol{u}_c, \tag{2.100}$$

where $\boldsymbol{\omega}_c = [\omega_{Rc}\ \omega_{Lc}]^T$. According to [11] (see also [2]), we propose the following scaling procedure, which guarantees fulfilling the configuration input limitations: $|\omega_R|,|\omega_L| \leqslant \omega_{max}$ and preserves the same direction of the computed control vector $\boldsymbol{u}_c$ and the rescaled (and limited) control vector[3] $\boldsymbol{u}_s$:

$$\boldsymbol{u}_s = \boldsymbol{\Phi}\boldsymbol{\omega}_s, \tag{2.101}$$

where $\boldsymbol{u}_s = [u_{1s}\ u_{2s}]^T$, $\boldsymbol{\omega}_s = [\omega_{Rs}\ \omega_{Ls}]^T$ and

$$s \overset{\Delta}{=} \max\left\{\frac{|\omega_{Rc}|}{\omega_{max}}, \frac{|\omega_{Lc}|}{\omega_{max}}\right\}, \qquad \boldsymbol{\omega}_s = \begin{cases} \boldsymbol{\omega}_c & \text{if } s \leqslant 1, \\ \frac{1}{s}\boldsymbol{\omega}_c & \text{if } s > 1. \end{cases} \tag{2.102}$$

Now, rescaled control vector $\boldsymbol{u}_s$ meets the control limitation $\Omega_{max} = f(V_{max})$ directly resulting from the value of ω_{max} and can be applied to system (2.1). In the next section the aforementioned control limitations will be taken into account in simulation tests. The performance of the presented controllers for the cases with and without limitations will be examined and compared.

2.5 Simulation Results

The effectiveness of the two proposed controllers will be illustrated by simulation results. The reference point is located at the origin: $\boldsymbol{q}_t = [0\ 0\ 0]^T$. Numerical simulations have been conducted within the time horizon of $T_h = 10[s]$ and for the following initial conditions: $\varphi(0) = 0$, $x(0) = 0$, and $y(0) = -3$ (parallel parking). The controllers have been tested for two cases: (A) without control signal limitations and (B) when practical limitations of wheel velocities have been imposed. In case (B) the following parameter values have been set[4]: wheel radius: $r = 0.026[m]$, axle length: $b = 0.066[m]$, and maximal wheel velocity: $\omega_{max} = 81[rad/s]$.

2.5.1 TVO Stabilizer

The parameters of the controller presented in Section 2.3.1 have been selected as: $k_1 = k_2 = 6$, $\boldsymbol{\xi}(0) = [0\ -1]^T$, $\psi_{10} = \pi$, $\psi_{20} = 4$, $\varepsilon_1 = \varepsilon_2 = 0.01$, $\alpha_1 = 2$ and $\alpha_2 = 3$.

---

[3] In the sense that $\boldsymbol{u}_s \,||\, \boldsymbol{u}_c$ or $\boldsymbol{u}_s = a\boldsymbol{u}_c$, where $0 < a \leqslant 1$.
[4] The real parameter values of the experimental mobile robot *MiniTracker 3* presented in [13] have been used.

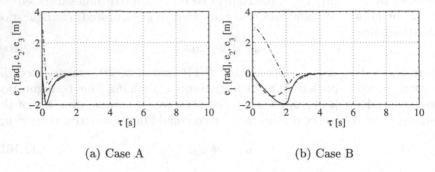

(a) Case A   (b) Case B

Fig. 2.4. Posture errors: e_1 (–), e_2 (- -), e_3 (-.-), TVO controller

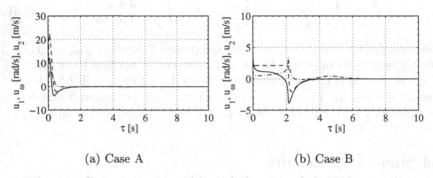

(a) Case A   (b) Case B

Fig. 2.5. Control signals: u_1 (–), u_2 (- -) and u_ω (-.-), TVO controller

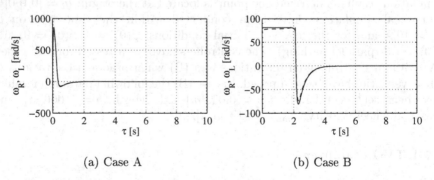

(a) Case A   (b) Case B

Fig. 2.6. Robot's wheel velocities: ω_R (–) and ω_L (- -), TVO controller

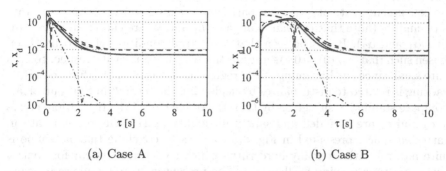

(a) Case A   (b) Case B

Fig. 2.7. Auxiliary signals – logarithmic value-scale: (black) x_1 (–), x_2 (- -), x_3 (-.-), (grey) x_{d1} (–), x_{d2} (- -), TVO controller

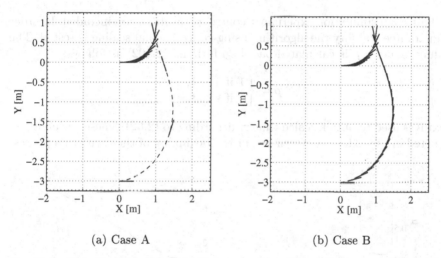

(a) Case A   (b) Case B

Fig. 2.8. Robot's path in the task space – parallel parking maneuvers, TVO controller

Case A

In Figs. 4(a)-8(a) the results of simulation concerning the TVO controller without control saturation are presented. From Fig. 4(a) it can be seen that the errors in Cartesian space converge exponentially to the neighborhood of the origin without significant overshoots. The lack of oscillatory behavior is ensured by making the initial values of the scaling functions ψ_1 and ψ_2 sufficiently high . In Fig. 5(a) the control signals: physical, $\boldsymbol{u}$, and virtual, u_ω, are presented. As one can see, although the initial values of u_ω are quite high, oscillations do not occur since the errors decrease fast. In Fig. 7(a) the evolution of the auxiliary signals $\boldsymbol{x}$ and $\boldsymbol{x}_d$ is depicted. It is interesting to note that vector x^* convergences to x_d while x_3 is driven to zero directly. Therefore, the convergence rate of $\|x\|$ is, in particular, related to the convergence rate of ψ_1 and ψ_2 (note that the values of gains k_1 and k_2 are chosen to be greater than α_1 and α_2). It can be seen that the initial condition of $\xi(0)$ has been chosen such that $x_{d1}(0) = 0$. As a result the tracking error with respect to z_1 is quite small since the beginning of the regulation process - note that no perfect tracking is related to disturbances by neglecting the term $\dot{\Psi}$ in the control law (compare stability result given by (2.58)). At the end of the regulation, errors e_1, e_2 and e_3 are bounded and satisfy inequality (2.41). The robot's path in Cartesian space, presented in Fig. 8(a), allows to conclude that its shape is quite natural without many hard turns which sometimes appear for control laws using time-varying feedback [3]. The presented strategy allows to avoid oscillatory behavior by proper tuning of the controller and is not very difficult.

Case B

In the second simulation experiment control saturations are included. In order to guarantee stability the algorithm using control signal scaling is tested. The scaling function μ is calculated based on formula (2.102) as follows:

$$\mu = \begin{cases} 1 & \text{if } s \leqslant 1, \\ \frac{1}{s} & \text{if } s > 1. \end{cases}$$

Next it is used to obtain scaled time τ_s according to (2.62). These variables are depicted in Fig. 2.9. As one can see, at the beginning of the regulation process

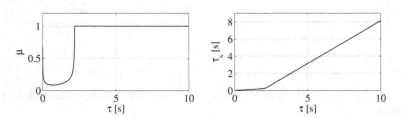

Fig. 2.9. Evolution of scaling function μ (left) and scaled time τ_s (right), TVO controller – **Case B**

the control signals are scaled significantly (about ten times with respect to the original values). After time 2.2[s] the signals are not scaled anywhere and the controller behaves as the original one. Comparing to the errors obtained in the previous case (see Figs. 4(a), 4(b), 7(a) and 7(b)) one can conclude that the presence of saturation makes the initial phase of the regulation slower. However, if the error values decrease and the control signal produced by nominal control law remains in a permissible range then the convergence rate is the same in both cases. It is very interesting to compare the paths of the robot. According to Figs. 8(a) and 8(b) one can see that the shapes of the trajectories are the same (independent on saturation), which is desirable from a practical point of view.

2.5.2 VFO Controller

The simulations with the VFO controller use the following set of parameters: $k_1 = 10, k_p = 5$, and $\eta = 4$. Since the continuous state variable $\varphi \in \mathbb{R}$ is not limited to the range $[-\pi, \pi)$, to avoid discontinuity in e_{1d} resulting from Definition (2.77), the continuous method of determining the auxiliary variable φ_d is applied. In the case of unlimited control signals (case (A)), this method can be treated as equivalent to the following formula: $\varphi_d(\tau) = \varphi_d(0) + \int_0^\tau \dot{\varphi}_d(\xi)d\xi$, where $\varphi_d(0)$ is computed by (2.77), $\dot{\varphi}_d$ is taken from (2.80), and the integral is computed numerically.

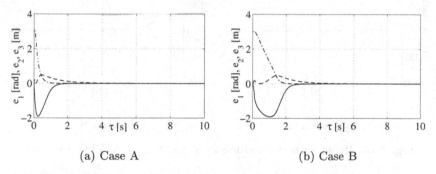

(a) Case A   (b) Case B

Fig. 2.10. Posture errors: e_1 (–), e_2 (- -), c_3 (-.-), VFO controller

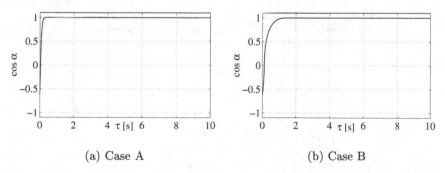

(a) Case A   (b) Case B

Fig. 2.11. Time plot of $\cos\alpha$, where $\alpha\angle(g_2^*, h^*)$, VFO controller

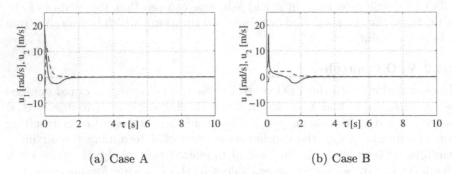

(a) Case A   (b) Case B

Fig. 2.12. Control signals: u_1 (–) and u_2 (- -), VFO controller

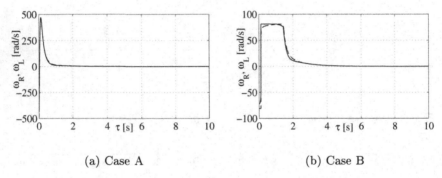

(a) Case A   (b) Case B

Fig. 2.13. Robot's wheel velocities: ω_R (–) and ω_L (- -), VFO controller

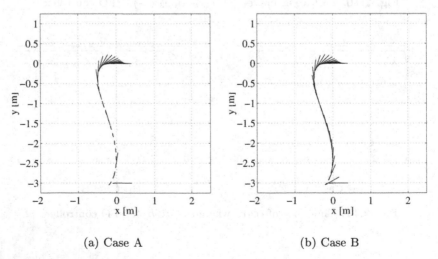

(a) Case A   (b) Case B

Fig. 2.14. Robot's path in the task space – parallel parking maneuvers, VFO controller

Case A

The control performance in the case without control signal limitations is presented in Fig. 10(a)-14(a). According to the time plots in Fig. 10(a) relatively fast error convergence can be seen. The orienting process is very effective since $\cos \alpha \approx +1$ after about $0.5[s]$. It should be noted that control signals u_1 and u_2 as well as ω_R and ω_L are bounded, non-oscillatory, and converge to zero (velocities ω_R and ω_L have been computed using (2.100), only for comparison with the time plots for case B). Figure 14(a) shows the resulted robot path in task space during the control process (the straight short lines denote the instantaneous vehicle orientation). One can find that the path is quite natural with only one switchback during the transient stage.

Case B

In this case the control signal limitation $\omega_{max} = 81[rad/s]$ is explicitly specified. The *nominal* equations of the VFO controller (2.81) and (2.85) are followed by the scaling procedure (2.100) (2.102) to limit the computed inputs[5]. The behavior of all signals in this case is presented in Figs. 10(b)-14(b). Error convergence is slower than in case A, but also relatively fast. The orienting process is less effective but $\cos \alpha \approx +1$ after about $1.5[s]$. Comparing Figs. 12(a) and 12(b) it is clear that control signals u_{1s} and u_{2s} are limited and explicit limitation regarding the value of ω_{max} can be observed in Fig. 13(b). It is interesting to note that the resulting path shape in task space is preserved in comparison with case A. Only the vehicle velocities along the path are rescaled (see the different densities of the short lines denoting the robot orientation in Figs. 14(a) and 14(b)). This feature seems to be very desirable from a practical point of view.

2.6 Conclusions

In this paper two point-stabilization strategies for a unicycle mobile robot have been presented and numerically tested. The first control strategy comes from the kinematic oscillator and transverse functions concept, which assures practical stability for the cart's posture error. The second control algorithm results from the simple geometrical interpretation and decomposition of a control task into orienting and pushing subtasks. The latter strategy leads to the time invariant discontinuous VFO controller, which guarantees asymptotic posture error convergence to zero. The practical aspect concerning the control signal limitations imposed directly on the vehicle wheel velocities has been considered. Simulations have been conducted for two cases: without limitations of input signals and with imposed limitations. A control signal

---

[5] Note that terms $\dot{h}_2$ and $\dot{h}_3$ which appear in Eq. (2.107) are determined also in case B with the *nominal* control $u_2 \equiv u_{2c}$ (not with the rescaled one u_{2s}). Hence, in case B, the feedforward term in Definition (2.81) should be denoted $\dot{\varphi}_{dc}$ rather than $\dot{\varphi}_d$. Note also that in case B $\dot{\varphi}_{dc} = \Gamma(\tau)\dot{\varphi}_d$, and $\Gamma(\tau) \equiv 1$ if $s \leqslant 1$.

scaling procedure has been proposed. The obtained performance has been compared for the two cases and two controllers. Future work will be focused on experimental validation of the presented strategies.

Appendix

────────── Derivation of relation (2.22). Recalling that J is a skew-symmetric matrix, $J^T = -J$, $\eta^T J \eta = 0$, where $\eta \in \mathbb{R}^2$, taking the time derivative of (2.13) and then using (2.20) results in

$$\dot{z}_3 = \left(x^{*T} + x_d^T\right) Jv + \dot{x}_d^T Jx^* = \left(x^{*T} + x_d^T\right) J\left[-k_1\left(x^* - x_d\right) + \dot{x}_d\right] + \dot{x}_d^T Jx^* =$$
$$= k_1\left(-x^{*T} Jx^* + x_d^T Jx_d\right) + k_1\left(-x_d^T Jx^* + x^{*T} Jx_d\right) +$$
$$+ x^{*T} J\dot{x}_d + \dot{x}_d Jx^* + x_d^T J\dot{x}_d = x_d^T J\dot{x}_d + 2k_1 x^{*T} Jx_d$$

────────── Derivation of (2.56).

Let us assume that a positive definite scalar function is written as

$$\dot{V} = -kV + \epsilon \exp\left(-\alpha\tau\right), \qquad (2.103)$$

where $k, \epsilon > 0$ are some scalars. Concerning the homogeneous ordinary differential equation $\dot{V} + kV = 0$ one can easily obtain that $V = C \exp\left(-k\tau\right)$, where C denotes some constant. Next, using the method of variation of parameters one can write that $V = C\left(\tau\right) \exp\left(-k\tau\right)$ and substitute this solution to (2.103), which yields

$$\dot{C}\left(\tau\right) = \epsilon \exp\left[\left(-\alpha + k\right)\tau\right]. \qquad (2.104)$$

Consequently, integrating term $\dot{C}\left(\tau\right)$ results in the following solution to (2.103)

$$V\left(\tau\right) = \begin{cases} V\left(0\right)\exp\left(-k\tau\right) + \dfrac{\epsilon}{k - \alpha}\left[\exp\left(-\alpha\tau\right) - \exp\left(-k\tau\right)\right] & \text{for } \alpha \neq k, \\ V\left(0\right)\exp\left(-k\tau\right) + \epsilon\tau \exp\left(-k\tau\right) & \text{for } \alpha = k. \end{cases} \qquad (2.105)$$

────────── Derivation of (2.92).

$$r \overset{(2.91)}{=} h^* - \dot{q}^* \overset{(2.85)}{=} \|h^*\| \begin{bmatrix} \dfrac{h_2}{\|h^*\|} - \cos\alpha\cos\varphi \\ \dfrac{h_3}{\|h^*\|} - \cos\alpha\sin\varphi \end{bmatrix}.$$

Now we can compute (assuming $\|h^*\| \neq 0$):

$$\|r\|^2 = \|h^*\|^2 \left[\frac{h_2^2}{\|h^*\|^2} - \frac{2h_2\cos\alpha\cos\varphi}{\|h^*\|} + \left(\cos\alpha\cos\varphi\right)^2 + \right.$$
$$+ \left.\frac{h_3^2}{\|h^*\|^2} - \frac{2h_3\cos\alpha\sin\varphi}{\|h^*\|} + \left(\cos\alpha\sin\varphi\right)^2\right] =$$
$$= \|h^*\|^2 \left[1 - 2\cos\alpha\frac{h_2\cos\varphi + h_3\sin\varphi}{\|h^*\|} + \cos^2\alpha\right] =$$
$$\overset{(2.86)}{=} \|h^*\|^2 \left(1 - 2\cos\alpha\cos\alpha + \cos^2\alpha\right) = \|h^*\|^2 \left(1 - \cos^2\alpha\right).$$

────────── Derivation of (2.93).

$$1 - \cos^2\alpha \overset{(2.86)}{=} 1 - \frac{\left(h_2\cos\varphi + h_3\sin\varphi\right)^2}{\|h^*\|^2 \|g_2^*\|^2} = \frac{h_2^2 + h_3^2 - \left(h_2\cos\varphi + h_3\sin\varphi\right)^2}{h_2^2 + h_3^2} =$$
$$= \frac{\left(h_2\sin\varphi - h_3\cos\varphi\right)^2}{h_2^2 + h_3^2}.$$

Assuming that $\varphi \to \varphi_d$, we get $\tan \varphi \overset{(2.77)}{\to} h_3/h_2$. Using this in the above formula we easily obtain (2.93).

───────────── Derivation of (2.97).

$$\begin{bmatrix} \dot{x} \\ \dot{y} \end{bmatrix} \overset{(2.71)}{=} g_2^*(\varphi)\, u_2 \overset{(2.87)}{=} \begin{bmatrix} \cos \varphi \\ \sin \varphi \end{bmatrix} g_2^{*T} h^* = \begin{bmatrix} h_2 \cos^2 \varphi + h_3 \cos \varphi \sin \varphi \\ h_2 \sin \varphi \cos \varphi + h_3 \sin^2 \varphi \end{bmatrix}.$$

At the limit $\varphi \to \varphi_d$ one gets:

$$\begin{bmatrix} \dot{x} \\ \dot{y} \end{bmatrix} = \begin{bmatrix} h_2 \cos^2 \varphi_d + h_3 \cos \varphi_d \sin \varphi_d \\ h_2 \sin \varphi_d \cos \varphi_d + h_3 \sin^2 \varphi_d \end{bmatrix}, \text{ and} \qquad (2.106)$$

$$\tan \varphi_d \overset{(2.77)}{=} \frac{h_3}{h_2} \;\Rightarrow\; \cos \varphi_d = \frac{h_2 \sin \varphi_d}{h_3}, \quad \sin \varphi_d = \frac{h_3 \cos \varphi_d}{h_2}.$$

Substituting formulas for $\cos \varphi_d$ and $\sin \varphi_d$ into the appropriate elements in (2.106) one gets (at the limit $\varphi \to \varphi_d$):

$$\begin{bmatrix} \dot{x} \\ \dot{y} \end{bmatrix} = \begin{bmatrix} h_2 \cos^2 \varphi_d + h_3 \frac{h_2 \sin \varphi_d}{h_3} \sin \varphi_d \\ h_2 \frac{h_3 \cos \varphi_d}{h_2} \cos \varphi_d + h_3 \sin^2 \varphi_d \end{bmatrix} = \begin{bmatrix} h_2 \\ h_3 \end{bmatrix}.$$

Recalling that $\dot{e}_2 = -\dot{x}$, $\dot{e}_3 = -\dot{y}$, relation (2.97) easily follows.

───────────── Now components $\dot{h}_2$ and $\dot{h}_3$ which appear in (2.80) will be described. Recalling Definitions (2.82) and (2.83) and assuming (2.69), simple computations give:

$$\dot{h}_2 = -k_p \dot{x} - \eta\, sgn(e_{20}) \frac{x\dot{x} + y\dot{y}}{\sqrt{x^2 + y^2}} \overset{(2.1)}{=} -u_2 \left(k_p \cos \varphi + \eta\, sgn(e_{20}) \frac{x \cos \varphi + y \sin \varphi}{\sqrt{x^2 + y^2}} \right),$$

$$\dot{h}_3 = -k_p \dot{y} \overset{(2.1)}{=} -u_2 (k_p \sin \varphi),$$

$$(2.107)$$

where u_2 is computed by (2.87).

References

1. M. Aicardi, G. Casalino, A. Bicchi, and A. Balestrino. Closed loop steering of unicycle-like vehicles via Lyapunov techniques. *IEEE Robotics and Automation Magazine*, 2:27–35, 1995.
2. G. Artus, P. Morin, and C. Samson. Tracking of an omnidirectional target with a unicycle-like robot: control design and experimental results. Technical Report 4849, INRIA, Sophia Antipolis, France, 2003.
3. A. Astolfi. *Asymptotic stabilization of nonholonomic systems with discontinuous control*. PhD thesis, Swiss Federal Institute of Technology, Zurich, 1996.
4. A. M. Bloch. *Nonholonomic mechanics and control*. Systems and Control. Springer, New York, 2003.
5. R. W. Brockett. Asymptotic stability and feedback stabilization. In R. W. Brockett, R. S. Millman, and H. H. Sussmann, editors, *Differential Geometric Control Theory*, pages 181–191. Birkhäuser, Boston, 1983.
6. R. W. Brockett, R. S. Millman, and H. J. Sussmann. *Differential Geometric Control Theory*. Birkhäuser, Boston, 1982.

7. C. Canudas de Wit, H. Khennouf, C. Samson, and O.J. Sørdalen. Nonlinear control design for mobile robots. In Y.F. Zheng, editor, *Recent Trends in Mobile Robots*, volume 11, chapter 5, pages 121–156. World Scientific, Singapore, 1993.
8. C. Canudas de Wit, B. Siciliano, and G. Bastin. *Theory of Robot Control*. Springer-Verlag, New York, 1996.
9. W. E. Dixon, D. M. Dawson, E. Zergeroglu, and A. Behal. *Nonlinear control of wheeled mobile robots*. Springer, London, 2001.
10. I. Dulęba. *Algorithms of motion planning for nonholonomic robots*. Wrocław University of Technology Publishing House, Wrocław, 1998.
11. P. Dutkiewicz, M. Michalski, and M. Michałek. Robust tracking with control vector constraints. In *Proceedings of the Second International Workshop On Robot Motion and Control*, pages 169–174, Bukowy Dworek, Poland, 2001.
12. M. Michałek and K. Kozłowski. Control of nonholonomic mobile robot with vector field orientation. In K. Tchoń, editor, *Advances of Robotics*, chapter 4, pages 235–246. Transport and Communication Publishers, Warsaw, 2005 (in Polish).
13. T. Jedwabny, M. Kowalski, M. Kiełczewski, M. Ławniczak, M. Michalski, M. Michałek, D. Pazderski, and K. Kozłowski. Nonholonomic mobile robot MiniTracker 3 for research and educational purposes. In *35th International Symposium on Robotics*, 2004.
14. I. Kolmanovsky and N. H. McClamroch. Developments in nonholonomic control problems. *IEEE Control Systems Magazine*, 15(6):20–36, 1995.
15. A. De Luca and G. Oriolo. Modeling and control of nonholonomic mechanical systems. In J. Angeles and A. Kecskementhy, editors, *Kinematics and Dynamics of Multi-Body Systems*, chapter 7, pages 277–342. Springer-Verlag, Wien, 1995.
16. P. Morin and C. Samson. Field oriented control of induction motors by application of the transverse function control approach. In *42nd IEEE Conference on Decision and Control*, pages 5921–5926, 2003.
17. P. Morin and C. Samson. Practical stabilization of driftless systems on Lie groups: the transverse function approach. *IEEE Transactions on Automatic Control*, 48(9):1496–1508, September 2003.
18. P. Morin and C. Samson. Trajectory tracking for non-holonomic vehicles: overview and case study. In *Proceedings of the 4th International Workshop On Robot Motion and Control*, pages 139–153, Puszczykowo, 2004.
19. G. Oriolo, A. De Luca, and M. Venditteli. WMR control via dynamic feedback linearization: design, implementation and experimental validation. *IEEE Transactions on Control System Technology*, pages 835–852, November 2002.
20. K. Kozłowski and D. Pazderski. Modelling and control of 4-wheel skid-steering mobile robot. *International Journal of Applied Mathematics and Computer Sciences*, 14(4):477–496, 2004.
21. D. Pazderski and K. Kozłowski. Practical stabilization of two-wheel mobile robot with velocity limitations using time-varying control law. In *Fith International Workshop on Robot Motion and Control*, pages 205–212, Dymaczewo, Poland, 2005.
22. C. Samson. Control of chained systems. Application to path following and time-varying point-stabilization of mobile robots. *IEEE Transactions on Automatic Control*, pages 64–77, January 1995.

3

Trajectory Tracking Control for Nonholonomic Mobile Manipulators

Alicja Mazur and Krzysztof Arent

Institute of Engineering Cybernetics, Wrocław University of Technology,
ul. Janiszewskiego 11/17, 50-372 Wrocław, Poland
{alicja|arent}@ict.pwr.wroc.pl

3.1 Introduction

In the paper we present new control algorithms for a special class of mobile manipulators, namely for nonholonomic mobile manipulators. A mobile manipulator is defined as a robotic system composed of a mobile platform and a manipulator mounted on the platform equipped with non–deformable wheels. Such a combined system is able to perform manipulation tasks in a much larger workspace than a fixed-base manipulator. Taking into account the type of mobility of their components, there are 4 possible configurations: type (h, h) – with both the platform and the manipulator holonomic, type (h, nh) – a holonomic platform with a nonholonomic manipulator, type (nh, h) – a nonholonomic platform with a holonomic manipulator, and finally type (nh, nh) – both the platform and the manipulator nonholonomic. The notion *doubly nonholonomic* manipulator was introduced in [13] for the type (nh, nh). The rigid manipulator can be a holonomic or a nonholonomic system depending on its construction.

In our considerations we focus on the latter two types of mobile manipulators, namely the mobile manipulator with nonholonomic platform only (nh, h) and the doubly nonholonomic mobile manipulator.

The problem of design of a control law for rigid robotic manipulators received much attention in late eighties and nineties of the last century. The algorithms were divided into two classes: the computed-torque algorithms, which came from the theory of linearization, see e.g. [8,14] and the passivity-based algorithms, which explored the passivity property of mechanical system, see e.g. [10,11]. So far in all considerations authors assumed that the every degree of freedom had independent linear engine – direct drive.

Recently, a new approach to the problem of robot drive has been proposed. In [7] Nakamura, Chung and Sørdalen presented a new nonholonomic mechanical gear, which is able to transmit velocities from the inputs to many passive joints (i.e. without actuators). In [7] the prototype of the nonholonomic manipulator was introduced and discussed. The nonholonomic constraints of the gear appear due to the rolling contact without slipping between balls of gear and special supporting wheels in the robot joints. Because the construction of the nonholonomic

K. Kozłowski (Ed.): Robot Motion and Control, LNCIS 335, pp. 55–71, 2006.
© Springer-Verlag London Limited 2006

manipulator is novel, only a few papers dealing with control problem for such an object have appeared, see e.g. [4, 6, 7].

This paper is devoted to the problem of trajectory tracking of nonholonomic mobile manipulators. Few authors have brought their attention to the control problem of such objects [5, 12, 13]. We present two versions of the tracking control algorithm – for mobile manipulator with nonholonomic platform only and for doubly nonholonomic mobile manipulator. We want to compare the control process of them using simulation and make some suggestion about a choice of drives for the robotic arm.

3.2 Nonholonomic Constraints

A mobile manipulator consists of two subsystems: a mobile platform and a rigid manipulator. The motion of the mobile platform is determined by nonholonomic constraints which describe a rolling-without-slipping assumption, while the rigid manipulator can be a holonomic or nonholonomic system – it depends on the construction.

In further considerations we focus on a selected mobile manipulator, namely a 3-pendulum mounted atop a mobile platform of the (2,0) class, called *unicycle*.

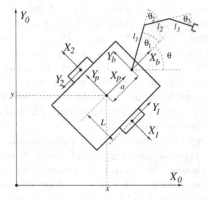

Fig. 3.1. Mobile manipulator: 3-pendulum on a platform of (2,0) class

We want to consider the influence of the nonholonomic gears designed by Nakamura, Chung and Sørdalen on the behavior of such a mobile manipulator and on the formulation of a control problem.

3.2.1 Kinematics of the Nonholonomic Mobile Platform of (2,0) Class

The motion of a mobile platform can be described using generalized coordinates $q_m \in R^n$ and generalized velocities $\dot{q}_m \in R^n$. The wheeled mobile platform should move without slipping of its wheels. This assumption implies the existence of l ($l < n$) independent nonholonomic constraints in the so-called Pfaff form

$$A_1(q_m)\dot{q}_m = 0, \tag{3.1}$$

where $A_1(q_m)$ is a full rank matrix of $(l \times n)$ size. Due to (3.1), since the platform velocity is always in the null space of A_1, it is always possible to find a vector of special auxiliary velocities $u \in R^m$, $m = n - l$, such that

$$\dot{q}_m = G_1(q_m)u, \tag{3.2}$$

where $G_1(q_m)$ is an $n \times m$ full rank matrix satisfying $A_1(q_m)G_1(q_m) = 0$.

Now we restrict our considerations to the nonholonomic mobile platform of $(2,0)$ class presented in Fig. 3.1. The nonholonomic constraints for such a platform can be expressed as follows

$$\dot{\xi} = \begin{pmatrix} \dot{x} \\ \dot{y} \\ \dot{\theta} \end{pmatrix} = \begin{pmatrix} v\cos\theta \\ v\sin\theta \\ w \end{pmatrix} = \begin{bmatrix} \cos\theta & 0 \\ \sin\theta & 0 \\ 0 & 1 \end{bmatrix} \begin{pmatrix} v \\ w \end{pmatrix}, \tag{3.3}$$

where (x, y) denote Cartesian coordinates of the mass center of the mobile platform in basic frame X_0Y_0, θ is the orientation of the platform – the angle between local frame X_pY_p and the basic frame – and L is a half of the platform width. Symbols v and w denote linear and angular velocity of the platform, respectively. We will call ξ the *posture coordinates* of the platform.

3.2.2 Kinematics of the Nonholonomic 3-pendulum

In [7] a new approach to the problem of robot drive has been presented. The authors have designed a new nonholonomic mechanical gear which is able to transmit velocities to many passive joints, see Fig. 3.2.

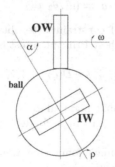

Fig. 3.2. Schematic of nonholonomic gear seen from above

The basic components of the gear are a ball and at least two wheels – an input wheel IW and an output wheel OW. The input wheel IW rotates about a fixed axis with a given angular velocity $\dot{\rho} = \eta_2$ and makes the ball rotate. The output wheel OW is driven by the ball and, using mechanical transmissions (e.g. shafts and gears or belts and pulleys) transmits velocities to the next joints of the manipulator which are not restricted to be planar. The nonholonomic constraints for the gears appear due to the rolling motion without slipping between the ball and the wheels in any gear in the robot joints.

The relationship between the joint velocities and driving velocities of the gears for the nonholonomic 3-pendulum presented in Fig. 3.1 can be described by the following matrix equation

$$\dot{q}_r = G_2(q_r)\eta,$$

or in more detail

$$\dot{q}_r = \begin{pmatrix} \dot{\theta}_1 \\ \dot{\theta}_2 \\ \dot{\theta}_3 \end{pmatrix} = \begin{bmatrix} 1 & 0 \\ 0 & a_2 \sin\theta_1 \\ 0 & a_3 \sin\theta_2 \cos\theta_1 \end{bmatrix} \begin{pmatrix} \eta_1 \\ \eta_2 \end{pmatrix}, \tag{3.4}$$

where θ_i is ith joint coordinate of the 3-pendulum, and constants a_2 and a_3 describe the gear ratio for transmission from the OW wheel in ith joint to the IW wheel in jth joint. The angular velocity of the first joint η_1 and the angular velocity of the driving input wheel IW η_2 play the role of control inputs to the kinematic equations (3.4).

3.3 Mathematical Model of a Nonholonomic Mobile Manipulator

3.3.1 Dynamics of a Mobile Manipulator with a Nonholonomic Platform

Let a vector of generalized coordinates of the mobile manipulator be denoted $q^T = (q_m^T, q_r^T)$, where q_m is a vector of mobile platform coordinates and q_r denotes a vector of joint coordinates of the rigid manipulator, i.e. the rigid 3-pendulum

$$q_r = (\theta_1, \theta_2, \theta_3)^T. \tag{3.5}$$

Because of the nonholonomy of constraints, to obtain the dynamic model of the mobile manipulator the d'Alembert Principle must be used

$$Q(q)\ddot{q} + C(q, \dot{q})\dot{q} + D(q) = A_{11}(q_m)\lambda + B\tau,$$

where:
$Q(q)$ – inertia matrix of the mobile manipulator,
$C(q, \dot{q})$ – matrix of Coriolis and centrifugal forces for the mobile manipulator,
$D(q)$ – vector of gravity,
$A_{11}(q_m)$ – matrix describing nonholonomic constraints,
λ – vector of Lagrange multipliers,
B – input matrix,
τ – vector of controls.

This model of dynamics can be expressed in more detail as

$$\begin{bmatrix} Q_{11} & Q_{12} \\ Q_{21} & Q_{22} \end{bmatrix} \begin{pmatrix} \ddot{q}_m \\ \ddot{q}_r \end{pmatrix} + \begin{bmatrix} C_{11} & C_{12} \\ C_{21} & C_{22} \end{bmatrix} \begin{pmatrix} \dot{q}_m \\ \dot{q}_r \end{pmatrix} + \begin{pmatrix} 0 \\ D_2 \end{pmatrix} = \begin{bmatrix} A_1^T \\ 0 \end{bmatrix} \lambda + \begin{bmatrix} B_{11} & 0 \\ 0 & I \end{bmatrix} \begin{pmatrix} \tau_m \\ \tau_r \end{pmatrix},$$

where A_1 is the Pfaff matrix for the mobile platform defined by (3.3), D_2 is a vector of gravity for the manipulator only, B_{11} describe which variables of the mobile platform are directly driven by the actuators, τ_m and τ_r are direct drives of the joints of the rigid manipulator. The above model of dynamics will be called a model in generalized coordinates.

Now we want to express the model of dynamics using auxiliary velocities (3.2) for the mobile platform. We compute

$$\ddot{q}_m = G_1(q_m)\dot{u} + \dot{G}_1(q_m)u,$$

and eliminate in the model of dynamics the vector of Lagrange multipliers (using the condition $G_1^T A_1^T = 0$) by left-sided multiplying the mobile platform equations by G_1^T matrix. After substituting for $\dot{q}_m$ and $\ddot{q}_m$ we get

$$\begin{bmatrix} G_1^T Q_{11} G_1 & G_1^T Q_{12} \\ Q_{21} G_1 & Q_{22} \end{bmatrix} \begin{pmatrix} \dot{u} \\ \ddot{q}_r \end{pmatrix} + \begin{bmatrix} G_1^T \left(C_{11} G_1 + Q_{11} \dot{G}_1 \right) & G_1^T C_{12} \\ C_{21} G_1 & C_{22} \end{bmatrix} \begin{pmatrix} u \\ \dot{q}_r \end{pmatrix} +$$

$$+ \begin{pmatrix} 0 \\ D_2 \end{pmatrix} = \begin{bmatrix} G_1^T B_{11} & 0 \\ 0 & I \end{bmatrix} \begin{pmatrix} \tau_m \\ \tau_r \end{pmatrix}, \tag{3.6}$$

or the same equations in a more compact form

$$Q^* \begin{pmatrix} \dot{u} \\ \ddot{q}_r \end{pmatrix} + C^* \begin{pmatrix} u \\ \dot{q}_r \end{pmatrix} + D^* = B^* \begin{pmatrix} \tau_m \\ \tau_r \end{pmatrix}. \tag{3.7}$$

The model (3.7) of the mobile manipulator dynamics expressed in auxiliary variables will be a point of departure for designing a new dynamic control algorithm.

3.3.2 Dynamics of a Doubly Nonholonomic Mobile Manipulator

The constraints (3.3) and (3.4) appearing in the doubly nonholonomic mobile manipulator are non-integrable; therefore to obtain a dynamical model of considered control system, we use the d'Alembert Principle

$$Q(q)\ddot{q} + C(q, \dot{q})\dot{q} + D(q) = A_{11}(q_m)\lambda_1 + A_{21}(q_r)\lambda_2 + B\tau, \tag{3.8}$$

where $A_{21}(q_r)$ is the matrix of nonholonomic constraints for the manipulator and B is a new input matrix defined in the following way

$$A_{21} = \begin{pmatrix} A_2^T(q_r) \\ 0 \end{pmatrix}, \qquad B = \begin{bmatrix} B_{11} & 0 \\ 0 & B_{22} \end{bmatrix}.$$

The submatrix B_{22} describes which coordinates of the nonholonomic manipulator are directly driven by the gear. Other elements of the model are defined similarly to symbols in section 3.3.1.

We express the kinematics (3.3) and (3.4) in one matrix equation as follows

$$\dot{q} = \begin{pmatrix} \dot{q}_m \\ \dot{q}_r \end{pmatrix} = \begin{bmatrix} G_1 & 0 \\ 0 & G_2 \end{bmatrix} \begin{pmatrix} u \\ \eta \end{pmatrix} = G\zeta \tag{3.9}$$

with $\zeta^T = (u^T, \eta^T)$. After substituting (3.9) into (3.8) we obtain

$$\bar{Q}\dot{\zeta} + \bar{C}\zeta + \bar{D} = \bar{B}\tau, \tag{3.10}$$

where

$$\bar{Q} = G^T Q G, \qquad \bar{C} = G^T (Q\dot{G} + CG), \qquad \bar{D} = G^T D, \qquad \bar{B} = G^T B.$$

We will refer to (3.10) as the model of dynamics in auxiliary coordinates.

3.4 Control Problem Statement

In the paper, our goal is to find control laws guaranteeing the proper cooperation between the mobile platform and the rigid manipulator mounted atop of it. We will assume that the desired task for the mobile manipulator can be decomposed into two independent parts: the rigid manipulator has to follow a desired admissible trajectory $q_{rd}(t)$, which defines a task of this subsystem, and the task of the platform is to follow a desired admissible trajectory $\xi_d(t)$. Our purpose will be to address the following control problem for mobile manipulators:

> Find control laws $\tau^T = (\tau_m^T, \tau_r^T)$ such that a mobile manipulator with fully known dynamics accomplishes the desired tasks for both subsystems, i.e. that trajectory tracking errors converge asymptotically to zero, in the presence of the nonholonomic constraints.

Additionally, we assume that the desired trajectories $q_{rd}(t), \xi_d(t)$ and their first and second time derivatives are bounded.

In order to design the tracking controller for the mobile manipulator it is necessary to observe that the complete mathematical model of the nonholonomic system is a cascade consisting of two subsystems. For this reason the structure of the controller is divided into two parts:

1. kinematic controller ζ_r – represents the embedded control inputs, which ensure realization of the task for the kinematics if the dynamics were not present. Such the controller generates 'velocity profiles' which can be executed in practice to realize the tracking task for nonholonomic subsystems.
2. dynamic controller – as a consequence of the cascaded structure of the system model, the system's velocities cannot be commanded directly, as it is assumed in the design of kinematic control signals, and instead they must be realized as the output of the dynamics driven by τ.

Because there exists a difference between the real velocities of the nonholonomic subsystem ζ and the embedded control input generated by the kinematic controller ζ_r, it is necessary to take into account the influence of the error $e_\zeta = \zeta - \zeta_r$ on the behaviour of the whole nonholonomic mobile manipulator.

3.5 Kinematic Control Algorithms

3.5.1 Kinematic Controller for the Mobile Platform – Samson & Ait-Abderrahim algorithm

In this work we use the concept of the trajectory tracking in the meaning of the posture tracking.

The desired trajectory for mobile platform has to satisfy nonholonomic constraints (3.3)

$$\dot{\xi}_d = \begin{pmatrix} \dot{x}_d \\ \dot{y}_d \\ \dot{\theta}_d \end{pmatrix} = \begin{pmatrix} v_d \cos \theta_d \\ v_d \sin \theta_d \\ w_d \end{pmatrix} . \tag{3.11}$$

We assume that $\xi_d(t)$ is time-variant ($\dot{\xi}_d \neq 0$); otherwise the task is reduced to the point stabilization and is considered as a separate task. Notice that non-reduction of the problem means that either v_d or w_d must remain different from zero.

Our goal is to achieve asymptotic convergence of standard tracking errors

$$e_b = \xi(t) - \xi_d(t) = \begin{pmatrix} e_x \\ e_y \\ e_\theta \end{pmatrix} = \begin{pmatrix} x - x_d \\ y - y_d \\ \theta - \theta_d \end{pmatrix}. \tag{3.12}$$

To simplify further considerations, we introduce other variables, so-called reference tracking errors e_m defined as

$$e_m := \begin{pmatrix} x_e \\ y_e \\ \theta_e \end{pmatrix} = \mathrm{Rot}(z, -\theta) \begin{pmatrix} e_x \\ e_y \\ e_\theta \end{pmatrix} = \begin{bmatrix} \cos\theta & \sin\theta & 0 \\ -\sin\theta & \cos\theta & 0 \\ 0 & 0 & 1 \end{bmatrix} \begin{pmatrix} e_x \\ e_y \\ e_\theta \end{pmatrix}. \tag{3.13}$$

It is obvious that due to non-singularity of the rotation matrix, asymptotic convergence of reference tracking errors e_m implies asymptotic convergence of standard tracking errors e_b.

Now we express the kinematics of the platform using e_m errors

$$\dot{e}_m = \begin{pmatrix} \dot{x}_e \\ \dot{y}_e \\ \dot{\theta}_e \end{pmatrix} = \begin{pmatrix} wy_e + v - v_d\cos\theta_e \\ -wx_e + v_d\sin\theta_e \\ w - w_d \end{pmatrix}, \tag{3.14}$$

where $u = [v, w]^T$ plays the role of a control input. As the input signals for the kinematics (3.14) we use the control law proposed in [1] by Samson and Ait-Abderrahim:

$$u_r = \begin{pmatrix} v_r \\ w_r \end{pmatrix} = \begin{pmatrix} -k_1 x_e + v_d\cos\theta_e \\ w_d - v_d y_e \dfrac{\sin\theta_e}{\theta_e} - k_2\theta_e \end{pmatrix}, \tag{3.15}$$

where $k_1, k_2 > 0$ are some control constants. This means that $u \equiv u_r$.

To prove the convergence of the presented kinematic control algorithm, the following Lyapunov-like function is selected

$$V_1(e_m) = \frac{1}{2}\left(x_e^2 + y_e^2 + \theta_e^2\right) \geqslant 0 \qquad \forall\, e_m. \tag{3.16}$$

After calculating the time derivative of V_1 along solutions of the closed-loop system (3.14)-(3.15) we obtain

$$\dot{V}_1 = x_e\dot{x}_e + y_e\dot{y}_e + \theta_e\dot{\theta}_e = -k_1 x_e^2 - k_2\theta_e^2 \leqslant 0. \tag{3.17}$$

From (3.17) we deduce that V_1 is non-increasing with time – it is bounded. Using Barbalat's lemma, it is easy to show that x_e and θ_e tend to zero.

On the other hand, the desired velocities realizing 'posture tracking' for the mobile platform v_d and w_d and their time derivatives are bounded by assumption. The main problem now is to prove that y_e tends to zero, too. To prove the convergence of y_e to zero, we use arguments of Barbalat's lemma applied to only one variable, namely to $\dot{x}_e$. We have shown that x_e is bounded and $\ddot{x}_e$ is bounded because it is a product of only bounded functions. This implies that $\dot{x}_e$ is uniformly continuous and from Barbalat's lemma we can conclude that $\dot{x}_e \to 0$ as $t \to \infty$. Variable $\dot{x}_e$ can be described by equation

$$\dot{x}_e = -k_1 x_e + w y_e \quad \longrightarrow \quad 0.$$

Variable x_e tends to zero and w (it is defined in fact by (3.15)) is a bounded function of time. This implies that y_e has to converge against zero. This completes the proof of the convergence of the kinematic control algorithm for trajectory tracking.

3.5.2 Kinematic Controller for the 3-pendulum – Jiang & Nijmeijer Algorithm

In order to control the doubly nonholonomic mobile manipulator, the kinematics (3.4) of the 3-pendulum will be converted into a chained form. It implies that all existing control laws for the chained form can be applied to the considered subsystem.

Conversion Into a Chained Form

As the kinematics of nonholonomic 3-pendulum we treat the extended equations (3.4), namely

$$\dot{\theta}_1 = \eta_1,$$
$$\dot{\theta}_2 = a_2 \sin\theta_1 \eta_2,$$
$$\dot{\theta}_3 = a_3 \sin\theta_2 \cos\theta_1 \eta_2, \qquad (3.18)$$
$$\dot{\phi} = \cos\theta_1 \cos\theta_2 \eta_2,$$

where ϕ is the orientation of the OW wheel mounted at the end of the second joint in the 3-pendulum. In [7] the authors have proven that it was impossible to transform the kinematics of the nonholonomic manipulator into a chained form, if the variable ϕ were not added to the state space. The coordinate transformation $z = h(\phi, q_r)$ and a state feedback $v = F(\phi, q_r)\eta$ proposed in [7] in general form is only local (it is valid only for $\theta_i \in (-\frac{\pi}{2}, \frac{\pi}{2}), i = 1, 2$), and it can be defined for example as

$$z = h(\phi, q_r) = \begin{pmatrix} z_1 \\ z_2 \\ z_3 \\ z_4 \end{pmatrix} = \begin{pmatrix} \phi \\ a_2 a_3 \dfrac{\tan\theta_1}{\cos^3\theta_2} \\ a_3 \tan\theta_2 \\ \theta_3 \end{pmatrix}, \qquad (3.19)$$

with new defined inputs

$$v_1 = \cos\theta_1 \cos\theta_2 \eta_2 = f_1 \eta_2, \qquad (3.20)$$

$$v_2 = a_2 a_3 \left(\frac{\eta_1}{\cos^3\theta_2 \cos^2\theta_1} + 3a_2 \frac{\sin^2\theta_1 \sin\theta_2 \eta_2}{\cos\theta_1 \cos^4\theta_2} \right) = f_{21}\eta_1 + f_{22}\eta_2. \qquad (3.21)$$

In the new coordinates the kinematics has the chained form

$$\dot{z}_1 = v_1,$$
$$\dot{z}_2 = v_2,$$
$$\dot{z}_3 = z_2 v_1,$$
$$\dot{z}_4 = z_3 v_1.$$

$$(3.22)$$

Kinematic Controller – Jiang & Nijmeijer Control Algorithm

For the kinematics in the chained form (3.22) we will apply a control algorithm proposed in [3] by Jiang and Nijmeijer. In the first step it is necessary to define basic tracking errors

$$x_{ie} = z_i - z_{id}, \qquad i = 1, \ldots, 4,$$

with z_{id} given by

$$z_d = h(\phi_d, q_{rd}) = \begin{pmatrix} \phi = f(\eta_{2d}, \theta_{1d}, \theta_{2d}) \\ a_2 a_3 \dfrac{\tan \theta_{1d}}{\cos^3 \theta_{2d}} \\ a_3 \tan \theta_{2d} \\ \theta_{3d} \end{pmatrix}. \qquad (3.23)$$

The kinematics will track a desired trajectory, if the tracking errors x_{ie} converge to 0. However, if we want to apply the Jiang & Nijmeijer control algorithm, we have to introduce modified tracking errors as follows

$$e_1 = x_{4e} - z_3 x_{1e},$$
$$e_2 = x_{3e} - z_2 x_{1e},$$
$$e_3 = x_{2e},$$
$$e_4 = x_{1e}.$$

$$(3.24)$$

It is obvious that asymptotic convergence of modified tracking errors e_i implies asymptotic convergence of basic tracking errors x_{ie}. Therefore in further considerations we will focus on a proof of convergence of the mentioned control algorithm, if the kinematics (3.4) is expressed in e_i errors

$$\dot{e}_1 = v_{1d} e_2 - z_2 e_4 (v_1 - v_{1d}),$$
$$\dot{e}_2 = v_{1d} e_3 - e_4 v_2,$$
$$\dot{e}_3 = v_2 - v_{2d},$$
$$\dot{e}_4 = v_1 - v_{1d},$$

$$(3.25)$$

where $v = [v_1, v_2]^T$ plays the role of a control input and v_{1d} and v_{2d} fulfil the nonholonomic constraints, i.e. $v_{1d} = \dot{z}_{1d}$ and $v_{2d} = \dot{z}_{2d}$. At the kinematic controller, i.e. $v \equiv v_r$ for the system (3.25) we take due to [3] the following control inputs

$$v_r = \begin{pmatrix} v_{1r} \\ v_{2r} \end{pmatrix} = \begin{pmatrix} v_{1d} + \dfrac{f}{\lambda - z_2(2e_1 + e_3)} \\ v_{2d} - k_3(e_1 + e_3) - 2 v_{1d} e_2 \end{pmatrix}, \qquad (3.26)$$

where

$$f = -k_4 e_4 + e_2 v_{2d} - k_3 e_2 (e_1 + e_3) - 2v_{1d} e_2^2$$

and $k_3, k_4 > 0$ are positive regulation parameters, whereas $\lambda > 0$ is some design parameter which locally defines a region of attraction for the presented control algorithm.

Along solutions of the closed-loop system (3.25)-(3.26) the time derivative of the following Lyapunov-like function, which is proper and positive definite,

$$V_2(e) = \frac{1}{2}e_1^2 + \frac{1}{2}e_2^2 + \frac{1}{2}(e_1 + e_3)^2 + \frac{\lambda}{2}e_4^2 \tag{3.27}$$

satisfies

$$\dot{V}_2 = -k_3(e_1 + e_3)^2 - k_4 e_4^2 \leqslant 0. \tag{3.28}$$

Thus the considered Lyapunov-like function is non-increasing. This implies that all e_i are bounded, and $V_2(t)$ converges to limit value $\overline{V}_2$. Using some variation of Barbalat's lemma, it can be shown that $e_i \to 0, i = 1, \ldots, 4$ asymptotically (see [3] for details).

3.6 Dynamic Control Algorithms

3.6.1 Dynamic Controller for a Mobile Manipulator with a Nonholonomic Platform

We consider the dynamics of a mobile manipulator given by (3.7). We assume that we know the solution u_r to the kinematic equation (3.3) which preserves 'posture trajectory tracking', namely a convergence of real coordinates ξ of the mobile platform to the desired trajectory ξ_d. Then we propose a new control algorithm guaranteeing asymptotic trajectory tracking for all coordinates of the mobile manipulator as follows

$$\begin{pmatrix} \tau_m \\ \tau_r \end{pmatrix} = (B^*)^{-1} \left\{ Q^* \begin{pmatrix} \dot{u}_r \\ \ddot{q}_{ref} \end{pmatrix} + C^* \begin{pmatrix} u_r \\ \dot{q}_{ref} \end{pmatrix} + D^* - \begin{pmatrix} K_1 e_u \\ K_2 s \end{pmatrix} - C_K \begin{pmatrix} e_u \\ s \end{pmatrix} \right\}, \tag{3.29}$$

where the symbols denote

$$\begin{pmatrix} e_u \\ s \end{pmatrix} = \begin{pmatrix} u - u_r \\ \dot{q}_r - \dot{q}_{ref} \end{pmatrix} = \begin{pmatrix} u - u_r \\ \dot{e}_r + \Lambda e_r \end{pmatrix}, \quad e_r = q_r - q_{rd}, \quad K_1, K_2, \Lambda > 0.$$

Variable q_{ref} is a special auxiliary reference trajectory defined as follows

$$\begin{pmatrix} \dot{q}_{ref} \\ \ddot{q}_{ref} \end{pmatrix} = \begin{pmatrix} \dot{q}_{rd} - \Lambda e_r \\ \ddot{q}_{rd} - \Lambda \dot{e}_r \end{pmatrix}.$$

It is easy to observe that variables q_{ref} and s are defined similar to sliding mode control proposed by Slotine and Li in [11], whereas K_1, K_2, Λ are symmetric and positive definite diagonal matrices. Matrix C_K is a correction matrix [2] necessary to obtain a skew-symmetry between inertia matrix Q^* and matrix C^* as follows

$$\frac{d}{dt}Q^* = (C^* + C_K) + (C^* + C_K)^T. \tag{3.30}$$

Remember that for mobile manipulators the skew-symmetry does not hold anymore. The closed-loop system (3.7)-(3.29) is described by error equation

$$Q^* \begin{pmatrix} \dot{e}_u \\ \dot{s} \end{pmatrix} = -(C^* + C_K) \begin{pmatrix} e_u \\ s \end{pmatrix} - \begin{bmatrix} K_1 & 0 \\ 0 & K_2 \end{bmatrix} \begin{pmatrix} e_u \\ s \end{pmatrix}. \tag{3.31}$$

The use of dynamic control algorithm to the whole equations of mobile manipulator introduces additional errors to the kinematics so that $u = u_r + e_u$ with u_r given by (3.15)

$$\begin{pmatrix} u_1 \\ u_2 \end{pmatrix} = \begin{pmatrix} -k_1 x_e + v_d \cos \theta_e + e_{u1} \\ w_d - v_d y_e \dfrac{\sin \theta_e}{\theta_e} - k_2 \theta_e + e_{u2} \end{pmatrix}. \tag{3.32}$$

To prove the trajectory tracking for all coordinates of the mobile manipulator, we choose the following Lyapunov-like function

$$V_3(e_m, e_u, s, q) = V_1(e_m) + \frac{1}{2} \begin{pmatrix} e_u & s \end{pmatrix}^T Q^*(q) \begin{pmatrix} e_u \\ s \end{pmatrix} \tag{3.33}$$

with V_1 defined by (3.16). Now we want to evaluate the time derivative of the Lyapunov-like function V_3 along the trajectories of the closed-loop system (3.14)-(3.32) and (3.31)

$$\dot{V}_3 = -k_1 x_e^2 - k_2 \theta_e^2 + x_e e_{u1} + \theta_e e_{u2} - K_{11} e_{u1}^2 - K_{12} e_{u2}^2 - s^T K_2 s. \tag{3.34}$$

The above expression can be rewritten as follows

$$\dot{V}_3 - -\left(k_1 - \frac{1}{4}\right) x_e^2 - \left(\frac{1}{2} x_e - e_{u1}\right)^2 - \left(\frac{1}{2}\theta_e - e_{u2}\right)^2 - \left(k_2 - \frac{1}{4}\right) \theta_e^2$$
$$-s^T K_2 s - (K_{11} - 1) e_{u1}^2 - (K_{12} - 1)e_{u2}^2 \leqslant 0. \tag{3.35}$$

The Lyapunov-like function V_3 is decreasing along any trajectory of a closed-loop system if the control parameters are greater than properly chosen numbers

$$k_1, k_2 > \frac{1}{4}, \qquad K_{11}, K_{12} > 1, \qquad K_{22} > 0.$$

From Barbalat's lemma, $\dot{V}_3$ tends to zero and, consequently, $x_e, \theta_e, s, e_{u1}, e_{u2}$ tend to zero. Because V_3 decreases, it means that all variables, namely $x_e, y_e, \theta_e, s, e_{u1}, e_{u2}$, are bounded. On the other hand, the desired velocities realizing 'posture tracking' for the mobile platform v_d and w_d and their time derivatives are bounded by assumption. The main problem now is to prove that y_e tends to zero, too. To prove convergence of y_e to zero, we invoke arguments similar to these presented in section 3.5.1 to show that $\dot{x}_e \to 0$ as $t \to \infty$. Variable $\dot{x}_e$ in a system disturbed by the dynamics of mobile manipulator can be described by the equation

$$\dot{x}_e = -k_1 x_e + e_{u1} + (e_{u2} + w_r) y_e \longrightarrow 0.$$

Variables x_e, e_{u1} and e_{u2} tend to zero and w_r is a bounded function of time. This implies that y_e has to converge to zero. This completes the proof.

3.6.2 Dynamic Controller for a Doubly Nonholonomic Mobile Manipulator

Consider the dynamics (3.10) of the doubly nonholonomic mobile manipulator. Corresponding to the transformation $z = h(\phi, q_r)$, the dynamic model (3.10) can be converted to the form [9]:

$$Q_1(z_e) \begin{pmatrix} \dot{u} \\ \dot{v} \end{pmatrix} + C_1(z_e, \dot{z}_e) \begin{pmatrix} u \\ v \end{pmatrix} + D_1(z_e) = B_1(z_e)\tau, \qquad (3.36)$$

where

$$z_e = \begin{pmatrix} \xi \\ z \end{pmatrix}, \qquad \begin{aligned} Q_1(z_e) &= \bar{Q}(q)|_{q=\xi, h^{-1}(z)}, \\ C_1(z_e, \dot{z}_e) &= \bar{C}(q)|_{q=\xi, h^{-1}(z)}, \\ D_1(z_e) &= \bar{D}(q)|_{q=\xi, h^{-1}(z)}, \\ B_1(z_e) &= \bar{B}(q)|_{q=\xi, h^{-1}(z)}. \end{aligned}$$

Then a dynamic control algorithm guaranteeing asymptotic trajectory tracking for all coordinates of the doubly nonholonomic mobile manipulator can be introduced

$$\tau = (B_1)^{-1} \left\{ Q_1 \begin{pmatrix} \dot{u}_r \\ \dot{v}_r \end{pmatrix} + C_1 \begin{pmatrix} u_r \\ v_r \end{pmatrix} + D_1 - K \begin{pmatrix} e_u \\ e_v \end{pmatrix} - C_K \begin{pmatrix} e_u \\ e_v \end{pmatrix} - \alpha \right\}, \quad (3.37)$$

where the symbols denote

$$\begin{pmatrix} e_u \\ e_v \end{pmatrix} = \begin{pmatrix} u_1 - u_{1r} \\ u_2 - u_{2r} \\ v_1 - v_{1r} \\ v_2 - v_{2r} \end{pmatrix}, \qquad \alpha = \begin{pmatrix} 0 \\ 0 \\ \alpha_1 \\ \alpha_2 \end{pmatrix} = \begin{pmatrix} 0 \\ 0 \\ \lambda e_4 - z_2 e_4 (2e_1 + e_3) \\ e_1 + e_3 - e_2 e_4 \end{pmatrix},$$

$$K = k \cdot I_4, \qquad k > 1$$

and C_K is a correction matrix defined by (3.30).

To prove the trajectory tracking for the doubly nonholonomic mobile manipulator, we choose the following Lyapunov-like function

$$V_4(e_m, e, e_u, e_v, z_e) = V_1(e_m) + V_2(e) + \frac{1}{2} \begin{pmatrix} e_u & e_v \end{pmatrix}^T Q_1(z_e) \begin{pmatrix} e_u \\ e_v \end{pmatrix}, \qquad (3.38)$$

where $V_1(e_m)$ is given by (3.16) and $V_2(e)$ is defined by (3.27). Before we start evaluating the time derivative of the Lyapunov-like function V_4 along the trajectories of the error equations, we have to calculate equations of the closed-loop system (3.36) with the control law (3.37)

$$Q_1 \begin{pmatrix} \dot{e}_u \\ \dot{e}_v \end{pmatrix} = -(C_1 + C_K) \begin{pmatrix} e_u \\ e_v \end{pmatrix} - K \begin{pmatrix} e_u \\ e_v \end{pmatrix} - \alpha. \qquad (3.39)$$

The time derivative of the Lyapunov-like function V_4 is equal to

$$\dot{V}_5 = \dot{V}_1 + \dot{V}_3 + \begin{pmatrix} e_u & e_v \end{pmatrix}^T Q_1 \begin{pmatrix} \dot{e}_u \\ \dot{e}_v \end{pmatrix} + \frac{1}{2} \begin{pmatrix} e_u & e_v \end{pmatrix}^T \dot{Q}_1 \begin{pmatrix} e_u \\ e_v \end{pmatrix}. \qquad (3.40)$$

Before we start evaluating $\dot{V}_4$ along trajectories of the closed-loop system, it is necessary to mention the influence of additional errors coming from dynamic control level and disturbing solutions to the kinematic equations (3.25). We will treat u_r, v_r as kinematic control signals in the ideal case (i.e. without dynamics), and then, as a kinematic control for the real case (with dynamics), we should take modified controls $u = u_r + e_u$ and $v = v_r + e_v$ as follows

$$u_1 = -k_1 x_e + v_d \cos \theta_e + e_{u1}, \tag{3.41}$$

$$u_2 = w_d - v_d y_e \frac{\sin \theta_e}{\theta_e} - k_2 \theta_e + e_{u2}, \tag{3.42}$$

$$v_1 = v_{1d} + \frac{f}{\lambda - z_2(2e_1 + e_3)} + e_{v1}, \tag{3.43}$$

$$v_2 = v_{2d} - k_3(e_1 + e_3) - 2v_{1d}e_2 + e_{v2}. \tag{3.44}$$

Having substituted the modified controls (3.41)–(3.42) into (3.14), (3.43)–(3.44) into (3.25), and (3.39) into (3.40) we obtain

$$\dot{V}_4 = -k_1 x_e^2 - k_2 \theta_e^2 + x_e e_{u1} + \theta_e e_{u2} - k_3(e_1 + e_3)^2 - k_4 e_4^2 + e_{v1}\alpha_1 + e_{v2}\alpha_2$$

$$+ \left(e_u\ e_v \right)^T \left\{ (-(C_1 + C_K + K) \begin{pmatrix} e_u \\ e_v \end{pmatrix} - \alpha \right\} + \frac{1}{2} \left(e_u\ e_v \right)^T \dot{Q}_1 \begin{pmatrix} e_u \\ e_v \end{pmatrix} =$$

$$= -(k_1 - \frac{1}{4})x_e^2 - (k_2 - \frac{1}{4})\theta_e^2 - (k-1)e_{u1}^2 - (k-1)e_{u2}^2 - (\frac{\theta_e}{2} - e_{u2})^2$$

$$- (\frac{x_e}{2} - e_{u1})^2 - k_3(e_1 + e_3)^2 - k_4 e_4^2 - k e_{v1}^2 - k e_{v2}^2 \leqslant 0. \tag{3.45}$$

Using arguments similar to [9] (some variation of Barbalat's lemma), we conclude that e_i, e_m, e_u and e_v tend to 0, if the control parameters are properly chosen

$$k_1, k_2 > \frac{1}{4}, \qquad k_3, k_4 > 0, \qquad k > 1.$$

Thus the asymptotic stability of the presented control algorithm has been proved.

3.7 Simulation Study

The simulations were run with the Matlab package and the Simulink toolbox[1]. As an object of simulations we have taken the 3-pendulum on the unicycle depicted in Fig. 3.1. The goal of the simulation study was to compare behavior of two types of nonholonomic mobile 3-pendulums: with nonholonomic constraints only in the mobile platform (only one nonholonomic subsystem) and with nonholonomic constraints in the mobile platform and in the gears designed by Nakamura, Chung and Sørdalen.

The desired trajectory for the manipulator is equal to $[\theta_{1d}(t), \theta_{2d}, \theta_{3d}] = [-0.1 \cos t, 0, 0.5]$. Variable ϕ is outside our interest because it plays only an auxiliary role in the transformation into the chained form. Such desired admissible trajectory can be defined by the desired velocities in the nonholonomic gears as $[\eta_{1d}(t), \eta_{2d}(t)] = [0.1 \sin t, 0]$. The desired trajectory for the mobile platform is a circle $[x_d(t), y_d(t), \theta_d(t)] = [10 \sin t, -10 \cos t, t]$. The gains of the kinematic

---

[1] Matlab package and the Simulink toolbox were available thanks to Wrocław Center of Networking and Supercomputing.

controllers are equal to $\lambda = 20$, $k_3 = 1$ and $k_4 = 10$ for the Jiang & Nijmeijer algorithm and $k_1 = 100$ and $k_2 = 10$ for the Samson algorithm. The gains of the dynamic controllers are equal to

$$K = \text{diag}\{80\}, \qquad K_1 = K_2 = \text{diag}\{80\}, \qquad \Lambda = \text{diag}\{20\}.$$

The initial position of the manipulator joints was equal to $(\theta_1(0), \theta_2(0), \theta_3(0)) = (0.5, -0.5, 0)$ and initial position of the platform $(x(0), y(0), \theta(0)) = (0, 0, 0.1)$. The trajectory tracking errors for joints of the 3-pendulum are depicted in Figs 3.3-3.5. The tracking of the desired trajectory for the mobile platform is presented in Fig. 3.6.

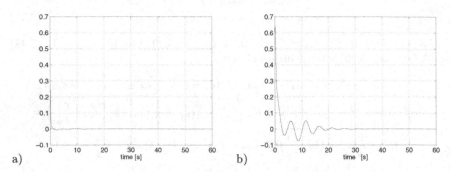

Fig. 3.3. Tracking error $\theta_1(t) - \theta_{1d}(t)$ [rad] for the first joint of the 3-pendulum: a) with direct drives, b) with nonholonomic gears

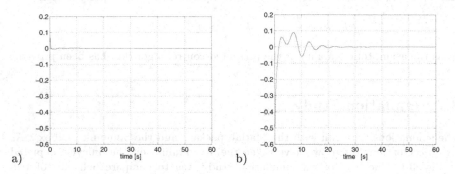

Fig. 3.4. Tracking error $\theta_2(t) - \theta_{2d}(t)$ [rad] for the second joint of the 3-pendulum: a) with direct drives, b) with nonholonomic gears

3.8 Conclusions

In this work we considered a problem of trajectory tracking of the nonholonomic mobile manipulator. Such a robotic system consists of the mobile platform with restricted mobility and the rigid manipulator mounted atop of it. The rigid manipulator can be treated as a holononomic or a nonholonomic subsystem depending on its construction. Nonholonomic drives are nonlinear and introduce to the description of the motion additional constraints, which come from the rolling-without-slipping assumption, similar to the rolling-without-slipping assumption for wheels of the mobile platform. Such assumptions imply the necessity of steering the nonholonomic mobile manipulator not only on the dynamical but on the kinematic level, too.

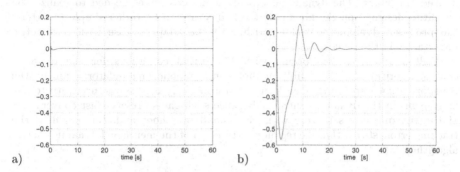

Fig. 3.5. Tracking error $\theta_3(t) - \theta_{3d}(t)$ [rad] for the third joint of the 3-pendulum: a) with direct drives, b) with nonholonomic gears

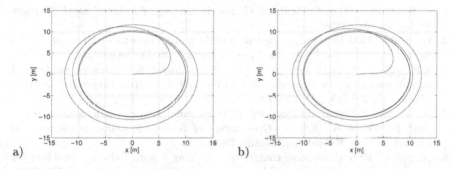

Fig. 3.6. Trajectory tracking for the mass center of the mobile platform equipped: a) with the holonomic 3-pendulum, b) with the nonholonomic 3-pendulum

If we use a mobile manipulator with the holonomic robotic arm, the control problem will be not easy, because the constraints appearing in only one subsystem are valid for the whole object although a coupled holonomic system can be selected from the mathematical description of the model. Dynamic control algorithm for such systems presented in this paper is general and can be used for a mobile platform with any robot mounted atop of it.

On the other hand, if we use doubly nonholonomic mobile manipulator, significant difficulties occur in the regulation process. First of them is the fact that nonholonomic constraints for every subsystem have a different form and need different kinematic controllers. Another problem is related to the selection of admissible trajectories for the robot joints. Because the kinematic controller realizes only trajectories which fulfil nonholonomic constraints, the set of properly chosen trajectories for the nonholonomic manipulator is more restricted than for the holonomic one (nonholonomic construction of a robotic arm is possible only for a robot with revolute joints). However, the nonholonomic manipulator needs only 2 engines to control its behavior.

The kinematic controllers generate velocities, which are necessary to move the nonholonomic subsystem (platform or manipulator) from its initial state to the desired trajectory. The dynamic controller generates forces needed to realize the velocities designed on the kinematic level. It seems to be rather simple to extend the proposed dynamic control algorithm to the adaptive case with parametric uncertainty in the dynamics.

Still another problem occurs with the initial conditions for the dynamic control algorithm for the doubly nonholonomic mobile manipulator. Namely, the transformation into the chained form is only local. It implies the necessity to use large gains in the dynamic controller because such gains preserve fast convergence of the tracking errors to zero with very small overshoot and, consequently, the tracking errors stay within the region of attraction of the mentioned dynamic control algorithm.

References

1. C. Canudas de Wit, B. Siciliano, G. Bastin *Theory of Robot Control.* Springer Verlag, London 1996.
2. I. Dulęba. Modeling and control of mobile manipulators. In: *Proc. 5th IFAC Symp. SYROCO'00*, Pergamon Press, Vienna 2000, pp. 687–692.
3. Z.P. Jiang, H. Nijmeijer. A recursive technique for tracking control of nonholonomic systems in chained form. *IEEE Trans. Automatic Control*, vol. 44, no. 2, 1999, pp. 265–279.
4. A. Mazur. Modeling and control of rigid manipulators with nonholonomic gear. In: Tchoń K. (Editor) *Cybernetics of Robotic Systems.* Transport and Communication Publishers, Warsaw 2004.
5. A. Mazur. Hybrid adaptive control laws solving a path following problem for non-holonomic mobile manipulators. *Int. J. Control*, vol. 77, no. 15, 2004, pp. 1297–1306.
6. M. Michałek, K. Kozłowski. Tracking controller with vector field orientation for 3-d nonholonomic manipulator. In: *Proc. Int. Workshop on Robot Motion and Control RoMoCo'04*, Puszczykowo 2004, pp. 181–189.
7. Y. Nakamura, W. Chung W, O.J. Sørdalen. Design and control of the nonholonomic manipulator. *IEEE Trans. Robotics & Automation*, vol. 17, no. 1, 2001, pp. 48–59.
8. H. Nijmeijer, A.J. Van der Schaft. *Nonlinear Dynamical Control Systems.* Springer Verlag, New York 1991.
9. M. Oya, C.-Y. Su, R. Katoh. Robust adaptive motion/force tracking control of uncertain nonholonomic mechanical systems. *IEEE Trans. Robotics & Automation*, vol. 19, no. 1, 2003, pp. 175–181.
10. N. Sadegh, R. Horowitz. Stability and robustness analysis of a class of adaptive controllers for robotic manipulators. *Int. J. Robotics Research*, vol. 9, no. 3, 1990, pp. 74–94.
11. J.J. Slotine, W. Li. Adaptive manipulator control: a case study. *IEEE Trans. Automatic Control* vol. 33, no. 11, 1988, pp. 995–1003.

12. K. Tchoń, J. Jakubiak. Acceleration-driven kinematics of mobile manipulators: an endogenous configuration space approach. In: J. Lenarčič, C. Galletti (Eds.) *On Advances in Robot Kinematics*. Kluwer Academic Publishers, The Netherlands 2004.
13. K. Tchoń, J. Jakubiak, K. Zadarnowska. Doubly nonholonomic mobile manipulators. In: *Proc. IEEE Int. Conf. on Robotics and Automation*, New Orleans 2004, pp. 4590–4595.
14. J.T. Wen, D.S. Bayard. New class of control laws for robotic manipulators: Part 1 – non-adaptive case. *Int. J. Control*, vol. 45, no. 5, 1988, pp. 1361–1386.

4

Bases for Local Nonholonomic Motion Planning

Ignacy Dulęba and Paweł Ludwików

Institute of Engineering Cybernetics, Wrocław University of Technology, ul. Janiszewskiego 11/17, 50-372 Wrocław, Poland
{iwd|pludwiko}@ict.pwr.wroc.pl

4.1 Introduction

Driftless nonholonomic systems are described by the equation

$$\dot{q} = \sum_{i=1}^{m} g_i(q) u_i \quad \dim q = n > m = \dim u, \tag{4.1}$$

where q is a configuration, g_i, $i = 1 \ldots, m$ are real analytical vector fields, called generators later on, and u is a vector of controls. The class of admissible controls steering the system (4.1) is composed of square integrable functions defined on the interval $[0, T]$, $u(\cdot) \in L_m^2[0, T]$, where time horizon $T > 0$ is fixed. Systems (4.1) are encountered in robotics while modeling, at the kinematic level, underactuated manipulators, floating space robots, underwater vehicles or mobile nonholonomic robots.

Local methods of motion planning based on a Lie-algebraic approach [6, 7] perform a sequence of the following steps in each local (around a current point in the state space) motion planning sub-task. At the beginning, in the first step of the procedure, a direction to the goal is determined, then this direction is expressed as a linear combination of basis vectors. The basis vectors are evaluated vector fields resulting from applying iteratively the Lie bracket to the generators of the system. In the third step, just derived coefficients of the linear combination are to be made control dependent. To this aim a generalized Campbell-Baker-Hausdorff-Dynkin formula (gCBHD) is usually applied. To avoid searching in the infinite-dimensional functional space of controls, it is a common approach to set a basis in the space of controls, to express each control as a series of the basis functions and to move the search into a parametric space of parameters of the series. For computational reasons the series are composed of a finite number of elements. For some classes of systems (4.1) (e.g. nilpotent systems [1]) this number can be determined. Controls' parameterizing is used by other techniques too [4, 10]. In the last step of local motion planning an inverse problem is solved to determine the coefficients of the controls that steer the system in the required direction. In local motion planning, presented above, bases are exploited twice. The first basis selects some vector fields while the second one sets admissible shapes of controls. Evidently, selection of the bases must influence the efficiency of local motion planning.

K. Kozłowski (Ed.): Robot Motion and Control, LNCIS 335, pp. 73–84, 2006.
© Springer-Verlag London Limited 2006

The outline of the paper is as follows. Section 2 recalls some necessary Lie-algebraic concepts and presents consequences of the gCBHD formula. In Section 3 three bases in the space of controls are defined and compared with each other, namely Fourier, Legendre and Haar bases, [2,14]. Section 4 provides evaluation of vector fields that satisfy the Jacobi identity condition. Section 5 concludes the paper.

4.2 Lie Algebraic Concepts and the gCBHD Formula for Driftless Nonholonomic Systems

We begin with introducing some basic concepts. For a pair of vector fields (vfs) X, Y their Lie bracket (another vector field) $[X, Y]$ can be defined. In coordinates

$$[X, Y] = \frac{\partial Y}{\partial q} X - \frac{\partial X}{\partial q} Y, \tag{4.2}$$

where q is the state vector. To each Lie bracket its degree can be assigned by counting the number of generators it is composed of. For example degree($[g_2, g_1]$) = 2, degree($[[g_2, g_1], g_2]$) = 3, degree(g_2) = 1.

The system (4.1) is fully described by its generators $g_1, \ldots, g_m$. The system has its associated Lie algebra, spanned by the generators. The Lie algebra is determined by its basis. One possible construction of the basis is due to Ph. Hall [12] and an effective method to generate the basis in presented in [5]. The basis can be used to check the small time controllability condition (LARC) for driftless nonholonomic systems (4.1), [3]. The basis H of a free Lie algebra is an ordered set of Lie monomials satisfying the following conditions [11]:

PH1 generators belong to the basis, $X_i \in H \mid i = 1, \ldots, m$,

PH2 if degree(B_i) < degree(B_j) then $B_i \overset{H}{<} B_j$,

PH3 $[B_i, B_j] \in H$ if and only if

   a) $B_i, B_j \in H$ and $B_i \overset{H}{<} B_j$ and

   b)   either $B_j = X_k$ for some k or $B_j = [B_l, B_r]$ with $B_l, B_r \in H$ and
        $B_l \overset{H}{\leqslant} B_i$.

In fact, PH2 is not a true membership condition, but it rather introduces a relation of precedence of vfs in the Ph. Hall basis. Clearly, $B_l \overset{H}{\leqslant} B_i$ is satisfied when either $B_l \overset{H}{<} B_i$ or $B_l = B_i$. Note that there is no stop condition in generating Ph. Hall basis, as the basis is infinite. For example, Ph. Hall basis spanned by two generators g_1, g_2 is composed of Lie brackets $g_1, g_2, [g_1, g_2], [g_1, [g_1, g_2]], [g_2, [g_1, g_2]] \ldots$. Ph. Hall basis elements can be ordered into layers composed of Lie brackets with the same degree. In the previous example the second layer is composed of one Lie bracket $[g_1, g_2]$ while the first layer contains two generators g_1, g_2.

Ph. Hall basis can be viewed as a sophisticated method of excluding redundancy from a set of all Lie monomials produced from generators by using the Lie bracket recursively. Two rules are applied: the antisymmetry

$$[X, Y] = -[Y, X], \tag{4.3}$$

and the Jacobi identity

$$[X, [Y, Z]] + [Z, [X, Y]] + [Y, [Z, X]] = 0, \tag{4.4}$$

where X, Y, Z are any Lie monomials (vfs). The first rule says that if $[X, Y]$ entered the basis then $[Y, X]$ cannot be included in the basis. The second rule excludes one Lie monomial from the set $[X, [Y, Z]], [Z, [X, Y]], [Y, [Z, X]]$.

Vector fields describe admissible directions of motion for the nonholonomic system (4.1). Now, it will be shown how they can be produced with controls with the use of the gCBHD formula. The formula describes (locally) a solution to a non-autonomous system of differential equations with a given initial condition [13]. When applied to the system (4.1), it determines (locally) a trajectory of the system initialized at $q(0)$

$$q(t) \simeq z(u(\cdot))(q(0)) + q(0), \quad u(\cdot) \in L_m^2[0, t]. \tag{4.5}$$

For $t \to 0$, the vector field $z(u(\cdot))$ takes a form of Lie series [9]. In sequel, the small time horizon t will be fixed and denoted T. In the paper [9] a pre-control form of the gCBHD formula has been derived for driftless nonholonomic systems (4.1). The state shift operator $z(t)$ takes the form of a Lie series composed of Ph. Hall basis elements premultiplied by control-dependent coefficients

$$z(u(\cdot)) = \sum_{i=1}^{\infty} \sum_{j=1}^{\#\{H_{ij} | degree(H_{ij}) = i\}} \alpha_{ij}(u(\cdot)) \cdot H_{ij}, \quad H_{ij} \in H. \tag{4.6}$$

In Eq. (4.6), the first summation is over layers while the second one is over items within layers. $H_{11} = g_1$, $H_{12} = g_2$, $H_{21} = [g_1, g_2]$, $H_{31} = [g_1, [g_1, g_2]]$, $H_{32} = [g_2, [g_1, g_2]]$. Expressions of α_{ij} control dependent coefficients premultiplying vector fields $g_1, g_2, [g_1, g_2], [g_1, [g_1, g_2]], [g_2, [g_1, g_2]]$ are collected below [9]

$$\alpha_{11}(T) = \int_{A_1(T)} u_1 \, ds^1, \qquad \alpha_{12}(T) = \int_{A_1(T)} u_2 \, ds^1$$

$$\alpha_{21}(T) = \frac{1}{2} \int_{A_2(T)} (u_{12} - u_{21}) \, ds^2, \quad \alpha_{31}(T) = \frac{1}{6} \int_{A_3(T)} (u_{112} - 2u_{121} + u_{211}) \, ds^3,$$

$$\alpha_{32}(T) = \frac{1}{6} \int_{A_3(T)} (-u_{122} + 2u_{212} - u_{221}) \, ds^3,$$

$$\tag{4.7}$$

where $A_r(T)$ is the r dimensional simplex, $A_r(T) = \{s \in R^r : 0 < s_1 < s_2 < \ldots < s_r < T\}$. $\int_{A_r(T)} = \int_{s_r=0}^{T} \int_{s_{r-1}=0}^{s_r} \cdots \int_{s_1=0}^{s_2}$, and dummy variables $ds^r = ds_1 \, ds_2 \ldots ds_r$. The controls in Eq. (4.7) are presented in abbreviated notation explained by the following example: $u_{2122} \to u_2(s_1) \cdot u_1(s_2) \cdot u_2(s_3) \cdot u_2(s_4)$. Note that generic (recursive) formulas for α_{ij} (Eq. (4.7)) can not be obtained as they are based on a basis of the Lie algebra and any basis (Lyndon, Ph. Hall) is determined algorithmically only (no explicit expressions for the basis elements are known).

Ph. Hall series (4.6) when evaluated at a given state q results in a vector of infinitesimal motion $z(t)(q)$ parametrized with controls. It has been proved [9] that, despite of a basis in the space of controls, each coefficient α_Z in Eq. (4.7) premultiplying vector field Z depends on the time horizon T according to the formula

$$\alpha_Z \sim T^{\mathrm{degree}(Z)/2}. \tag{4.8}$$

4.3 Evaluation of Bases in Space of Controls

Three bases in the space of controls will be compared. They represent different classes of control signals: harmonic, polynomial and piece-wise constant, respectively. Their very first normalized elements are presented below:
the Fourier basis ($\omega = 2\pi/T$)

$$F_0(t) = \sqrt{\frac{1}{T}}, \qquad F_2(t) = \sqrt{\frac{2}{T}} \cdot \cos(\omega t),$$

$$F_1(t) = \sqrt{\frac{2}{T}} \cdot \sin(\omega t), \quad F_3(t) = \sqrt{\frac{2}{T}} \cdot \sin(2 \cdot \omega t),$$

the Legendre basis

$$L_0(t) = \sqrt{1} \cdot T^{-1/2},$$
$$L_1(t) = \sqrt{3} \cdot T^{-3/2}(2t - T),$$
$$L_2(t) = \sqrt{5} \cdot T^{-5/2}(6t^2 - 6tT + T^2),$$
$$L_3(t) = \sqrt{7} \cdot T^{-7/2}(20t^3 - 30t^2T + 12tT^2 - T^3),$$

and the Haar basis

$$A_0(t) = \sqrt{\frac{1}{T}}, \quad t \in [0, T] \qquad A_1(t) = \sqrt{\frac{1}{T}} \cdot \begin{cases} 1 & t \in [0, \frac{T}{2}) \\ -1 & t \in [\frac{T}{2}, T] \end{cases}$$

$$A_2(t) = \sqrt{\frac{2}{T}} \cdot \begin{cases} 1 & t \in [0, \frac{T}{4}) \\ -1 & t \in [\frac{T}{4}, \frac{T}{2}] \end{cases} \qquad A_3(t) = \sqrt{\frac{2}{T}} \cdot \begin{cases} 1 & t \in [\frac{T}{2}, \frac{3T}{4}) \\ -1 & t \in [\frac{3T}{4}, T], \end{cases}$$

where at non-determined points in the interval $[0, T]$ the basis functions take the value of 0.

To fix attention, two-input systems ($m = 2$) will be considered and evaluation will be carried out for the first three layers of the Ph. Hall basis. It is natural to expect that bases are comparable when the same number of their first elements are considered. In all cases the first four elements of each basis are taken into account. We begin our considerations with some remarks on the computational complexity of deriving coefficients (4.7) for different bases. In order to obtain the coefficients, the integration over a multi-dimensional simplex has to be performed. With the use of symbolic packages, computation of the coefficients is a time consuming task. Computational time is the longest for the Fourier basis, medium for the Legendre basis and the shortest for the Haar basis. However, data preparation for the integration of Haar basis elements is the hardest compared to the other two bases (for the Haar basis the integration period should be decomposed into sub-intervals with constant controls).

Table 4.1. Coefficients of Ph. Hall series for selected pairs of basis elements $(u_i, u_j) = (B_i, B_j)$ in the space of controls

the pair	g_1	g_2	$[g_1, g_2]$	$[g_1, [g_1, g_2]]$	$[g_2, [g_1, g_2]]$
the Haar basis					
(A_0, A_0)	1	1	0	0	0
(A_0, A_1)	1	0	-0.25	0	-0.042
(A_0, A_2)	1	0	-0.088	0.022	-0.01
(A_0, A_3)	1	0	-0.088	-0.022	-0.01
(A_1, A_2)	0	0	-0.088	-0.022	-0.01
(A_1, A_3)	0	0	0.088	0.022	0.01
(A_2, A_3)	0	0	0	0	0
the Legendre basis					
(L_0, L_0)	1	1	0	0	0
(L_0, L_1)	1	0	0.29	0	-0.05
(L_0, L_2)	1	0	0	0.037	-0.012
(L_0, L_3)	1	0	0	0	-0.0056
(L_1, L_2)	0	0	0.13	-0.03	0
(L_1, L_3)	0	0	0	0	0
(L_2, L_3)	0	0	0.085	0	0.0011
the Fourier basis					
(F_0, F_0)	1	1	0	0	0
(F_0, F_1)	1	0	-0.26	0	-0.038
(F_0, F_2)	1	0	0	0.036	-0.012
(F_0, F_3)	1	0	-0.11	0	-0.0095
(F_1, F_2)	0	0	-0.16	-0.036	0
(F_1, F_3)	0	0	0	0	0
(F_2, F_3)	0	0	0	0	0

In Eq. (4.8) the preference of low degree Ph. Hall basis elements over high degree ones is clearly visible as T is small. In Table 4.1 the coefficients of Ph. Hall series generated with pairs of controls (u_1, u_2) that belong to the Fourier, Legendre and Haar bases are presented. It can be deduced that, despite a basis used, absolute values of coefficients of low degree vector fields are larger than those of higher degrees. It is useless to generate first degree vector fields with harmonics (A_0, A_i), (L_0, L_i), (F_0, F_i), $i = 0 - 3$, when the same effect can be obtained by applying piece-wise constant controls, either $u_1 = 1, u_2 = 0$ or $u_1 = 0, u_2 = 1$. Therefore, we compare the bases with respect to their ability to generate especially useful vectors that form a natural basis in the space of directions, namely Ph. Hall directions of motion. The directions can be defined as vectors having only one non-zero component in a given layer and vanishing coefficients for all lower degree layers. Values of coefficients from higher degree layers are not so important, as they can be reduced by decreasing T, cf. Eq. (4.8). Examples of Ph. Hall directions follow $g_1 \rightarrow (1, 0, *, *, *, \ldots)$, $g_2 \rightarrow (0, 1, *, *, *, \ldots)$, $[g_1, g_2] \rightarrow (0, 0, 1, *, *, \ldots)$, $[g_1, [g_1, g_2]] \rightarrow (0, 0, 0, 1, 0, *, \ldots)$, $[g_2, [g_1, g_2]] \rightarrow (0, 0, 0, 0, 1, *, \ldots)$, where $*$ stands for any value, and 1 means any non-zero value. In comparing the ability of the bases to generate Ph. Hall directions, the Fourier basis is better than the Legendre basis, cf. Table 4.1 as the $[g_1, g_2]$ Ph. Hall direction is generated by (F_1, F_2) with the amplitude 0.16 and by (L_1, L_2) with the amplitude 0.13 (while the Haar basis can produce this direction with amplitude 0.088). Finally, the bases will be compared with respect to the maximal amplitude of basic controls. The Fourier basis elements F_i, $i \geqslant 1$ give $\sqrt{2/T}$ and $\sqrt{1/T}$ for F_0. For the Haar basis, $A_0, A_1 \to \sqrt{1/T}$, and consecutive packs of 2^i, $i = 1, \ldots$ elements result in the maximal amplitude $\sqrt{2^i} \cdot \sqrt{1/T}$. For the Legendre basis the amplitude equal to $L_i \to \sqrt{i+1}\sqrt{1/T}$. According to the criterion, the Fourier basis is the best while the Legendre and the Haar bases are equivalent as the maximal amplitude of their elements increase with the index of the basis element (in fact, slightly slower for the Haar basis). In [8] some analytic measures to compare bases were proposed based on evaluation of vector fields.

4.4 Evaluation of Vector Fields Constrained with the Jacobi Identity

Selection of vector fields to the Ph. Hall basis is somewhat arbitrary. Among three vector fields subordinated to the Jacobi identity (4.4), one is excluded according to rule PH3. As it has been proved in [9], all vector fields that belong to the same layer can be generated with the same amount of energy. It is reasonable to state the question which vector field should be excluded (possibly violating rule PH3) to make the other two the most energy efficient in producing a motion. It will be assumed later on that the three co-planar vector fields are considered locally, at a given state, and two of them must generate a given vector of motion optimizing energy expenditure. One simplification is assumed. For vector $\mathbf{x}$, expressed as a linear combination of equi-energy costly vectors $\mathbf{y}, \mathbf{z}$: $\mathbf{x} = \xi_1 \mathbf{y} + \xi_2 \mathbf{z}$, its energy cost is equal $\xi_1^2 + \xi_2^2$ (motion along each vector $\mathbf{y}, \mathbf{z}$ costs one unit of energy).

Let $\mathbf{a}$, $\mathbf{b}$, $\mathbf{c}$ be non-zero co-planar vectors (vector fields evaluated at a given state) that satisfy the Jacobi identity condition

$$\mathbf{a} + \mathbf{b} + \mathbf{c} = 0. \tag{4.9}$$

Without loosing generality, possibly using rotation and scaling, the vectors can be located on the plane xy and parameterized as follows

$$\mathbf{a} = (1, 0), \quad \mathbf{b} = (l\cos\theta, l\sin\theta), \quad \mathbf{c} = (-1 - l\cos\theta, -l\sin\theta), \tag{4.10}$$

where $\theta \in (0, \pi)$. Eq. (4.9) states that the three vectors are clearly redundant to span a plane. In order to form a basis, one of the vectors has to be removed. There are three possible bases constituted by the pairs of vectors $\mathbf{a}\&\mathbf{b}$, $\mathbf{a}\&\mathbf{c}$ and $\mathbf{b}\&\mathbf{c}$. Each among them allows to express any point $\boldsymbol{p} = (x, y)$ (equivalently a vector connecting the origin with the point) as a linear combination of the basis vectors

$$\boldsymbol{p} = \alpha_1 \mathbf{a} + \alpha_2 \mathbf{b} = \beta_1 \mathbf{a} + \beta_2 \mathbf{c} = \gamma_1 \mathbf{b} + \gamma_2 \mathbf{c}. \tag{4.11}$$

Eq. (4.11) allows to express coefficients of the combinations as functions of coordinates of $\boldsymbol{p}$ and parameters l, θ

$$\alpha_1 = x - y\cot\theta, \quad \alpha_2 = \frac{y}{l\sin\theta}, \tag{4.12}$$

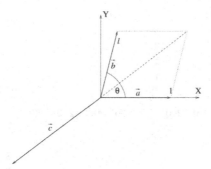

Fig. 4.1. Vectors $\mathbf{a}, \mathbf{b}, \mathbf{c}$ on a plane

$$\beta_1 = x - y\left(\cot\theta + \frac{1}{l\sin\theta}\right), \qquad \beta_2 = -\frac{y}{l\sin\theta}, \tag{4.13}$$

$$\gamma_1 = -x + y\left(\cot\theta + \frac{1}{l\sin\theta}\right), \qquad \gamma_2 = -x + y\cot\theta, \tag{4.14}$$

while energy expenditure to generate $\mathbf{p}$ for each basis equals

$$E_\alpha(x,y) = \sum_{i=1}^{2}\alpha_i^2, \qquad E_\beta(x,y) = \sum_{i=1}^{2}\beta_i^2, \qquad E_\gamma(x,y) = \sum_{i=1}^{2}\gamma_i^2. \tag{4.15}$$

Expressing $\mathbf{p}$ in polar coordinates, $x = r\cos\psi$, $y = r\sin\psi$, allows to concentrate on direction of motion ψ rather than distance r

$$E_\alpha(r,\psi) = r^2\cdot\left[\frac{\sin^2\psi}{l^2\sin^2\theta} + (\cos\psi - \cot\theta\sin\psi)^2\right], \tag{4.16}$$

$$E_\beta(r,\psi) = \frac{r^2}{l^2}\left[\frac{\sin^2\psi}{\sin^2\theta} + \left(-l\cos\psi + l\cot\theta\sin\psi + \frac{\sin\psi}{\sin\theta}\right)^2\right], \tag{4.17}$$

$$E_\gamma(r,\psi) = \frac{r^2}{l^2\sin^2\theta}\left[2l^2\sin^2(\theta-\psi) - 2l\sin(\theta-\psi)\sin\psi + \sin^2\psi\right]. \tag{4.18}$$

Regions where pairs of vectors form the optimal basis are determined by the equation

$$\min(E_\alpha(\psi), E_\beta(\psi), E_\gamma(\psi)) \tag{4.19}$$

to be solved for $\psi \in [0, 2\pi]$. Solution of Eq. (4.19) only depends on one variable ψ (l, θ are parameters and r does not influence the solution — it rather expands it radially). Simulations were performed for angles $\theta = \{45°, 90°, 120°\}$ and lengths $l = \{1/4, 1, 4\}$. Energy expenditure for each pair of the base vectors are visualized in Fig. 4.2, while the best selection of the base vectors is depicted in Fig. 4.3.

To simplify determining the regions, it is enough to find critical values of angle ψ where switches between optimal bases occur. For example the critical value of ψ for the switch from set $\mathbf{a\&b}$ to $\mathbf{a\&c}$ (cf. Eqns. (4.16) (4.17)) is derived from the equation

$$0 = E_\alpha(r,\psi) - E_\beta(r,\psi) = E_\alpha(\psi) - E_\beta(\psi). \tag{4.20}$$

In Fig. 4.4 the optimal range for vectors $\mathbf{a}, \mathbf{b}, \mathbf{c}$ is presented as the function of θ (l is the parameter), while in Fig. 4.5 the variables are interchanged and l serves as the

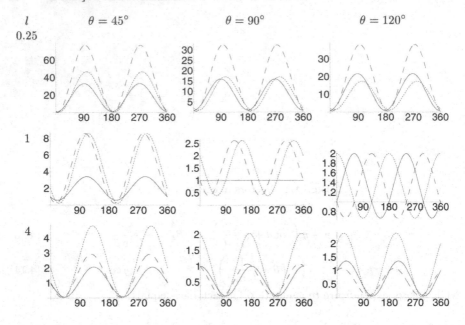

Fig. 4.2. Energy required to produce a given direction. $E_\alpha, E_\beta, E_\gamma$ are drawn with solid/dashed/dotted line, respectively

running variable and θ is the parameter. It is worth noting that these vectors are likely to be included in the basis which form the angle as close to the right angle as possible. The lengths of the vectors, although they influence this characteristics, are as important as the angles.

The presented approach to select two among the three vector fields subordinated to the Jacobi identity can be extended to many vector fields forming a set of independent vectors when evaluated at the current state. In this case more than one Jacobi triple may appear and the optimization problem becomes complex due to many variables involved in trigonometric expressions. However, the problem may arise rarely. For example, the first selection due to the Jacobi identity for the system (4.1) described by three generators g_1, g_2, g_3 is valid for $k \geqslant 3$ degree vector fields. As low degree vector fields from the Ph. Hall basis $g_1, g_2, [g_1, g_2], [g_1, g_3], [g_1, g_3], [g_2, g_3]$ are numerous (it is assumed that they are independent at the current state), the selection really matters for $n \geqslant 6$ dimensional state spaces.

4.5 Conclusions

In this paper bases for local nonholonomic motion planning based on Lie-algebraic approach were evaluated at the current state. The basis for a Lie algebra of vector fields and the basis for a space of controls were examined. Attributive bases for various classes of controls were considered: harmonic functions represented by the

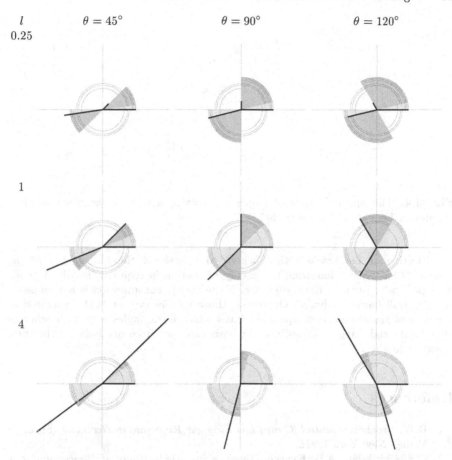

Fig. 4.3. The optimal choice of basis vectors for selected values of parameters θ and l. White/lightgray/darkgray background fill areas where the optimal pair of vectors are **a**&**b**/**a**&**c**/**b**&**c**, respectively. Bold lines indicate the locations and lengths of the base vectors. Single/double/triple arcs circle areas where **a**, **b** and **c** contribute to the optimal basis, respectively

Fourier basis, polynomials represented by the Legendre basis and piece-wise constant controls represented by the Haar basis. From the control practice point of view, it is important to evaluate the influence of basis elements in the space of controls on resulting motion. It is known that any motion can be composed of basic pieces and the pieces can be computed in off-line mode (before any control process is initialized). The Fourier basis seems to be the best among the bases tested. The basis effectively generates pure motions along Ph. Hall directions. It does not suffer as much as the other two bases from frequency band-limit on controls (the Haar basis suffers the most as its Fourier expansion is composed of many high frequency components). The maximal amplitudes of the Fourier basis elements are smaller than the amplitudes of the Haar and Legendre bases.

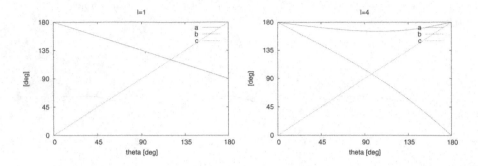

Fig. 4.4. The optimal range of angles for vectors **a**, **b**, **c** as the function of θ. Parameter l equals 1 (left), 4 (right)

Selection of Lie brackets that belong to a basis of the Lie algebra of the system (4.1), when minimizing the energy of motion is required, hardly depends on a particular point in the state space. If the energy consumption is not an issue, the Ph. Hall basis is adviced. Otherwise, those vectors (vector fields evaluated at the current state) should compose the basis which form angles with each other as close to the right angle as possible. The basis can vary from one point in the state space to another.

References

1. R.W. Brockett. *Control Theory and Singular Riemannian Geometry.* Springer Verlag, New York 1981.
2. G.S. Chirikijan, A.B. Kyatkin. *Engineering Applications of Noncommutative Harmonic Analysis.* CRC Press, Boca Raton 2001.
3. W.L. Chow. Über Systeme von linearen partiellen Differentialgleichungen erster Ordnung. *Math. Ann.*, vol. 117, no. 1, 1939, pp. 98–105 (in German).
4. A.W. Divelbiss, J.T. Wen. Nonholonomic path planning with inequality contraints. In: *Proc. IEEE Conf. on Robotics and Automation*, 1994, pp. 52–57.
5. I. Dulęba. Checking controllability of nonholonomic systems via optimal generation of Ph. Hall basis. In: *Proc. IFAC Symp. on Robot Control*, 1997, Nantes, France, pp. 485–490.
6. I. Dulęba. Locally optimal motion planning of nonholonomic systems. *Journal of Robotic Systems*, 1997, pp. 767–788.
7. I. Dulęba. Nonholonomic motion planning based on nonholonomic spheres. In: *Proc. 10th Conf. on Advanced Robotics ICAR*, Budapest 2001, pp. 321–326.
8. I. Dulęba. Bases comparison in control space of nonholonomic systems. In: *Proc. 4th Workshop on Robot, Motion and Control*, Puszczykowo 2004, pp. 161–166.
9. I. Dulęba, W. Khefifi. Pre-control form of the generalized Campbell-Baker-Hausdorff-Dynkin formula for affine nonholonomic systems. *Syst. Contr. Lett.*, 2005 (in print).

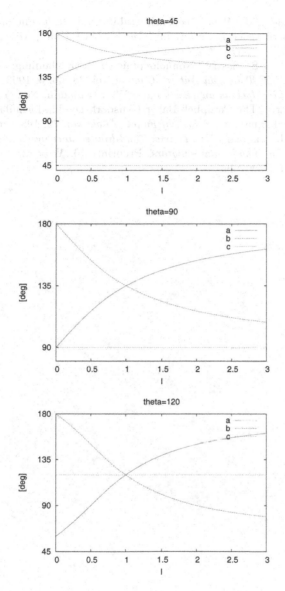

Fig. 4.5. The optimal range of angles for vectors $\mathbf{a}, \mathbf{b}, \mathbf{c}$ as the function of l. Parameter θ equals $45°$, $90°$, $120°$

10. F. Lizarralde, J.T. Wen. Feedback stabilization of nonholonomic systems based on path space iteration. In: *Proc. MMAR Conf.*, Międzyzdroje 1995, pp. 485–490.
11. R.M. Murray, S. Sastry. Nonholonomic motion planning: steering using sinusoids. *IEEE Trans. on Autom. Control*, vol. 38, no. 5, 1993, pp. 700–716.
12. J.P. Serre. *Lie Algebras and Lie groups*. W.J. Benjamin, New York 1965.
13. R.S. Strichartz. The Campbell-Baker-Hausdorff-Dynkin formula and solutions of differential equations. *Journ. of Funct. Anal.*, vol. 72, 1987, pp. 320–345.
14. K. Tchoń, J. Jakubiak. *Non-Fourier band-limited Jacobian inverse kinematics algorithms for mobile manipulators*. Preprints 10, Wrocław Univ. of Techn. Reports, Wrocław 2003.

On Drift Neutralization of Stratified Systems

István Harmati, Bálint Kiss, and Emese Szádeczky-Kardoss

Department of Control Engineering and Information Technology
Budapest University of Technology and Economics,
Magyar Tudósok krt. 2, H-1117 Budapest, Hungary
harmati@iit.bme.hu, bkiss@iit.bme.hu, kardoss@seeger.iit.bme.hu

5.1 Introduction

Several mechanical systems possess stratified configuration space [1], which means that each contact combination implies a distinct nonholonomic constraint dividing the configuration space into intersecting submanifolds called strata. The equations of motion may be different on each submanifold and change discontinuously when the system steps across to another submanifold. The submanifold with the lowest dimension (highest number of constraints) is called the bottom stratum and it plays a distinguished role in the MPP, because the subsystem defined on it is not controllable (in the sense that the Lie Algebra Rank Condition is not satisfied). Therefore, one has to switch to other strata to find feasible path between any two points of the bottom stratum. To achieve this goal, successive or cyclic switching between strata is required. Such a strategy supports legged robots in the gaiting [2].

Point-to-Point (PTP) stratified motion planning algorithm (MPA) in [1] steers the driftless system into a desired configuration. The procedure exploits the results of an MPA used for smooth systems, reported in [3], which finds a flow sequence of control vector fields connecting the initial and final configurations. The piecewise constant inputs provided achieve the goal with exact precision if the Lie algebra generated by the control vector fields of the system is nilpotent, otherwise the solution is a distance dependent approximation. In order to solve the obstacle avoidance problem by path planning, one may apply Lie bracket averaging by highly oscillatory sequence (HOS) of inputs [4] for smooth driftless system and its extended version to driftless SKS [2]. This paper does not focus on time issues, therefore the term trajectory is used as the synonym of the term path.

Path planning (and tracking) can be easily solved for a restricted class of smooth kinematic systems which are called differentially flat [5]. For such systems, the MPP is reduced to an interpolation problem in the space of the flat output since there is a one-to-one correspondence between sufficiently smooth trajectories of the flat output and feasible trajectories of the system. The method can be applied to a subsystem of driftless SKS [2].

The MPAs presented in the paper consider SKSs with drift. The idea is to neutralize drift from the model and trace back the problem to recently developed

K. Kozłowski (Ed.): Robot Motion and Control, LNCIS 335, pp. 85–96, 2006.
© Springer-Verlag London Limited 2006

approach [6]. The idea of drift neutralization emerges also in flat MPA. The simulation results are illustrated on a robotic rowboat model.

The remaining part of the paper is organized as follows. The next section describes the rowboat model (without drift). Section 5.3 gives some background material for SKSs. Sections 5.4 and 5.5 detail the MPAs for the rowboat model with drift including simulation results. Section 5.6 gives some concluding remarks.

5.2 The Robotic Rowboat Model

The robotic rowboat model is adopted from legged robotics, a typical example of SKSs, where a six-legged robot corresponds to a six-paddled rowboat. Both paddles and legs are the tool of making and breaking contact with the environment (terrain or river) and they change the kinematics discontinuously when transition is performed between the two states. Let x, y denote the position of the boat, and let θ denote its orientation. Paddles are divided into two groups. A paddle group consists of paddles from the same side. Paddles from the same group move in the same manner and paddle angles (ϕ_1, ϕ_2) belonging to paddle groups do not leave an admissible range given by (ϕ_{min}, ϕ_{max}). The distances of the two sets of paddles from the water surface are represented by h_1 and h_2, respectively.

Let the length of the paddles be $l = 1$. The values of h_1 and h_2 define the following strata: $S_0 : \mathbb{R}^5 \times [\phi_{min}, \phi_{max}]^2$, $S_1 : h_1 < 0$ (paddle group 1 is in the water), $S_2 : h_2 < 0$ (paddle group 2 is in the water), and $S_{12} := S_1 \cap S_2 : h_1 < 0, h_2 < 0$ (all paddles are in the water). Note that S_{12} is referred to as the bottom stratum. We assume a typical paddling motion such that the angles ϕ_1 and ϕ_2 change always at constant depth h_1, h_2 in the water with paddles just immersed, i.e. paddles are active at negative small values of h_1 and/or h_2.

The system has $s = 3$ strata, whose equations of motion can be separated as follows. On the stratum S_{12}, all paddles are in the water, the state vector is $x_c = \left(x, y, \theta, \phi_1, \phi_2 \right)^T$ and the equation of motion is given by

$$S_{12} : \dot{x}_c = \left(\cos\theta, \sin\theta, 1, 1, 0 \right)^T u_1 + \left(\cos\theta, \sin\theta, -1, 0, 1 \right)^T u_2 =$$
$$= \left(g_{12,1} \; g_{12,2} \right) \left(u_1 \; u_2 \right)^T . \tag{5.1}$$

On the stratum S_1, the paddle group 1 is in the water, the state vector is $x_c = \left(x, y, \theta, \phi_1, \phi_2, h_2 \right)^T$ and the equation of motion is given by

$$S_1 : \dot{x}_c = \left(\cos\theta, \sin\theta, 1, 1, 0, 0 \right)^T u_1 + \left(0, 0, 0, 0, 1, 0 \right)^T u_2$$
$$+ \left(0, 0, 0, 0, 0, 1 \right)^T u_4 = \left(g_{1,1} \; g_{1,2} \; g_{1,1}^{off} \right) \left(u_1 \; u_2 \; u_4 \right)^T . \tag{5.2}$$

On the stratum S_2, the paddle group 2 is in the water, the state vector is $x_c = \left(x, y, \theta, \phi_1, \phi_2, h_1 \right)^T$ and the equation of motion is given by

$$S_2 : \dot{x}_c = \left(\cos\theta, \sin\theta, -1, 0, 1, 0 \right)^T u_1 + \left(0, 0, 0, 1, 0, 0 \right)^T u_2$$
$$+ \left(0, 0, 0, 0, 0, 1 \right)^T u_3 = \left(g_{2,1} \; g_{2,2} \; g_{2,1}^{off} \right) \left(u_1 \; u_2 \; u_3 \right)^T . \tag{5.3}$$

Observe that $g_{1,1}^{off}$ and $g_{2,1}^{off}$ control only the paddle heights and hence they are only responsible for the contacts and the switches between strata. The vector fields equipped with these properties are called moving off vector fields while remaining vector fields are called moving on vector fields.

5.3 Background Material

Our method in the next section applies intensively some previous methods solving the smooth and stratified MPP. In this section, the philosophy of these algorithms is illustrated on the robotic rowboat. It is assumed (and the robotic rowboat model actually has the property) that the distributions defined within the strata are nonsingular. For more detail due to the algorithms, the reader is referred to [1], [3], [4], [2].

The first problem emerged is that the bottom system (5.1) is not controllable, i.e. it does not satisfy the Lie Algebra Rank Condition (LARC) : $\bar{\Delta}|_x = T_x S_{12}$, $x \in S_{12}$, $\Delta = \mathrm{span}\{g_{12,1}, g_{12,2}\}$, where $\bar{\Delta}$ is the involutive closure of the distribution Δ and $T_x S_{12}$ denotes the tangent space of S_{12} at x. The stratified MPA of [1] claims that one may insert any vector field $g_i \in S_1$ or $g_i \in S_2$ into the distribution Δ if it commutes with moving off vector fields $g_{1,1}^{off}$ and $g_{2,1}^{off}$. This distribution is denoted by Δ_e. Commutation of two vector fields in our context means that their Lie bracket is zero. In other words, a flow along one vector field followed by a flow along another one is equivalent to the same flow sequence in the reverse order. Distribution Δ_e constructed by the commutation property plays an important role in the next theorem.

Theorem 1. *If the extended distribution Δ_e satisfies LARC then the system is stratified controllable.*

Proof: The proof can be found in [1].

If Δ_e spanned by those moving on vector fields that satisfy commutation property, ensures stratified controllability, one traces back the stratified MPP to a smooth MPP. Note also that the moving off vector fields play no role furthermore in motion planning since they are responsible only for strata switching. The way to find an appropriate sequence of flows along the moving on vector fields of Δ_e is reported in [3]. First of all, the procedure requires $n = dim(x)$ independent vector fields along which the system can move. Although the method intends to reach only the desired final configuration x_f without any path tracking, determination of so-called P. Hall coordinates is required. In order to solve this issue, a trajectory between the initial configuration x_i and x_f is to be defined. For this, let $g_{12,3}$ be the Lie bracket of $g_{12,1}$ and $g_{12,2}$, i.e. $g_{12,3} = [g_{12,1}, g_{12,2}]$ which, together with $g_{12,1}$ and $g_{12,2}$, spans $\bar{\Delta}_e$. The Lie bracket $g_{12,3} = [g_{12,1}, g_{12,2}]$ plays an important role because if there is a system with control vector fields $g_{12,1}$ and $g_{12,2}$ then any admissible trajectory of the system evolves on that manifold which is generated by the integration of $\bar{\Delta}_e$. Let $G = [g_{12,1}, g_{12,2}, g_{12,3}, g_{1,2}, g_{2,2}]$. If all the vector fields of $\bar{\Delta}_e$ were controlled by an independent control v_i then the equations of motion of the rowboat would be $\dot{x}(t) = G(x)(v_1(t), \ldots, v_5(t))^T = G(x)v(t)$ and a straight line trajectory between x_f and x_i would imply $\dot{x} = (x_f - x_i)/T = \lambda$ and a fictitious input $v = G^{-1}(x)\lambda$.

However, $g_{12,3}$ has no real input, therefore the Philip Hall coordinates are obtained by recursive integrals

$$\tilde{h}_1(T) = \int\limits_0^T v_1(s)ds \qquad \tilde{h}_2(T) = \int\limits_0^T v_2(s)ds$$

$$\tilde{h}_3(T) = \int\limits_0^T (-v_1(s)\tilde{h}_2(s) + v_3(s))ds \qquad (5.4)$$

$$\tilde{h}_4(T) = \int\limits_0^T v_4(s)ds \qquad \tilde{h}_5(T) = \int\limits_0^T v_5(s)ds$$

where the form of $\tilde{h}_3(T)$ is different because it is defined by the Lie bracket $g_{12,3} = [g_{12,1}, g_{12,2}]$. Choosing $T = 2$sec, $x_f = (-0.3,\ 0.4,\ 0,\ 0,\ 0)^T$ (x_i is the origin), the absolute value of P. Hall coordinates $\tilde{h}_1 = -0.15$, $\tilde{h}_2 = -0.15$, $\tilde{h}_3 = 0.1888$, $\tilde{h}_4 = 0.15$, $\tilde{h}_5 = 0.15$ are to be applied and the flow sequence solution is

$$x_s = \Phi(g_{2,2}, 0.15) \circ \Phi(g_{1,2}, 0.15) \circ \Phi(-g_{12,2}, 0.43) \circ \Phi(-g_{12,1}, 0.43)$$
$$\circ\Phi(g_{12,2}, 0.43) \circ \Phi(g_{12,1}, 0.43) \circ \Phi(g_{12,2}, 0.15) \circ \Phi(g_{12,1}, 0.15)(x_i)$$
$$= (-0.3,\ 0.39,\ 0,\ 0,\ 0,\ )^T \approx x_f, \qquad (5.5)$$

where the Campbell-Baker-Hausdorff formula

$$\Phi([g_{i_1}, g_{i_2}], \epsilon^2) \approx \Phi(-g_{i_2}, \epsilon) \circ \Phi(-g_{i_1}, \epsilon) \circ \Phi(g_{i_2}, \epsilon) \circ \Phi(g_{i_1}, \epsilon)(x) \qquad (5.6)$$

is applied to approximate the flow along the Lie bracket direction. The flow $\Phi(g_{2,2}, 0.15)$ means that the system moves in the direction of vector field $g_{2,2}$ with time duration 0.15. The exact reach of x_f is not accomplished because the system is not nilpotent (i.e. recursive Lie brackets do not vanish, see [3]). As a consequence, the rowboat arrives to x_f not at T but $\sum_{j=1}^5 \left|\tilde{h}_j\right|$. If the arrival time $T = 2$sec is required to achieve, one may change the system input from 1 to $\sum_{j=1}^5 \left|\tilde{h}_j\right|/T$ and $\tilde{h}_j = \left|\tilde{h}_j\right| \sum_{j=1}^5 \left|\tilde{h}_j\right|/T$.

One way to improve considerably the algorithm is to replace PTP smooth MPA by smooth path planning. It is shown in [4] that a single Lie bracket direction such as $g_{12,3} = [g_{12,1}, g_{12,2}]$ of the robotic rowboat can be approximated by control input u_1 (of $g_{12,1}$) and control input u_2 (of $g_{12,2}$). More exactly, if $j \to \infty$ then

$$u_1^j(t) = 2\sqrt{j}\cos(jt) \qquad u_2^j(t) = 2\sqrt{j}\sin(jt) \qquad (5.7)$$

implies a Lie bracket averaging, i.e. a trajectory along the flow of $g_{12,3} = [g_{12,1}, g_{12,2}]$. Unfortunately, the method cannot be applied directly on stratified systems [6] and the complexity of the appropriate highly oscillatory input such as (5.7) becomes very complicated if the algorithm requires recursive Lie brackets.

5.4 Drift Neutralization in Stratified Framework

5.4.1 MPA with Drift Neutralization for Noninvolutive SKSs

In this subsection, we provide an MPA that performs path planning while drift is present.

Step 1. Create a multiple stratified system. As the first step, equations of motion of drifted SKS are to be determined due to strata, i.e.

$$S_0 \; : \; \dot{x} = f_0 + g_{0,1}u_{0,1} + \cdots + g_{0,m_0}u_{0,m_0}$$
$$S_1 \; : \; \dot{x} = f_1 + g_{1,1}u_{1,1} + \cdots + g_{1,m_1}u_{1,m_1}$$
$$\vdots$$
$$S_I \; : \; \dot{x} = f_I + g_{I,1}u_{I,1} + \cdots + g_{I,m_I}u_{I,m_I} \tag{5.8}$$

where S_0 denotes the bottom stratum, $S_1, \ldots, S_I$ are the higher strata (with less constraints). Note that for the sake of easier description, strata are numbered by integer in the following. Except for the drift terms f, (5.8) is similar to (5.1)-(5.3).

Step 2. Create a driftless extended bottom stratified system (DEBSS). In the next couple of steps, the algorithm omits drifts $f_0, \ldots, f_I$ from (5.8) and, according to Section 5.3, creates the so called driftless bottom stratified system by means of generating vector fields of $\bar{\Delta}_e$:

$$\dot{x} = g_{0,1}u_{0,1} + \cdots + g_{0,m_0}u_{0,m_0} + [\cdot, \cdot]_{S_0}$$
$$+ g_{1,1}|_{S_0} u_{1,1} + \cdots + g_{1,m_1}|_{S_0} u_{1,m_1} + [\cdot, \cdot]_{S_1}$$
$$\vdots$$
$$+ g_{I,1}|_{S_0} u_{I,1} + \cdots + g_{I,m_I}|_{S_0} u_{I,m_I} + [\cdot, \cdot]_{S_I} \tag{5.9}$$

where the notation $|_{S_0}$ refers to vector fields restricted to bottom stratified system but defined originally in another stratum. The term 'restricted' is due to the operation that projects vector fields into the bottom stratified system. In the model of a robotic rowboat it means that extra dimensions beyond the 5th dimension are 'cut off' from the vector fields. Notation $[\cdot, \cdot]_{S_i}$ denotes those (recursive) Lie brackets of vector fields $g_{i,j}$, $j = 1, \ldots, m_i$ which are independent from each other and $g_{i,j}$, $j = 1, \ldots, m_i$. From now on, the planning is performed in the bottom stratum, hence the notation $|_{S_0}$ is omitted. Recall that by definition, the generating vector fields of $\bar{\Delta}_e$ satisfy the commutation condition of Theorem 1.

Step 3. Construction of the Decoupled Noninvolutive Representation (DNIR) that satisfies the DLARC (see forthcoming definitions). In the next steps, we assume that DEBSS is noninvolutive and controllable.

Definition 1. *DEBSS is noninvolutive, if $\dim(\Delta_e) < \dim(\bar{\Delta}_e)$.*

Definition 2. *DEBSS is controllable, if $\dim(\bar{\Delta}_e) = n$, where n is the dimension of the bottom stratum.*

An appropriate order of vector fields manages to achieve a better path tracking. Hence, it is useful to define a representation of DEBSS that reflects the order of chosen vector fields.

Definition 3. Let $\Xi(\Delta) = k$, $k \in \{-1, 0, 1, \ldots, I\}$ if the vector fields that generate Δ belong to kth stratum. If Δ is generated by vector fields from different strata then $\Xi(\Delta) = -1$.

Definition 4. Consider a DEBSS and let n be the dimension of its bottom stratum. A sequence of involutive closures $\bar{\Delta}^l = \{\bar{\Delta}_1, \ldots, \bar{\Delta}_l\}$, $l \leqslant n$ is said to be a Decoupled Noninvolutive Representation (DNIR) of the DEBSS if $\Xi(\Delta_i) \geqslant 0$ for all $i = 1, \ldots, l$.

If $\Xi(\Delta_i) = k$ then the vector fields that span Δ_i will be denoted $\{g_1^i, \ldots, g_{n_i}^i\}$, where n_i denotes the number of vector fields taking part in Δ_i from the kth stratum. If $g_{p,q}$ of stratum S_p in (5.8) corresponds to g_j^i of Δ_i then $u_{p,q}$ in S_p is equivalent to u_j^i. Similarly, if $\Xi(\bar{\Delta}_i) = k$ then the vector fields that span $\bar{\Delta}_i$ will be denoted $\{g_1^i, \ldots, g_{h_i}^i\}$, where h_i denotes the number of vector fields taking part in $\bar{\Delta}_i$ from the kth stratum in the DEBSS. If $g_{p,q}$ of stratum S_p in DEBSS corresponds to g_j^i of Δ_i in DNIR then $u_{p,q}$ in S_p is equivalent to u_j^i in DNIR.

The most advantageous DNIRs bundle moving on vector fields into the shortest length of sequence which satisfies stratified controllability. For the test of stratified controllability one may use DLARC property.

Definition 5. Consider a DNIR $\bar{\Delta}^l = \{\bar{\Delta}_1, \ldots, \bar{\Delta}_l\}$ $l \leqslant n$. A DNIR satisfies the Decomposed Lie Algebra Rank Condition (DLARC) if $\sum_{j=1}^{l} \bar{\Delta}_j = T_x\mathcal{C}_0$, where $\mathcal{C}_0$ is the bottom stratum and $T_x\mathcal{C}_0$ is the tangent space at $x \in \mathcal{C}_0$.

The following result assists to find a DNIR that allows to reach a final point.

Theorem 2. DLARC is a sufficient condition of stratified controllability.

Proof: The proof can be found in [6].

DLARC is a strong condition. It is possible that DLARC does not hold but the system is stratified controllable. This comes from the fact that Theorem 1 allows using the Lie brackets of two control vector fields from different strata while Theorem 2 does not. In compensation, one can provide an MPA to this restricted class of driftless stratified systems.

Step 4. Determine base points. Starting from DEBSS and Δ_e, a PTP stratified MPA is accomplished (see Section 5.3). The order of generating vector fields in the flow sequence is defined by the DNIR constructed in the previous step. The solution of the PTP (smooth) MPA provides a trajectory between x_i and x_f. Note again that the flow sequence terminates exactly at x_f only if the system is nilpotent. By choosing x_f close enough to x_i, arbitrary precision can be achieved.

Definition 6. The configurations $b_0 = x_i$, $b_1 = \Phi(g_{h_1}^1, t_{h_1}^1) \circ \cdots \circ \Phi(g_1^1, t_1^1)b_0$, $b_2 = \Phi(g_{h_2}^2, t_{h_2}^2) \circ \cdots \circ \Phi(g_1^2, t_1^2)b_1$, $\ldots$, $b_l = x(T) = \Phi(g_{h_l}^l, t_{h_l}^l) \circ \cdots \circ \Phi(g_1^l, t_1^l)b_{l-1}$, namely the points that separate the flows of different strata in the solution of the PTP stratified MPA are called base points.

Note that some vector fields in the definition of base points are obtained as Lie brackets of the original control vector fields. Base points are fixed and every reference trajectory has to go through them.

Step 5.Compute control inputs to decomposed reference trajectory. Every stratum contains two neighboring base points which can be considered as $x_{i,k}$ initial and $x_{f,k}$ final state of a smooth MPP, where $j = 1, \ldots, l$. It is possible to approximate any desired path between $x_{i,k}$ and $x_{f,k}$ on the integral manifold of $\bar{\Delta}_k$. Let the generating moving on vector fields of $\bar{\Delta}_k$ define $G^k = [g_1^k, \ldots, g_{h_k}^k]$. Then, for any desired trajectory segment $x_d^k(t)$, it holds that $x_d^k(0) = x_i^k$, $t_k = \sum_{z=1}^{h_k} t_{i_k, z}$, $x_d(t_k) = x_f^k$, $\dot{x}_d^k(t) \in \bar{\Delta}_k \left(x_d^k(t) \right)$, $t < t_k$. Such a trajectory is called a decomposed trajectory. The extended input of $x_d^k(t)$ is computed by $v(t) = (G^k)^{-1} \dot{x}_d^k(t)$. Since the Lie bracket direction cannot be realized by an existing control input $u_k(t)$, it is approximated by Lie bracket averaging (5.7). In fact, $u_k(t)$ has to be shifted in time because trajectory segments are shifted in time according to the order of vector fields in the DNIR. Connecting the control inputs in time, one obtains the input which approximates the decomposed reference trajectory.

Step 6.Drift neutralization. Let $u_k(t)$ be the control input in the kth path segment. (If k is in subscript then it is the input in $S_{\Xi(\bar{\Delta}_k)}$, if k is in superscript then it is the input in the DNIR of the integral manifold of $\bar{\Delta}_k$.) The system then moves on the integral manifold of $\bar{\Delta}_k$, where $\bar{\Delta}_k$ is the kth (involutive) distribution in the DNIR of the DEBSS. The drift appearing in the kth segment is $f_k(x)$ from (5.8), which has been neglected so far in the planning algorithm. Based on the previous steps, we obtained the inputs for a driftless subsystem of $S_{\Xi(\bar{\Delta}_k)}$ in (5.8) that approximates the desired decomposed path in the kth path segment. In order to neutralize the effect of the drift, consider first the subsystem generated by $\bar{\Delta}_k$ in the kth subsegment with its corresponding drift:

$$\dot{x} = f_{\Xi(\bar{\Delta}_k)} + g_1^k u_1^k + \cdots + g_{h_k}^k u_{h_k}^k = f_{\Xi(\bar{\Delta}_k)} + G^k u^k \qquad (5.10)$$

with $u^k = \left(u_1^k, \ldots, u_{h_k}^k \right)^T$, where $\Xi(\bar{\Delta}_k)$ defines the stratum whose drift is valid in the kth subsegment. Extend now u^k with $u_f^k = (u_{f,1}^k, \ldots, u_{f,h_k}^k)^T$, i.e. $u^k := u^k + u_f^k$, where $u_{f,s}^k = - \langle g_s^k, f_{\Xi(\bar{\Delta}_k)} \rangle / \|g_s^k\|$, $s = 1, \ldots, h_k$. As before, the input belonging to any Lie bracket direction can be approximated by real control input components if one applies the Lie bracket averaging (5.7). The computation of u^k is carried out similarly for every path segment $k = 1, \ldots, l$. Since the kth subsegment identifies its stratum via the function $\Xi(\bar{\Delta}_k)$, input u^k in the kth subsegment determines the real input u_k also in the stratum $S_{\Xi(\bar{\Delta}_k)}$.

Step 7.Insertion of moving off vector fields at the base points. The control of the drifted SKS for path planning is available.

The following results identify the applicability of the algorithm.

Proposition 1. *Let there be given a stratified system with drift and consider the algorithm defined in this subsection. The error of the trajectory planning coincides with the error of the Lie bracket averaging if*

C1) $j \to \infty$ in Lie bracket averaging (5.7)
C2) the reference trajectory is decomposed due to the DNIR structure defined in the algorithm
C3) the drift is independent from the strata i.e. $f_1 = f_2 = \cdots = f_l =: f$
C4) $f \in \bar{\Delta}_1 \cap \bar{\Delta}_2 \cap \cdots \cap \bar{\Delta}_l$.

Proof: Consider first a system without drift. It is known from Section 5.3 and also from [3] that PTP motion planning does not reach exactly the final point if the system is not nilpotent. This error does not have contribution to trajectory planning since the decomposed trajectory has to go through the reached final point and not the desired final point. Since by condition $C2$, the reference trajectory is decomposed, it evolves on the integral manifold of the actual distribution $\bar{\Delta}_k$ in the kth subsegment. Additionally, the Lie bracket averaging in the kth subsection is a smooth trajectory planning problem, and since $C1$ holds, the theory of [4] (see Section 5.3) ensures that the trajectory generated by u^k converges to the decomposed reference trajectory.

We have to show only that the additional drift term does not induce any additional error except the one coming from the Lie bracket averaging. Consider any point of the desired path, say, in the kth subsegment and disregard for a moment that u^k steering the driftless system along the decomposed reference trajectory has already been computed. In other words, assume that $u^k = u_{f,s}^k$ substituting $u_{f,s}^k = -\left\langle g_s^k, f_{\Xi(\bar{\Delta}_k)} \right\rangle$, $s = 1, \ldots, h_k$ into (5.10) yields

$$\dot{x} = f_{\Xi(\bar{\Delta}_k)} - \sum_{s=1}^{h_k} g_s^k \left\langle g_s^k / \left\| g_s^k \right\|, f_{\Xi(\bar{\Delta}_k)} \right\rangle. \tag{5.11}$$

Condition $C3$ ensures that $q(x) = \left(\left\langle g_1^k / \left\| g_1^k \right\|, f_{\Xi(\bar{\Delta}_k)} \right\rangle, \ldots, \left\langle g_{h_k}^k / \left\| g_{h_k}^k \right\|, f_{\Xi(\bar{\Delta}_k)} \right\rangle \right)^T$ completes a projection of drift on the base vectors i.e.

$$f_{\Xi(\bar{\Delta}_k)} = \sum_{s=1}^{h_k} q_s g_s^k = G^k(x) q(x) \tag{5.12}$$

in the kth subsegment, hence (5.11) is replaced by $\dot{x} = 0$. If condition $C4$ is satisfied, this argument can be applied for all subsegments, which means that $u_{f,s}^k$ eliminates drift $f_{\Xi(\bar{\Delta}_k)}$. The only active component which remains in u^k steers the driftless system along the decomposed reference trajectory. Only the conversion of u^k to u_k, which applies the Lie bracket averaging, generates error. This completes the proof.

Proposition 2. *Let there be given a stratified system with drift and consider the algorithm defined in this subsection. The error of the trajectory tracking coincides with the error of the Lie bracket averaging if*

C1) $j \to \infty$ in the Lie bracket averaging (5.7)
C2) the reference trajectory is decomposed using the DNIR structure defined in the algorithm
C3) $f_{\Xi(\bar{\Delta}_k)} \in \bar{\Delta}_k$, for every $k = 1, \ldots, l$.

Proof: The proof is similar to that of Proposition 1. The only difference is that here the drift is not the same in every stratum. Although the drift may be different in strata, the argument in the proof of Proposition 1 to the validation of (5.11) and (5.12) can be applied again. Equations (5.11) and (5.12) are valid mutatis mutandis in every subsegment, i.e. neglecting the error of the Lie bracket averaging, the drift is neutralized, which completes the proof.

5.4.2 Simulation Results on a Robotic Rowboat

As before, the kinematic model of a robotic rowboat is used again for simulation. The MPP consists of following straight-line trajectory towards $x_f =$

$(-0.3,\ 0.4,\ 0,\ 0,\ 0)^T$ from the origin of the reference frame. The rowboat tries to accomplish its task in a river with a constant drift $f_r = (0.3,\ 0.3,\ 0,\ 0,\ 0)^T$. We assume that the drift appears only when paddles are in the river. It is a good approximation if paddles move fast enough in the air when they are lifted from the water. It can be verified that Condition $C3$ of Proposition 2 is fulfilled.

The simulation results of the robotic rowboat in the river with drift are shown in Figure 5.1. The figure shows the path provided by PTP stratified MPA, and the path provided by the previously proposed algorithm with $j = 34000$, step size $5 \cdot 10^{-4}$ sec and step size 10^{-4} sec using a standard ODE solver. It can be observed that in spite of the drift in S_{12}, both algorithms achieve the final point $(-0.3,\ 0.39,\ 0,\ 0,\ 0,\ )^T$ given to the PTP stratified MPA. Step size $5 \cdot 10^{-4}$ sec implies observable oscillation in the path while 10^{-4} sec does not. In general, one concludes that smaller step size reduces oscillation but results in longer computation time. Long computation (around one hour on a 2.8GHz P4 computer) is due to the high frequency in inputs, which is illustrated in Figure 5.1 and means higher resolution integration of the ODE with Matlab. It is also observed that small j in the Lie bracket averaging does not generate a path towards the desired final point. Moreover, the acceptable value of j depends also on the step size chosen.

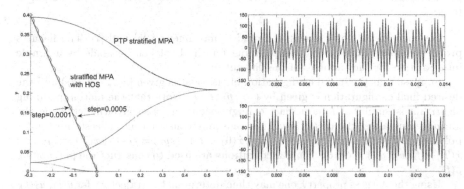

Fig. 5.1. *Left:* The paths provided by stratified MPA with PTP and stratified trajectory tracing by HOS inputs ($j = 34000$, step size is $5 \cdot 10^{-4}$ sec [oscillating path] and step size is 10^{-4} sec ['straight-line' path]). *Right:* The HOS inputs at the beginning of simulation period. ($j = 34000$, step size is 10^{-4} sec)

5.5 Exact Reaching along Smooth Curves in the xy Plane

Consider now the same rowboat model given by (5.1) such that the configuration variables ϕ_1 and ϕ_2 are omitted. Moreover, consider a drift in the bottom stratum such that it depends on the boat position (x, y). This gives the following dynamical system:

$$\dot{x} = f_x(x, y) + \cos\theta(u_1 + u_2) \qquad (5.13)$$

$$\dot{y} = f_y(x, y) + \sin\theta(u_1 + u_2) \qquad (5.14)$$

$$\dot{\theta} = f_\theta(x, y, \theta) + u_1 - u_2, \qquad (5.15)$$

where f_x, f_y, and f_θ are smooth functions of their arguments and may be interpreted as the position dependent drift and vortex of the river. This allows to model any kind of flow profile to a given river section.

Observe that this system with drift is such that the knowledge of the sufficiently smooth trajectory $t \to (x(t), y(t))$ allows to determine the time functions of the remaining variables, namely $\theta(t)$, $u_1(t)$, and $u_2(t)$ such that (5.13)-(5.15) are identically satisfied. Note first that

$$\theta = \arctan 2\left(\dot{y} - f_y(x, y)\right), \left(\dot{x} - f_x(x, y)\right), \qquad (5.16)$$

where $\arctan 2$ gets its value in the $(-\pi, \pi]$ interval in radians. Next, one gets $u_1 + u_2 = (\dot{x} - f_x(x, y))\cos\theta + (\dot{y} - f_y(x, y))\sin\theta$, which, together with (5.15), allows to obtain u_1 and u_2. We have just shown the following.

Proposition 3. *The system* (5.13)-(5.15) *is differentially flat. The flat output is* (x, y).

For more details on flatness, the reader may refer to [5] and [7]. The flatness property implies that the motion planning for the boat can be made by a simple interpolation for the variables of the flat output.

Suppose that the initial configuration of the boat is given by (x_I, y_I, θ_I) and the desired final configuration is given by (x_F, y_F, θ_F). (Note that the motion planning algorithm is similar if intermediate configurations are also given). Let T be the journey time along the trajectory. Then one may construct an at least third-order polynomial for $x(t)$ such that $x(0) = x_I$, $\dot{x}(0) - f_x(x_I, y_I) = \cos\theta_I$, $x(T) = x_F$, and $\dot{x}(T) - f_x(x_F, y_F) = \cos\theta_F$. Similar conditions are used to construct the polynomial $y(t)$.

Using the flatness property, one may then determine the trajectory for $\theta(t)$, $u_1(t)$, and $u_2(t)$ as described above. If one also wants to have the trajectory of ϕ_1 and ϕ_2, one has to integrate the corresponding equations, namely $\dot{\phi}_1 = u_1$ and $\dot{\phi}_2 = u_2$. This allows detecting the back or forward switch points for the paddles (i.e. paddle gaits). Each time when one paddle angle goes out of range one inserts a gait which lifts off and rotates back the corresponding paddle group. It is important to recall that these gaits are considered to be instantaneous.

An example trajectory created using the above described method is presented in Figure 5.2. The drift of the river depends on the distance from the shoreline and reaches its maximal value at the river centerline. The initial orientation of the boat is $\pi/2$ rad, its desired final orientation is $-\pi/2$ rad. The journey time along the trajectory is 50 sec. Note that the boat axis is no longer tangent to the path since there is a drift.

Recall that similar procedure may apply to other systems such that the control of the restricted bottom stratum is of interest and the corresponding dynamical system is differentially flat.

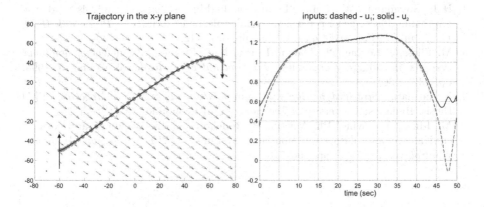

Fig. 5.2. Flatness based motion planning. *Left:* path of the boat COG; *Right:* input signals

5.6 Conclusions

Two MPAs were presented for stratified systems with drift. The first method is able to construct a so-called decomposed reference trajectory and neutralize special class of drifts in a sense that path planning and drift neutralization error is due only to the Lie bracket averaging. The precision depends highly on the parameter choices and the computational capacity available. The method is not able to handle general drift but it inspires to extend the results. The second method presented is specific to the example system considered and provides exact reaching using the flatness property of the system.

Acknowledgments

The research was partly supported by the Hungarian Science Research Fund under grant OTKA T 042634 and by the János Bolyai Research Scholarship of the Hungarian Academy of Sciences.

References

1. B. Goodwine. *Control of stratified systems with robotic applications.* PhD thesis, California Institute of Technology, 1998.
2. I. Harmati, B. Kiss, B. Lantos. Two motion planning approaches for six-legged robots. In: *Proc. of Int. Workshop on Robot Motion and Control RoMoCo,* Bukowy Dworek 2002, pp. 87-92.
3. G. Lafferriere, H.J. Sussmann. Motion planning for controllable systems without drift. In: *Proc. IEEE Int. Conf. on Robotics and Automation,* 1991, pp. 1148–1153.

4. H.J. Sussmann, W. Liu.   Limits of highly oscillatory controls and the approximation of general paths by admissible trajectories. In: *Proc. 30th IEEE Conf. on Decision and Control*, 1991, pp. 437–442.
5. M. Fliess, J. Lévine, Ph. Martin, P. Rouchon. Flatness and defect of nonlinear systems: introductory examples. *Int. Journal of Control*, vol. 61, no. 6, 1995, pp. 1327–1361.
6. I. Harmati. *Stratified motion and manipulation algorithms in robotics.* PhD thesis, Budapest University of Technology and Economics, 2004.
7. M. Fliess, J. Lévine, Ph. Martin, P. Rouchon. A Lie-Bäcklund approach to equivalence and flatness of nonlinear systems. *IEEE Trans. on Automatic Control*, vol. 38, 1999, pp. 700–716.

Control and Mechanical Systems

Part II

Control and Mechanical Systems

6

Novel Adaptive Control of Partially Modeled Dynamic Systems

József K. Tar[1], Imre J. Rudas[1], Ágnes Szeghegyi[1], and Krzysztof Kozłowski[2]

[1] Budapest Tech Polytechnical Institution
   H-1034 Budapest, Bécsi út 96/b, Hungary
   Tar.Jozsef@nik.bmf.hu, Rudas@bmf.hu, Szeghegyi.Agnes@kgk.bmf.hu
[2] Poznań University of Technology, ul. Piotrowo 3a, 60-965 Poznań, Poland
   Krzysztof.Kozlowski@put.poznan.pl

6.1 Introduction

The basic components of Soft Computing were almost completely developed by the sixties. In our days SC means either separate or integrated application of Neural Networks (NN) and Fuzzy Systems (FS) enhanced with high parallelism of operation and supported by several deterministic, stochastic or combined parameter-tuning methods (learning). The main advantage of using FS is evading the development of intricate analytical system models.

Regarding the use of NNs, typical problem classes have been identified for the solution of which typical uniform architectures (e.g. multilayer perceptron, Kohonen-network, Hopfield-network, Cellular Neural Networks, CNN Universal Machine, etc.) have been elaborated. For instance, a typical application of NNs is linearization of nonlinear sensor signals [1].

Fuzzy systems have the great advantage of providing mathematically rigorous and systematic representation of vague or imprecise information in a form similar to human languages [2]. They use membership functions of typical shapes (e.g. trapezoidal, triangular or step-like, etc.), and the fuzzy relations can also be utilized in a standardized way by using different classes of fuzzy operators.

The first phase of applying traditional SC, that is the identification of the problem class and finding the appropriate structure for dealing with it, is usually easy. The next phase, i.e. determining the necessary size of the structure and fitting its parameters via machine learning is far less easy. For neural networks certain solutions start from quite a large initial network and apply dynamic pruning for getting rid of the dead nodes [3]. An alternative method starts with a small network, and the number of nodes is increased step by step (e.g. [4,5]). Due to the possible existence of local optima, for a pure backpropagation training, inadequacy of a given number of neurons cannot be concluded simply. To evade this difficulty, learning methods considerably improved in the last decade (e.g. [6]). Inclusion of Genetic Algorithms (GA) [7] or other stochastic or semi-stochastic elements [8] were of great significance. In spite of this development it can be stated that for strongly coupled non-linear Multiple Input - Multiple Output (MIMO) systems traditional

K. Kozłowski (Ed.): Robot Motion and Control, LNCIS 335, pp. 99–111, 2006.
© Springer-Verlag London Limited 2006

SC still has several drawbacks. The number of the necessary fuzzy rules strongly increases with the degree of freedom and the intricacy of the problem. To reduce modeling complexity fuzzy interpolation methods were developed and checked [9]. Similarity relations can also be utilized in the design of fuzzy diagnostic systems [10]. However, conventional fuzzy modeling techniques also need further investigation and development [11]. Similar problems arise regarding the necessary number of neurons in a neural network approach. External dynamic interactions on which normally no satisfactory information is available influence the system's behavior in a dynamic manner. Both the big size of the necessary structures, the huge number of parameters to be tuned, as well as the goal varying in time are still serious problems.

Realizing that generality and uniformity of the traditional SC structures exclude the application of plausible simplifications gave rise to the idea that by addressing narrower problem classes a novel branch of soft computing could be developed by use of far simpler and far more lucid uniform structures and procedures than the classical ones.

The first steps in this direction were made in the field of Classical Mechanical Systems (CMSs) [12], based on the Hamiltonian formalism detailed e.g. in [13]. This approach used the internal symmetry of CMSs, the Symplectic Group (SG) of Symplectic Geometry in the tangent space of the physical states of the system. The result of the *situation-dependent system identification* was a symplectic matrix compensating the effects of the inaccuracy of the rough dynamic model initially used as well as the external dynamic interactions not modeled by the controller. By use of perturbation calculus it was proved that under certain restrictions this new approach could be successful in the control of the whole class of classical mechanical systems [14]. It is interesting that the method of Taylor series extension combined with the Hamiltonian formalism is widely used in our days for problem solution (e.g. [15, 16]).

Later the problem was considered from a purely mathematical point of view. It became clear that all the essential steps used in the control could be realized by other mathematical means than the symplectic matrices related to some phenomenological interpretation. Other Lie groups are defined in a similar manner by some basic quadratic expressions like in the case of the Generalized Lorentz Group [17], the Stretched and the Partially Stretched Orthogonal Matrices [18], or Symplectic Matrices of Special Structure [19]. In these approaches the Lie group used in the control does not describe any internal physical symmetry of the system to be controlled.

The next essential step was to turn from the inaccurate modeling and not modeled external perturbations to the control of partially modeled physical systems containing internal degrees of freedom that were not modeled by the controller. The first results belonged to a special case in which the not modeled parts corresponded to very stiff but flexible joints in a robot arm [20]. The great stiffness of the not modeled joints made this situation special in the sense that the motion belonging to these degrees of freedom was very much restricted.

The *cart plus inverted pendulum* or the *cart plus double inverted pendulums* systems are popular paradigms in control technology due to their certain characteristics that make their control difficult or non-trivial. In [21] simulation and experimental results were published on the application of the energy method for the stabilization of the inverted pendulum around its homoclinic orbit. In [22] the badly conditioned inertia matrix of the cart plus double pendulum system was considered. In the present paper the same system is chosen as a paradigm from a quite different point of view. Its linear degree of freedom meaning the horizontal translation of the whole system and the axis of one of the pendulums are driven only. The other pendulum can move freely according to the nonlinear coupling present in the system. By use of numerical simulation it is shown that the novel adaptive approach is able to control this system.

6.2 Formulation of the Control Task

From a purely mathematical point of view the control task can be formulated as follows. There is given some imperfect model of the system on the basis of which some excitation is calculated to obtain a desired system response $\mathbf{i}^d$ as $\mathbf{e} = \varphi(\mathbf{i}^d)$. The system has its inverse dynamics described by the unknown function $\mathbf{i}^r = \psi(\varphi(\mathbf{i}^d))$ that results in a realized response $\mathbf{i}^r$ instead of the desired one, $\mathbf{i}^d$. Normally one can obtain information via observation only on the function $\mathbf{f}()$ considerably varying in time, and no possibility exists to directly manipulate the nature of this function: only $\mathbf{i}^d$ as the input of $\mathbf{f}()$ can be deformed to $\mathbf{i}^{d*}$ to achieve and maintain the $\mathbf{i}^d = \mathbf{f}(\mathbf{i}^{d*})$ state. (Only the model function $\mathbf{f}()$ can directly be manipulated.) On the basis of the modification of the method of renormalization widely applied in physics the following *scaling iteration* was suggested for finding the proper deformation:

$$S_n \mathbf{f}(\mathbf{i}_{n-1}) = \mathbf{i}_0 \qquad (6.1)$$

in which $\mathbf{i}_0$ denotes some initial trial, and the S_n matrices describe certain linear transformations to be specified later. It is evident that if $S_n \to \mathsf{I}$, and $S_n S_{n-1} \ldots S_1 \to S$ i.e. the series of the linear transformations so converge to the identity operator that the products of the matrices also converges to a finite matrix, just the proper deformation is approached, therefore the controller "learns" the behavior of the observed system via step-by-step amendment and maintenance of the initial model. The satisfactory condition for such convergence is outlined and explained in Section Theorem 1. It is worth noting that these considerations are independent of the particular algebraic details that are discussed in the forthcoming part of the paper.

Since (6.1) ambiguously determines the possible applicable quadratic matrices we have additional freedom in choosing the appropriate ones. The most important points of view are fast and efficient computation, and the ability for remaining as close to the identity operator as possible.

For making the problem mathematically unambiguous (6.1) can be transformed into a matrix equation by placing the values of $\mathbf{f}$ and $\mathbf{i}$ into well-defined blocks of bigger matrices. Via computing the inverse of the matrix containing $\mathbf{f}$ in (6.1) the problem can be made mathematically well defined. Since the calculation of the inverse of one of the matrices is needed in each control cycle it is expedient to choose special matrices of fast and easy invertibility. Within the block matrices the response arrays may be extended by adding to them a "dummy", that is physically not interpreted dimension of constant value, in order to evade the occurrence of the mathematically dubious $0\to 0$, $0\to$finite, and finite$\to 0$ transformations. In the present paper the special symplectic matrices announced in [19] were applied for this purpose. In general, the Lie group of the Symplectic Matrices (SM) is defined by the equations

$$S^T \Im S = \Im \equiv \begin{pmatrix} 0 & 1 \\ -1 & 0 \end{pmatrix} \tag{6.2}$$

in which $\det S = 1$. The inverse of such matrices can be calculated in a computationally very cost-efficient manner as $S^{-1} = \Im^T S^T \Im$. If q_1 and q_3 are the controlled and modeled generalized coordinates ($\mathbf{q} = [q_1, q_3]^T$), and q_2 is the not modeled/uncontrolled one of the 3 Degree Of Freedom (DOF) mechanical system under consideration then the symplectic matrices are constructed from the *desired* and the *observed* joint coordinate accelerations being the response of the mechanical system to the excitation of torque and force. The columns of the matrix defined in (6.3):

$$M(\ddot{\mathbf{q}}) \equiv [m_1 \ m_2 \ m_3 \ \dots] \equiv \begin{bmatrix} \ddot{q}_1 & -\ddot{q}_1 & e_1^{(3)} & \dots \\ \ddot{q}_3 & -\ddot{q}_3 & e_2^{(3)} & \dots \\ d & -d & e_3^{(3)} & \dots \\ D & \frac{\|\ddot{\mathbf{q}}\|^2 + d^2}{D} & e_4^{(3)} & \dots \end{bmatrix} \tag{6.3}$$

can be used for constructing the "candidate" symplectic matrix

$$S(\ddot{\mathbf{q}}) \equiv \begin{bmatrix} \mathbf{0} & \mathbf{0} & \mathbf{0} & \dots & \frac{-\mathbf{m}^{(1)}}{s} & \frac{-\mathbf{m}^{(2)}}{s} & -\mathbf{e}^{(3)} & \dots \\ \mathbf{m}^{(1)} & \mathbf{m}^{(2)} & \mathbf{e}^{(3)} & \dots & \mathbf{0} & \mathbf{0} & \mathbf{0} & \dots \end{bmatrix}, \tag{6.4}$$

in which the symbols $\mathbf{e}^{(3)}$, $\mathbf{e}^{(4)}$ denote two pairwisely orthogonal unit vectors that lie in the orthogonal subspace of the first two columns of the block matrix, and d is the dummy parameter used for avoiding singular transformations, furthermore

$$D^2 \equiv \ddot{\mathbf{q}}^T \ddot{\mathbf{q}} + d^2, s = 2D^2. \tag{6.5}$$

Equations (6.3) and (6.4) cannot be "deduced". They are mere algebraic means for creating simple symplectic matrices the structure of which is suggested by $\Im$. They are used in (6.1) to uniquely define the S_n matrices as $S_n S(\mathbf{f}(\mathbf{i}_{n-1})) = S(\mathbf{i}_0)$. This construction satisfies (6.1) since the first two elements of the lower half of the first column of $S(\mathbf{f}(\mathbf{i}_{n-1}))$ are just mapped to those of $\mathbf{i}_0$ being in similar position within $S(\mathbf{i}_0)$. The unit vectors can be created e.g. by using El Hini's algorithm [18], which, while rotates vector $\mathbf{b}$ into the direction of vector $\mathbf{a}$, leaves their orthogonal subspace invariant. So if the operation starts with an orthonormal set $\{\mathbf{e}^{(1)}, \dots\}$ that is rigidly rotated until $\mathbf{e}^{(1)}$ becomes parallel with the first column of M, its second column will lie in the orthogonal subspace of the first one spanned by the transformed $\{\mathbf{e}^{*(2)}, \dots\}$ set. In the next step this whole set can be rigidly rotated until the new $\mathbf{e}^{**(2)}$ becomes parallel with the second column of M. (This operation leaves the previous set $\mathbf{e}^{*(1)}$ invariant because it is orthogonal to the two vectors determining this special rotation.) With the above completion the appropriate operation in (6.1) evidently equals to the identity operator if the desired response is equal to the observed one, and remains in the close vicinity of the unit matrix if the non-zero desired and realized responses are very close to each other.

Since amongst the conditions for which the convergence of the method was proved near-identity transformations were supposed in the perturbation theory, a parameter ξ measuring the "extent of the necessary transformation", a "shape factor" σ, and a "regulation factor" λ can be introduced in a linear interpolation with small positive ε_1 and ε_2 values as

$$\xi \equiv \frac{\|\mathbf{f} - \mathbf{i}^d\|}{\max(\|\mathbf{f}\|, (\|\mathbf{i}^d\|)}, \lambda \equiv 1 + \varepsilon_1 + (\varepsilon_1 - 1 - \varepsilon_2)\frac{\xi\sigma}{1 + \xi\sigma}, \tag{6.6}$$

$$\hat{\mathbf{i}}^d \equiv \lambda \left(\mathbf{i}^d - \mathbf{f} \right). \tag{6.7}$$

This interpolation reduces the task of the adaptive control in the more critical sessions and helps to keep the necessary linear transformation in the vicinity of the identity operator. For considerable relative transformation ξ is big resulting in $\lambda \approx \varepsilon_2$, while for small ξ $\lambda \approx 1 + \varepsilon_1$ that means a little "extrapolation". Other important fact concerning the details of the numerical calculations is the ratio of $\|\ddot{\mathbf{q}}\|$ and d in (6.3). The controller has *a priori* information only on the nominal accelerations, but in the error-relaxation process much higher desired accelerations may occur. For this, a slowly forgetting integrating filter was introduced to create a weighting factor for $0 < \beta < 1$ as

$$w(t_i) \equiv \frac{\sum_{j=0}^{\infty} \beta^j \|\ddot{\mathbf{q}}^{\mathbf{Des}}(t_{i-j})\|}{\sum_{s=0}^{\infty} \beta^s} \tag{6.8}$$

and in (6.3) instead of the actual values of $\ddot{\mathbf{q}}$ the actual weighted ones were taken into account (β determines the time-constant of forgetting that has to be determined experimentally via simulation). The numerical realization of such a filter is very easy: the content of a buffer has to be multiplied by β in each control cycle, and the new $\|\ddot{\mathbf{d}}^{Des}\|$ value has to be added to it. It is also easy to calculate the sum of the weights in the denominator of (6.8): $\Sigma = 1/(1 - \beta)$. If (6.1) is not "statically" interpreted and in the $(n + 1)^{th}$ control cycle a new $S_n S_{n-1} \dots S_1$ matrix is applied for the desired accelerations slowly varying in time, the control law is obtained.

6.3 Description of the System to Be Controlled

The cart considered consisted of a body and wheels of negligible momentum and inertia having the overall mass of M [kg]. The pendulums were assembled on the cart by parallel shafts and arms of negligible masses and lengths L_1 and L_2[m], respectively. At the end of the arms balls of negligible sizes and considerable masses of m_1 and m_2 [kg] were attached, respectively. The Euler-Lagrange equations of motion of this system are given as follows:

$$\begin{bmatrix} Q_1 \\ Q_2 \\ Q_3 \end{bmatrix} = \begin{bmatrix} m_1 L_1^2 & 0 & -m_1 L_1 \sin q_1 \\ 0 & m_2 L_2^2 & -m_2 L_2 \sin q_2 \\ -m_1 L_1 \sin q_1 & -m_2 L_2 \sin q_2 & (M + m_1 + m_2) \end{bmatrix} \begin{bmatrix} \ddot{q}_1 \\ \ddot{q}_2 \\ \ddot{q}_3 \end{bmatrix} + \tag{6.9}$$

$$+ \begin{bmatrix} -m_1 L_1 \cos q_1 \dot{q}_1 \dot{q}_3 - m_1 g \cos q_1 \\ -m_2 L_2 \cos q_2 \dot{q}_2 \dot{q}_3 - m_2 g \cos q_2 \\ -m_1 L_1 \cos q_1 \dot{q}_1^2 - m_2 L_2 \cos q_2 \dot{q}_2^2 \end{bmatrix}$$

in which g denotes the gravitational acceleration [m/s^2], Q_1 and Q_2 ($N \times m$) denote the driving torque at shaft 1 and 2, respectively, and Q_3 [N] stands for the force moving the cart in the horizontal direction. The appropriate rotational angles are q_1 and q_2 [rad], and the linear degree of freedom belongs to q_3 [m]. In our case Q_2 was set to be zero because it played the role of the not modeled degree of freedom without driving. The positive definite nature of the inertia matrix M guarantees that

$$if \begin{bmatrix} \ddot{q}_1 \\ 0 \\ \ddot{q}_3 \end{bmatrix} \neq \mathbf{0} \ then \ [\ddot{q}_1 \ 0 \ \ddot{q}_3] \ \mathbf{M} \begin{bmatrix} \ddot{q}_1 \\ 0 \\ \ddot{q}_3 \end{bmatrix} > 0, \tag{6.10}$$

therefore if (6.9) is so rearranged that $\ddot{q}_2$ is moved to the right hand side of the equation and the equation for $Q_2 = 0$ is dealt with separately we again obtain a positive definite inertia matrix for the modeled subsystem q_1 and q_3, therefore the complete stability of the control is guaranteed. The forthcoming simulation confirms this statement.

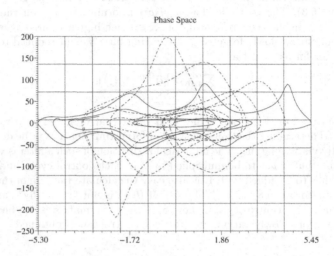

Fig. 6.1. Nonadaptive control: the nominal and computed phase trajectories of the controlled axes $\dot{q}_1$ (rad/s) vs. q_1 [rad] (solid lines) and $\dot{q}_3$ [m/s] vs. q_3 [m] (dash-dot lines)

6.4 Simulation Results

While (6.9) described the state propagation of the physical system, for the desired relaxation of the trajectory tracking error of the two controlled joints a simple PID-type rule was prescribed by the use of purely kinematic terms. This error relaxation could be achieved exactly only in the possession of the exact dynamic model of the physical system to be controlled. Instead of the exact actual dynamic model the constant $10 \times I$ (I = unit matrix of the dimensions 2×2) was used as the inertia matrix belonging to q_1 and q_3, and the Coriolis and gravitational terms were modeled by the constant vector $[10, 10]^T$. This evidently corresponds to a very rough approximation of the reality in which $m_1 = m_2 = 10$ kg, $L_1 = L_2 = 2$m, and for the cart mass $M = 2.5$ kg were chosen. In the forthcoming simulations the following numerical data were used: $d = 80$, $\beta = 0.92$, $\sigma = 0.5$, $\varepsilon_1 = 0.2$, $\varepsilon_2 = 10^{-5}$.

Figure 6.1 exemplifies the inadequacy of the rough initial model during the whole control session via displaying the phase trajectories of the controlled axes in the case of the nonadaptive control (the identical "canonical" curves belong to the nominal motion). The adaptive counterpart of this motion can be seen in Fig. 6.2 that belongs to the same rough initial model updated step-by-step. Figure 6.3 gives the trajectory tracking error for the controlled axes. The improvement seems to be

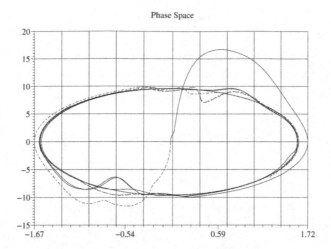

Fig. 6.2. Adaptive control: the nominal and computed phase trajectories of the controlled axes $\dot{q}_1$ [rad/s] vs. q_1 [rad] (solid lines) and $\dot{q}_3$ [m/s] vs. q_3 [m] (dash-dot lines)

essential. The sessions of increased phase trajectory errors may be related to the motion of the coupled second pendulum the phase trajectory of which is described in Fig. 6.4. In this case there is no prescribed "nominal" trajectory. This axis suffers very hectic variation in its rotational velocity that is coupled to and disturbs the motions of the controlled axes. However, due to adaptivity, the phase trajectories of these latter ones remain almost canonical.

To reveal some details of the operation of the adaptive control the *regulating factor* λ and the *weighting factor* w vs. time are described in Figs. 6.5 and 6.6.

Their fine structure also indicates that besides the imprecision of the initial model its incompleteness i.e. the effect of the coupled and not modeled axis is quite considerable. This effect is even more enhanced in the diagram displaying the norm of the (S-I) matrix vs. time (Fig. 6.7). While in general this norm is relatively very small in comparison with that of the unit matrix of appropriate size ($\sqrt{8}$) as it was expected on the basis of perturbation calculus, its local peaks are in coincidence with that of the regulating and the weighting factors, and they are significant.

6.5 Conclusions

In this paper the operation of an adaptive control based on a novel branch of soft computing was investigated via simulation in controlling a system of incomplete and partial model. The joint not modeled by the controller was left freely without any actuation and was in strong nonlinear coupling with the controlled and actuated ones. It was found that the novel method could successfully compensate for the effect of the excitation of the unconstrained joint of unbounded motion, in spite of the fact that it can obtain quite considerable kinetic and potential energy. It seems to be expedient to make similar investigations for other physical systems of not

Joint Coordinate Errors vs. Time [0.001 s]

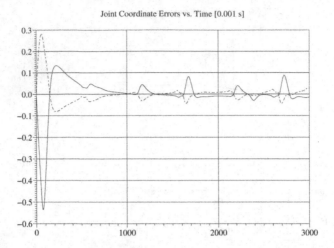

Fig. 6.3. Adaptive control: the trajectory tracking error for the controlled axes q_1 [rad] (solid line) and q_3 [m] (dash-dot line) vs. time [ms]

Phase Space of the Unmodeled DOF

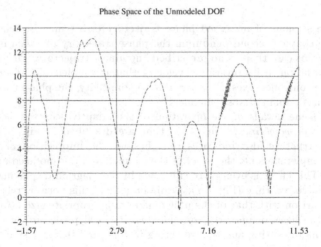

Fig. 6.4. Adaptive control: the phase trajectory of the uncontrolled axis $\dot{q}_2$ [rad/s] vs. q_2 [rad] (dashed line)

modeled and unactuated degrees of freedom in strong nonlinear coupling with the actuated and controlled ones.

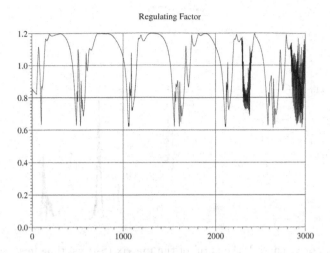

Fig. 6.5. Adaptive control: the regulating factor λ (dimensionless) vs. time [ms] (solid line)

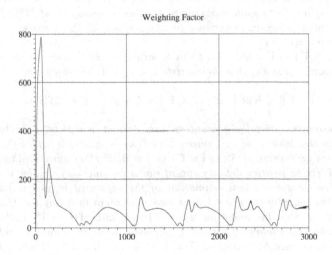

Fig. 6.6. Adaptive control: the weighting factor w (dimensionless) vs. time [ms] (solid line)

Theorem 1. *The forthcoming considerations serve rather as geometrically interpreted lucid explanations than some rigorous proofs. However, they help us outline the possible sets of physical systems for the control of which the presented method can work. The concept of **Complete Stability** was invented for externally driven physical systems as e.g. Cellular Neural Networks described by the state propagation equation $\dot{x} = f(x, u)$. The system is completely stable if for constant external excitation u the state variable $x \to x_\infty$ as $t \to \infty$, i.e. for constant input excitation the*

Norm of S–UNIT

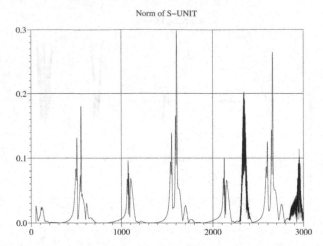

Fig. 6.7. Adaptive control: the norm of the matrix (S-I) vs. time [ms] (solid line)

system's response asymptotically converges to a constant output. If the variation of the input is far slower than the system's dynamics it provides us with a continuous response corresponding to some mapping of the time-varying input [23]. The same concept can evidently be used in our case for the series of the S_n matrices at constant or slowly varying vector $\mathbf{i}^d$.

The $\mathbf{f}(S_n...S_1\hat{\mathbf{i}}^d) \to \mathbf{i}^d$ requirement can be expressed in a more or less restricted form. For instance, assume, that there exists $0 < K < 1$ for which

$$\left\|\mathbf{f}\left(S_n \dots S_1 \mathbf{i}^d\right) - \mathbf{i}^d\right\| \leqslant K\|\mathbf{f}\left(S_{n-1} \dots S_1 \mathbf{i}^d\right) - \mathbf{i}^d\| \leqslant \cdots \leqslant K^n \|\mathbf{f}(\mathbf{i}^d) - \mathbf{i}^d\|. \quad (6.11)$$

*This requirement trivially guarantees the desired complete stability with a convergence to the desired value. Suppose that $\mathbf{f}(\mathbf{x})$ is invertible and differentiable, and there exists an inverse of $\mathbf{i}^d$ as $\hat{\mathbf{i}} = \mathbf{f}^{-1}(\mathbf{i}^d) \neq 0$. Furthermore, let the Jacobian of $\mathbf{i}$ that is $\partial\mathbf{f}/\partial\mathbf{x}$ be positive definite and of norm considerably smaller than 1. Let us suppose, too, that the actual estimation of the deformed input $\mathbf{i}$ is quite close to $\mathbf{f}^{-1}(\mathbf{i}^d)$. Consequently there must exist two near-identity linear transformations T and S in the chosen Lie group for which $T\hat{\mathbf{i}} = \mathbf{i}$, and $S\mathbf{f}(\mathbf{i}) = \mathbf{i}^d$. Following the classic **perturbation theory**, if variable ξ is chosen to be the "small variable" then the above operators can be written as $T = I + \xi G$, and $S = I + \xi H$ where G and H are certain generators of the group, i.e. the elements of the tangent space at the unit element. Taking into account only the 0th and the 1st order terms in ξ the following estimations can be obtained:*

$$\mathbf{f}(\mathbf{i}) = \mathbf{f}\left((I + \xi G)\hat{\mathbf{i}}\right) \approx \mathbf{f}(\hat{\mathbf{i}}) + \xi\frac{\partial\mathbf{f}}{\partial\mathbf{i}}G\hat{\mathbf{i}} + O\left(\xi^2\right) = \mathbf{i}^d + \xi\frac{\partial\mathbf{f}}{\partial\mathbf{i}}G\hat{\mathbf{i}} + O\left(\xi^2\right), \quad (6.12)$$

$$\mathbf{i}^d = S\mathbf{f}(\mathbf{i}) \approx (I + \xi H)\left(\mathbf{i}^d + \xi\frac{\partial\mathbf{f}}{\partial\mathbf{i}}G\hat{\mathbf{i}}\right) = \mathbf{i}^d + \xi\left(H\mathbf{i}^d + \frac{\partial\mathbf{f}}{\partial\mathbf{i}}G\hat{\mathbf{i}}\right) + O\left(\xi^2\right). \quad (6.13)$$

This implies that $H\mathbf{i}^d \approx -\frac{\partial\mathbf{f}}{\partial\mathbf{i}}G\hat{\mathbf{i}}$. Next approximation can be estimated as

$$\mathbf{f}(S\mathbf{i}) = \mathbf{f}(ST\hat{\mathbf{i}}) = \mathbf{f}\left([I + \xi H]\,[I + \xi G]\hat{\mathbf{i}}\right) \approx \mathbf{i}^d + \xi\frac{\partial\mathbf{f}}{\partial\mathbf{i}}(G + H)\hat{\mathbf{i}} + O\left(\xi^2\right). \quad (6.14)$$

For convergence $\|\mathbf{f}(\mathbf{Si}) - \mathbf{i}^d\| \leqslant K\|\mathbf{f}(\mathbf{i}) - \mathbf{i}^d\|$ *is required:*

$$|\xi| \, \|\frac{\partial \mathbf{f}}{\partial \mathbf{i}}(\mathsf{G} + \mathsf{H})\hat{\mathbf{i}}\| \leqslant K|\xi| \, \|\frac{\partial \mathbf{f}}{\partial \mathbf{i}}\mathsf{G}\hat{\mathbf{i}}\|. \tag{6.15}$$

With the above approximation this means that

$$\|\frac{\partial \mathbf{f}}{\partial \mathbf{i}}\mathsf{H}\hat{\mathbf{i}} - \mathsf{H}\mathbf{i}^d\| \equiv \|\mathsf{H}\mathbf{i}^d - \frac{\partial \mathbf{f}}{\partial \mathbf{i}}\mathsf{H}\hat{\mathbf{i}}\| \leqslant K\|\mathsf{H}\mathbf{i}^d\|. \tag{6.16}$$

Now take into account that in the case of the possession of a perfect dynamic model $\hat{\mathbf{i}} = \mathbf{i}^d$. *For an approximate one* $\hat{\mathbf{i}} \approx \mathbf{i}^d$, *that is both the directions and the norms of these vectors are approximately equal to each other. This implies that the angle between these vectors is acute. Since matrix multiplication is continuous operation, for a finite generator* $\mathsf{H}\,\mathsf{H}\hat{\mathbf{i}} \approx \mathsf{H}\mathbf{i}^d$, *therefore the angle between these vectors is acute, too. Due to the positive definite nature of* $\partial \mathbf{f}/\partial \mathbf{i}$ *multiplication with it also can result in an acute angle between* $\mathsf{H}\mathbf{i}^d$ *and* $(\partial \mathbf{f}/\partial \mathbf{i})\mathsf{H}\hat{\mathbf{i}}$. *Compare this with the requirement given in (6.16)! In the expression at the left hand side* $\mathsf{H}\mathbf{i}^d$ *obtains a correction approximately in the direction opposite to its own one. The small norm of the Jacobian* $\partial \mathbf{f}/\partial \mathbf{i}$ *together with the approximately equal norms of* $\mathsf{H}\mathbf{i}^d$ *and* $\mathsf{H}\mathbf{i}^d$ *guarantee that the norm of the corrected vector will be smaller than that of the original one and will not exceed it in the approximately opposite direction. With this observation the proof is completed.*

It is expedient to offer a few words to the practical significance of these requirements. For instance, in the wide class of Classical Mechanical Systems in the role of this Jacobian the inertia matrix appears. It is well known that this matrix and its inverse are positive definite, but their norm in general can be arbitrary. However, in the Computed Torque control based on a very approximate model in Section 6.2 in $\mathbf{i}^r = \psi(\varphi(\mathbf{i}^d))$ *the rough model of the inertia matrix can be a small positive scalar times the unit matrix* I *that determines the exerted torque. In the joint coordinate accelerations this scalar is multiplied with the inverse of the inertia matrix providing the Jacobian. It is evident that in general this Jacobian will be positive definite, and its norm can be made arbitrarily small by properly setting the scalar value in the rough model. This implies that the proposed method can work in the control of the quite wide class of Classical Mechanical and other physical systems. The simulation presented in this paper well exemplifies this capacity.*

6.6 Acknowledgment

The authors gratefully acknowledge the aid by the Polish-Hungarian Bilateral Technology Co-operation Project for PL-2/01, and the support by the Hungarian National Research Fund OTKA T 034651 supporting their work.

References

1. I. Kováčová, L. Madarász, D. Kováč, J. Vojtko. Neural network linearization of pressure force sensor transfer characteristic. *Proc. 8th IEEE Int. Conf. on Intelligent Engineering Systems INES*, ISBN 973-662-120-0, 2004, pp. 79-82.

2. J. Vaščák, L. Madarász. The design of the evaluated linguistic approximation. *Bulletins for Applied Mathematics*, Technical University of Budapest and Technical University of Košice, October 1995, pp. 247-250.
3. R. Reed. Pruning algorithms – a survey. *IEEE Trans. on Neural Networks*, vol. 4, no. 5, 1993, pp. 740–747.
4. S. Fahlmann, C. Lebiere. The cascade-correlation learning architecture. *Advances in Neural Information Processing Systems*, vol. 2, 1990, pp. 524–532.
5. T. Nabhan, A. Zomaya. Toward generating neural network structures for function approximation. *Neural Networks*, vol. 7, 1994, pp. 89–91.
6. G. Magoulas, N. Vrahatis, G. Androulakis. Effective backpropagation training with variable stepsize. *Neural Networks*, vol. 10, 1997, pp. 69-82.
7. W. Kinnenbrock. Accelerating the standard backpropagation method using a genetic approach. *Neurocomputing*, vol. 6, 1994, pp. 583–588.
8. A. Kanarachos, K. Geramanis. Semi-stochastic complex neural networks. *Proc. of IFAC-CAEA '98 Control Applications and Ergonomics in Agriculture*, 1998, pp. 47–52.
9. D. Tikk, P. Baranyi, T.D. Gedeon, L. Muresan. Generalization of the rule interpolation method resulting always in acceptable conclusion. *Tatra Mountains Math. Publ.*, vol. 21, 2001, pp. 73-91.
10. J. Vaščák, L. Madarász. Similarity relations in diagnosis fuzzy systems. *Journal of Advanced Computational Intelligence*, vol. 4, no. 4, 2000, pp. 246–250.
11. D. Tikk, Gy. Biró, T.D. Gedeon, L.T. Kóczy, J.D. Yang. Improvements and critique on Sugeno's and Yasukawa's qualitative modeling. *IEEE Trans. on Fuzzy Systems*, vol. 10, no. 5, 2002, pp. 596-606.
12. J.K. Tar, I.J. Rudas, J.F. Bitó. Group theoretical approach in using canonical transformations and symplectic geometry in the control of approximately modeled mechanical systems interacting with unmodelled environment. *Robotica*, vol. 15, 1997, pp. 163-169.
13. V.I. Arnold. *Mathematical Methods of Classical Mechanics*. Műszaki Könyvkiadó, Budapest 1985 (in Hungarian).
14. J.K. Tar, I.J. Rudas, J.F. Bitó, K.R. Kozłowski. A modified renormalization algorithm based adaptive control guaranteeing complete stability for a wide class of physical Systems. In: Elmenreich W., Tenreiro Machado J.A., Rudas I.J. (Eds.) *Intelligent Systems at the Service of Mankind*, UBOOKS, Augsburg 2004.
15. E. Miletics, G. Molnárka. Taylor series method with numerical derivatives for initial value problems. *Journal of Computational Methods in Sciences and Engineering*, vol. 3, no. 3, 2003, pp. 319-329.
16. E. Miletics. Energy conservative algorithm for numerical solution of initial value problems of hamiltonian systems. *Proc. IEEE Int. Conf. on Computational Cybernetics ICCC*, ISBN 963 7154 18 3, 2003, pp. 1-4.
17. J.K. Tar, I.J. Rudas, J.F. Bitó, K. Jezernik. A generalized Lorentz group-based adaptive control for DC drives driving mechanical components. *Proc. 10th Int. Conf. on Advanced Robotics ICAR*, ISBN 963 7154 05 1, 2001, pp. 299-305.
18. Y. El Hini. Comparison of tha application of the sympletic and the partially stretched orthogonal transformations in a new branch of adaptive control for mechanical devices. *Proc. 10th Int. Conf. on Advanced Robotics ICAR*, ISBN 963 7154 05 1, 2001, pp. 701-706.

19. J.K. Tar, A. Bencsik, J.F. Bitó, K. Jezernik. Application of a new family of symplectic transformations in the adaptive control of mechanical systems, *Proc. 28th Annual Conf. of the IEEE Industrial Electronics Society*, ISBN 0-7803-7474-6, 2002, pp. 1499–1504.

20. J.K. Tar, I.J. Rudas, L. Horváth, K. Kozłowski. Analysis of the effect of backlash and joint acceleration measurement noise in the adaptive control of electro-mechanical systems. *Proc. IEEE Int. Symp. on Industrial Electronics ISIE*, ISBN 0-7803-7912-8 (CD issue), file BF-000965.pdf, 2003.

21. R. Lozano, I. Fantoni, D.J. Block. Stabilization of the inverted pendulum around its homoclinic orbit. *Systems & Control Letters*, vol. 40, no. 3, 2000, pp. 197-204.

22. J.K. Tar, I.J. Rudas, L. Horváth, S.G. Tzafestas. Adaptive control of the double inverted pendulum based on novel principles of soft computing. *Proc. Int. Conf. in Memoriam John von Neumann*, ISBN 963 7154213, 2003, pp. 257–268.

23. T. Roska. *Development of kilo real-time frame rate TeraOPS computational capacity topographic microprocessors*. Plenary Lecture at 10th Int. Conf. on Advanced Robotics ICAR, 2001, Budapest, Hungary, August 22-25, 2001.

7

Example Applications of Fuzzy Reasoning and Neural Networks in Robot Control

Waldemar Wróblewski

Chair of Control and Systems Engineering, Poznań University of Technology
ul. Piotrowo 3a, 60-965 Poznań, Poland waldemar.wroblewski@put.poznan.pl

7.1 Introduction

Most typical tasks met in the area of robotics – control of robot manipulators and mobile robots – are performed with use of mathematical models of these devices and their environment. For some time, there have appeared attempts of implementation of artificial intelligence methods in robot control. The paper presents two such applications.

First, an implementation of neural networks in control of a robot manipulator is presented. Many neural network controllers for robot arms have been proposed in literature [2, 7, 9, 10]. Usually they are based on full state measurements, where joint positions and velocities are required. Joint position measurements with use of encoders usually give accurate results, while joint velocity measurements with use of tachometers are disturbed by noise. This may reduce the performance of the manipulator. Therefore the idea of replacing velocity measurements by joint velocity estimation with use of an observer has been developed in several papers. Here, the effectiveness of two alternative neural controllers presented in [9] and [8] is compared. In the first one, by Kim and Lewis [9], the neural network is used as an observer to estimate the joint velocity. The other feedforward neural network is used as a controller. The weights of both networks are tuned on-line. On the other side, Jungbeck and Madrid in [8] introduced an optimal controller, also implemented as a feedforward neural network.

The next part of the paper is devoted to an implementation of fuzzy logic reasoning in control of mobile robots. The idea of using fuzzy logic to control is not new and comes from attempts of building pseudo-intelligent inference and decision systems. Fuzzy inference resembles human decision process – input signals are estimated in a fuzzy, subjective way. The results affect control signals by inference rules, which are assumed to be a set of simple fuzzy rules resulting from human experience in solving the considered problem.

Mobile robots are generally nonholonomic systems due to the constraints imposed on their kinematics. Namely, the equations describing the constraints cannot be integrated symbolically to obtain explicit relations between robot positions in local and global coordinate frames. There are many algorithms for control of mobile robots described in literature [1]. However, control of nonholonomic

K. Kozłowski (Ed.): Robot Motion and Control, LNCIS 335, pp. 113–127, 2006.
© Springer-Verlag London Limited 2006

systems is a difficult problem. In tasks such as trajectory tracking or following a predefined trajectory, it is necessary to control simultaneously the position and orientation of the robot, as well as its velocity.

The paper is organized as follows. In Section 7.2 the equations of the robot manipulator dynamics and of the neural network observer are outlined. Then, in Section 7.3, both considered controllers are presented and example simulation results for a double pendulum are presented. Section 7.4 presents a kinematics model of type (3,0) mobile robot [1] and the structure of its control system. In Section 7.5, a fuzzy controller for the trajectory tracking task is presented with example simulation results for this system compared to those of a classical PD controller [4,5].

7.2 Mathematical Models of the Manipulator and of the Neural Network Observer

7.2.1 Manipulator Dynamics

The mathematical model of the dynamics of an N-link robot manipulator is described by the following equation [4,5]:

$$\boldsymbol{\tau} = \mathbf{M}(\boldsymbol{q})\ddot{\boldsymbol{q}} + \mathbf{C}(\boldsymbol{q}, \dot{\boldsymbol{q}})\dot{\boldsymbol{q}} + \boldsymbol{g}(\boldsymbol{q}) + \mathbf{F}_v\dot{\boldsymbol{q}} + \boldsymbol{f}_c(\dot{\boldsymbol{q}}) + \boldsymbol{\tau}_d, \qquad (7.1)$$

where $\mathbf{M}(\boldsymbol{q}) \in \mathbb{R}^{N \times N}$ is a symmetric positive definite uniformly bounded matrix, $\mathbf{C}(\boldsymbol{q}, \dot{\boldsymbol{q}})\dot{\boldsymbol{q}} \in \mathbb{R}^{N \times 1}$ is the vector of Coriolis/centrifugal forces, $\mathbf{F}_v \in \mathbb{R}^{N \times N}$ is the diagonal matrix of viscous friction coefficients, $\boldsymbol{f}_c(\boldsymbol{q}) \in \mathbb{R}^{N \times 1}$ is the vector of Coulomb friction forces, and $\boldsymbol{g}(\boldsymbol{q}) \in \mathbb{R}^{N \times 1}$ is the vector of gravitational forces. The ($N \times 1$) vectors $\boldsymbol{q}$, $\dot{\boldsymbol{q}}$, $\ddot{\boldsymbol{q}}$ denote generalized position, velocity, and acceleration, respectively. The vectors $\boldsymbol{\tau}$, $\boldsymbol{\tau}_d \in \mathbb{R}^{N \times 1}$ denote control input torque and bounded unknown disturbance torque, respectively.

7.2.2 Approximating Neural Networks

A two-layer feedforward neural network [3] is shown in Fig. 7.1. Activation functions of the hidden layer have been denoted as $\sigma_i(\cdot)$ (they are combined as vector $\boldsymbol{\sigma}$), while the output layer is linear. There are N and N_o neurons in the hidden and the output layer, respectively. The weight matrix of the output layer is $\mathbf{W} \in \mathbb{R}^{(N+1) \times N_o}$ – its vertical size has been extended by 1 to take into account the bias input '1'. Thus, the extended vector of the input layer is $\bar{\boldsymbol{\sigma}} = \begin{bmatrix} \boldsymbol{\sigma}^T & 1 \end{bmatrix}^T$.

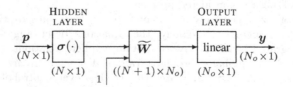

Fig. 7.1. Two-layer feedforward neural network

This network may be used as an approximator [3] when the functions $\sigma_i(\boldsymbol{p})$ form a basis. It approximates a nonlinear function $\boldsymbol{y}(\boldsymbol{p}) \in \mathbb{R}^{N_o \times 1}$:

$$\boldsymbol{y}(\boldsymbol{p}) = \mathbf{W}^T \bar{\sigma}(\boldsymbol{p}) + \varepsilon(\boldsymbol{p}) \tag{7.2}$$

with a reconstruction error $\varepsilon(\boldsymbol{p})$. So, an estimate $\hat{\boldsymbol{y}}(\boldsymbol{p})$ of $\boldsymbol{y}(\boldsymbol{p})$ is calculated by the neural network:

$$\hat{\boldsymbol{y}}(\boldsymbol{p}) = \widehat{\mathbf{W}}^T \bar{\sigma}(\boldsymbol{p}), \tag{7.3}$$

where $\widehat{\mathbf{W}}$ is an estimate of the ideal neural network weight matrix $\mathbf{W}$ and can be obtained off-line or on-line by learning algorithms.

7.2.3 Neural Network Observer

In this section the concept and model of neural implementation of the observer introduced in [8, 9] is briefly presented. The link position and velocity estimation errors are defined as:

$$\tilde{\boldsymbol{q}} = \boldsymbol{q} - \hat{\boldsymbol{q}}, \qquad \dot{\tilde{\boldsymbol{q}}} = \dot{\boldsymbol{q}} - \dot{\hat{\boldsymbol{q}}}, \tag{7.4}$$

where $\hat{}$ denotes estimated quantities. Assuming state variable vectors $\boldsymbol{x}_1 = \boldsymbol{q}$, $\boldsymbol{x}_2 = \dot{\boldsymbol{q}}$, the robot dynamics equation (7.1) may be rewritten as:

$$\begin{cases} \dot{\boldsymbol{x}}_1 = \boldsymbol{x}_2, \\ \dot{\boldsymbol{x}}_2 = \boldsymbol{h}_o(\boldsymbol{x}_1, \boldsymbol{x}_2) + \mathbf{M}^{-1}(\boldsymbol{x}_1)\boldsymbol{\tau}, \end{cases} \tag{7.5}$$

where $\boldsymbol{h}_o(\boldsymbol{x}_1, \boldsymbol{x}_2)$ is the following nonlinear function:

$$\boldsymbol{h}_o(\boldsymbol{x}_1, \boldsymbol{x}_2) = -\mathbf{M}^{-1}(\boldsymbol{x}_1)\left[\mathbf{C}(\boldsymbol{x}_1, \boldsymbol{x}_2)\boldsymbol{x}_2 + \boldsymbol{y}(\boldsymbol{x}_1) + \mathbf{F}_v \dot{\boldsymbol{x}}_1 + \boldsymbol{f}_c(\boldsymbol{x}_1) + \boldsymbol{\tau}_d\right]. \tag{7.6}$$

It is assumed in [8, 9], that the inertia matrix $\mathbf{M}(\boldsymbol{x}_1)$ is known but, due to uncertainties of Coriolis/centrifugal terms and a complicated form of friction terms, $\boldsymbol{h}_o(\boldsymbol{x}_1, \boldsymbol{x}_2)$ is represented by the neural network which shows approximation properties. This network is described by the following formula:

$$\boldsymbol{h}_o(\boldsymbol{x}_1, \boldsymbol{x}_2) = \mathbf{W}_o^T \boldsymbol{\sigma}_o(\boldsymbol{x}_1, \boldsymbol{x}_2) + \varepsilon_0(\boldsymbol{x}_1, \boldsymbol{x}_2), \tag{7.7}$$

where $\boldsymbol{\sigma}_o(\boldsymbol{x}_1, \boldsymbol{x}_2) \in \mathbb{R}^{N_o \times 1}$ is the input vector of the network, N_o is the number of input basic functions, and $\mathbf{W}_o \in \mathbb{R}^{N_o \times N}$ is the weight matrix of the neural network.

The functional estimate of (7.7) in terms of $\hat{\boldsymbol{x}}_1$, $\hat{\boldsymbol{x}}_2$ is given by

$$\hat{\boldsymbol{h}}_o(\hat{\boldsymbol{x}}_1, \hat{\boldsymbol{x}}_2) = \widehat{\mathbf{W}}_o^T \boldsymbol{\sigma}_o(\hat{\boldsymbol{x}}_1, \hat{\boldsymbol{x}}_2) \tag{7.8}$$

with the current values of the elements of the weight matrix $\widehat{\mathbf{W}}_o$ calculated by the tuning algorithm.

The proposed observer has the form:

$$\begin{cases} \dot{\hat{\boldsymbol{z}}}_1 = \hat{\boldsymbol{x}}_2 + k_D \tilde{\boldsymbol{x}}_1, \\ \dot{\hat{\boldsymbol{z}}}_2 = \widehat{\mathbf{W}}_o^T \boldsymbol{\sigma}_o(\hat{\boldsymbol{x}}_1, \hat{\boldsymbol{x}}_2) + \mathbf{M}^{-1}(\boldsymbol{x}_1)\boldsymbol{\tau} + \mathbf{K}\tilde{\boldsymbol{x}}_1, \end{cases} \tag{7.9}$$

where $\tilde{\boldsymbol{x}}_1 = \boldsymbol{x}_1 - \hat{\boldsymbol{x}}_1$, $k_D > 0$, $\mathbf{K} = \mathbf{K}^T$ and all non-zero elements of $\mathbf{K}$ are > 0. The estimates $\hat{\boldsymbol{x}}_1$, $\hat{\boldsymbol{x}}_2$ of the system state (7.5) are [9]:

$$\begin{cases} \hat{\boldsymbol{x}}_1 = \hat{\boldsymbol{z}}_1, \\ \hat{\boldsymbol{x}}_2 = \hat{\boldsymbol{z}}_2 + k_P \tilde{\boldsymbol{x}}_1, \end{cases} \tag{7.10}$$

(not $\hat{\boldsymbol{z}}_2 = \hat{\boldsymbol{z}}_2 + k_P \tilde{\boldsymbol{x}}_1$ as erroneously written in [9]).

Therefore, the dynamic equations of the observer may be rewritten in terms of $\hat{\boldsymbol{x}}_1$, $\hat{\boldsymbol{x}}_2$ as follows:

$$\begin{cases} \dot{\hat{\boldsymbol{x}}}_1 = \hat{\boldsymbol{x}}_2 + k_D \tilde{\boldsymbol{x}}_1, \\ \dot{\hat{\boldsymbol{x}}}_2 = \dot{\hat{\boldsymbol{x}}}_2 + k_P \dot{\tilde{\boldsymbol{x}}}_1 = \widehat{\mathbf{W}}_o^T \boldsymbol{\sigma}_o(\hat{\boldsymbol{x}}_1, \hat{\boldsymbol{x}}_2) + \mathbf{M}^{-1}(\hat{\boldsymbol{x}}_1)\boldsymbol{\tau} + \mathbf{K}\tilde{\boldsymbol{x}}_1 + k_P \dot{\tilde{\boldsymbol{x}}}_1. \end{cases} \tag{7.11}$$

This observer is not implementable as it requires the value of the error $\dot{\tilde{\boldsymbol{x}}}_1$, which is not available. The proposed observer [9] implies particular structure of the control system with a feedforward neural network inserted in the feedback path.

Finally, the observer dynamics model expressed in terms of the estimate error is as follows:

$$\begin{cases} \dot{\tilde{\boldsymbol{x}}}_1 = \tilde{\boldsymbol{x}}_2 - k_D \tilde{\boldsymbol{x}}_1, \\ \dot{\tilde{\boldsymbol{x}}}_2 = -k_P \tilde{\boldsymbol{x}}_2 - (\mathbf{K} - k_P k_D \mathbf{U})\tilde{\boldsymbol{x}}_1 + \widehat{\mathbf{W}}_o^T \boldsymbol{\sigma}_o(\hat{\boldsymbol{x}}_1, \hat{\boldsymbol{x}}_2), \end{cases} \tag{7.12}$$

where $\mathbf{U} \in \mathbb{R}^{N \times N}$ is a unit matrix. This form is useful for deriving the weight update rule of the neural network observer in terms of the known position estimation error.

It is proven in [9] that if the observer satisfies the following conditions:

$$k_P > k_D^2 - N_o/2, \tag{7.13}$$

$$\lambda_m(\mathbf{K}) > (k_P^2 + N_o k_D^2)/(2k_D), \tag{7.14}$$

where $\lambda_b(\mathbf{K})$ is the smallest eigenvalue of $\mathbf{K}$, then the weight tuning

$$\dot{\widehat{\mathbf{W}}}_o = -k_D \mathbf{F}_o \boldsymbol{\sigma}_o(\hat{\boldsymbol{x}}_1, \hat{\boldsymbol{x}}_2)\tilde{\boldsymbol{x}}_1^T - \varkappa_o \mathbf{F}_o \|\tilde{\boldsymbol{x}}_1\| \widehat{\mathbf{W}}_o - \varkappa_o \mathbf{F}_o \widehat{\mathbf{W}}_o, \tag{7.15}$$

where $\varkappa_o > 0$ is a sufficiently large parameter, $\mathbf{F}_o = \mathbf{F}_o^T$ and all non-zero elements of $\mathbf{F}_o$ are > 0, leads to uniformly ultimately bounded state estimation error $\tilde{\boldsymbol{x}}_1$, $\tilde{\boldsymbol{x}}_2$, and weight estimation error $\widetilde{\mathbf{W}}_o$.

7.3 Comparison of Alternative Controllers

7.3.1 Neural Controllers

Let us assume that the desired trajectory $\boldsymbol{q}_d(t) \in \mathbb{R}^N$ (link position vector) and its time derivatives are known. The manipulator link tracking errors $\boldsymbol{e}(t)$, $\dot{\boldsymbol{e}}(t)$ are defined as:

$$\boldsymbol{e}(t) = \boldsymbol{q}_d(t) - \boldsymbol{q}(t) = \boldsymbol{q}_d(t) - \boldsymbol{x}_1(t), \tag{7.16}$$

$$\dot{\boldsymbol{e}}(t) = \dot{\boldsymbol{q}}_d(t) - \dot{\boldsymbol{q}}(t) = \dot{\boldsymbol{q}}_d(t) - \dot{\boldsymbol{x}}_1(t), \tag{7.17}$$

and the mixed tracking error:

$$\hat{\boldsymbol{r}}(t) = \dot{\boldsymbol{e}}(t) + \boldsymbol{\Lambda}\boldsymbol{e}(t), \tag{7.18}$$

where $\boldsymbol{\Lambda}$ is a constant positive definite matrix. The manipulator dynamics model can be written in terms of $\hat{\boldsymbol{r}}(t)$ as follows:

$$\mathbf{M}(\boldsymbol{x}_1)\dot{\hat{\boldsymbol{r}}}(t) = -\mathbf{C}(\boldsymbol{x}_1, \boldsymbol{x}_2)\hat{\boldsymbol{r}}(t) - \boldsymbol{\tau}(t) + \boldsymbol{h}_c(\boldsymbol{x}_1, \boldsymbol{x}_2), \qquad (7.19)$$

where

$$\boldsymbol{h}_c(\boldsymbol{x}_1, \boldsymbol{x}_2) = \mathbf{M}(\boldsymbol{x}_1)(\ddot{\boldsymbol{q}}_d + \dot{\tilde{\boldsymbol{x}}}_2 + \boldsymbol{\Lambda}\dot{\boldsymbol{e}}) + \mathbf{C}(\boldsymbol{x}_1, \boldsymbol{x}_2)(\dot{\boldsymbol{q}}_d + \tilde{\boldsymbol{x}}_2 + \boldsymbol{\Lambda}\boldsymbol{e}) +$$
$$+ \boldsymbol{g}(\boldsymbol{x}_1) + \mathbf{F}_v\boldsymbol{x}_2 + \boldsymbol{f}_c(\boldsymbol{x}_2) + \boldsymbol{\tau}_d \qquad (7.20)$$

represents the unknown dynamics of the robot arm.

Kim and Lewis in [9] construct the output feedback neural controller:

$$\boldsymbol{\tau}(t) = \widehat{\mathbf{W}}_{c1}^T \boldsymbol{\sigma}_{c1}(\hat{\boldsymbol{x}}_1, \hat{\boldsymbol{x}}_2) + \mathbf{K}_v\hat{\boldsymbol{r}}(t) - k_z \frac{\hat{\boldsymbol{r}}(t)}{\|\hat{\boldsymbol{r}}(t)\|}, \qquad (7.21)$$

where the last component is a robustifying term, with the weight tuning rule:

$$\dot{\widehat{\mathbf{W}}}_{c1} = \mathbf{F}_{c1}\boldsymbol{\sigma}_c(\hat{\boldsymbol{x}}_1, \hat{\boldsymbol{x}}_2)\hat{\boldsymbol{r}}^T - \varkappa_{c1}\mathbf{F}_{c1}\|\hat{\boldsymbol{r}}\|\widehat{\mathbf{W}}_{c1}, \qquad (7.22)$$

where $\varkappa_{c1} > 0$, $\mathbf{F}_{c1} = \mathbf{F}_{c1}^T$ and all non-zero elements of $\mathbf{F}_{c1}$ are > 0. It has been shown in [9] that $\tilde{\boldsymbol{x}}_1$, $\tilde{\boldsymbol{x}}_2$, $\hat{\boldsymbol{r}}(t)$, and $\widehat{\mathbf{W}}_{c1}$ are uniformly ultimately bounded.

Alternatively, Jungbeck and Madrid in [8] propose another neural controller assuming that the input control torque is

$$\boldsymbol{\tau}(t) = \boldsymbol{h}_c(\boldsymbol{x}_1, \boldsymbol{x}_2) - \boldsymbol{u}(t), \qquad (7.23)$$

where $\boldsymbol{u}(t) \in \mathbb{R}^{N \times 1}$ is a temporary control input to be optimized. They obtain the following system:

$$\dot{\tilde{\boldsymbol{z}}} = \begin{bmatrix} \dot{\tilde{\boldsymbol{e}}} \\ \dot{\hat{\boldsymbol{r}}} \end{bmatrix} = \begin{bmatrix} -\boldsymbol{\Lambda} & \mathbf{U} \\ \mathbf{0} & -\mathbf{M}^{-1}\mathbf{C} \end{bmatrix} \begin{bmatrix} \boldsymbol{e} \\ \hat{\boldsymbol{r}} \end{bmatrix} + \begin{bmatrix} \mathbf{0} \\ \mathbf{M}^{-1} \end{bmatrix} \boldsymbol{u}(t) = \mathbf{A}(\boldsymbol{x}_1, \boldsymbol{x}_2)\tilde{\boldsymbol{z}} + \mathbf{B}(\boldsymbol{x}_1)\boldsymbol{u}(t) \quad (7.24)$$

with $\mathbf{A}(\boldsymbol{x}_1, \boldsymbol{x}_2) \in \mathbb{R}^{2N \times 2N}$, $\mathbf{B}(\boldsymbol{x}_1) \in \mathbb{R}^{2N \times N}$, $\tilde{\boldsymbol{z}}(t) \in \mathbb{R}^{2N \times 1}$. To optimize $\boldsymbol{u}(t)$, the quadratic index defined as

$$J(\boldsymbol{u}) = \frac{1}{2} \int_{t_0}^{\infty} \left(\tilde{\boldsymbol{z}}^T \mathbf{Q}\tilde{\boldsymbol{z}} + \boldsymbol{u}^T \mathbf{R}\boldsymbol{u} \right) dt, \qquad (7.25)$$

which is subject to constraints (7.24), should be minimized. For the optimal control $\boldsymbol{u}^*(t)$ there exists a function

$$V = \frac{1}{2}\tilde{\boldsymbol{z}}^T \mathbf{P}(\boldsymbol{x}_1)\tilde{\boldsymbol{z}} = \frac{1}{2}\tilde{\boldsymbol{z}}^T \begin{bmatrix} \mathbf{K}_2 & \mathbf{0} \\ \mathbf{0} & \mathbf{M}(\boldsymbol{x}_1) \end{bmatrix} \tilde{\boldsymbol{z}}, \qquad (7.26)$$

where $\boldsymbol{\Lambda}$ in (7.18) and $\boldsymbol{K}_2$ in (7.26) can be found by solving the Riccati differential equation:

$$\mathbf{PA} + \mathbf{A}^T\mathbf{P} - \mathbf{PBR}^{-1}\mathbf{B}^T\mathbf{P} + \dot{\mathbf{P}} + \mathbf{Q} = 0. \qquad (7.27)$$

Then, the optimal control $\boldsymbol{u}^*(t)$ minimizing (7.25) subject to (7.24) is

$$\boldsymbol{u}^*(t) = -\mathbf{R}^{-1}\mathbf{B}^T\mathbf{P}(\boldsymbol{x}_1)\tilde{\boldsymbol{z}} = -\mathbf{R}^{-1}\hat{\boldsymbol{r}}. \qquad (7.28)$$

Taking into account this optimal control to the model (7.19) we obtain [8]:

$$\dot{\tilde{z}} = (A + BR^{-1}B^T P)\tilde{z} + B \left[\widetilde{W}_{c2}^T \sigma_c(t) + \tau_d(t) + k_z \frac{\hat{r}(t)}{\|\hat{r}(t)\|} \right], \qquad (7.29)$$

where k_z is bounded and the last component is a robustifying term. The adaptive learning rule for the neural network weights:

$$\dot{\widehat{W}}_{c2} = F_{c2}\sigma_c(\hat{p})(B^T P(x_1)\tilde{z})^T - \varkappa_{c2}\|\tilde{z}\|\widehat{W}_{c2}, \qquad (7.30)$$

where $\varkappa_{c2} > 0$, $F_{c2} = F_{c2}^T$ and all non-zero elements of F_{c2} are > 0, leads to uniformly ultimately bounded $e(t)$, $\hat{r}(t)$, and $\widehat{W}_{c2}$.

7.3.2 Simulation Results

Both output feedback controllers have been tested using the mathematical model of the double pendulum, which resembles a 2-link planar manipulator. It is assumed that the lengths of both links are $l_1 = l_2 = 1$m and the masses are $m_1 = 1$kg, $m_2 = 0.3$kg. The Coriolis/centrifugal terms are not taken into account. The matrix of viscous friction coefficients is $F_v = \text{diag}\{0.85\ 0.85\}$; Coulomb friction forces are equal to zero. Both joints of the manipulator have to follow the reference trajectories

$$q_{d1} = 1 + 0.5\sin(5\pi t), \qquad q_{d2} = 1 + 0.5\cos(5\pi t)$$

in 2 seconds and the sampling time is $T_p = 0.005$s.

All neurons in hidden layers are sigmoidal, but their curvature can be adjusted individually. The neural observer has 10 hidden neurons and the input vector is:

$$p_o = \begin{bmatrix} \tilde{q}^T & \hat{q}^T & \dot{\hat{q}}^T & \|\tilde{q}_1\| & \|\tilde{q}_2\| & \|\hat{q}_1\| & \|\hat{q}_2\| \end{bmatrix}.$$

Each neural controller has 14 hidden neurons and their input vector is:

$$p_c = \begin{bmatrix} q^T & \dot{q}^T & e^T & \dot{e}^T & r^T & q_d^T & \dot{q}_d^T \end{bmatrix}.$$

The learning process of all networks has been conducted on-line.

The simulation results are shown in Fig. 7.2. The time transients of the angular positions for the first and second (optimal) controller are drawn as solid and dotted lines, respectively. The reference trajectories are drawn as dashed lines.

Both controllers make the pendulum follow the reference trajectories. The learning process is faster for the second link – after some time the current trajectory follows precisely the reference one. It is interesting that, after the initial phase of learning, the results for both controller implementations are practically the same. The first controller, though non-optimal, works similarly to the other one.

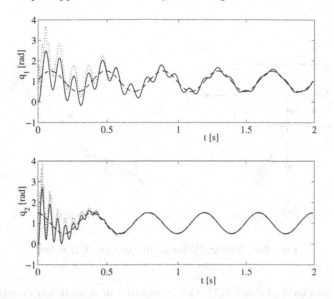

Fig. 7.2. Simulation results for output feedback controllers

7.4 Wheeled Platform and its Control Scheme

7.4.1 Kinematics Model of the Mobile Robot

The position and orientation of a mobile robot moving on a plane, expressed in the base coordinate frame $0X_0Y_0$ (see: Fig. 7.3), is described by the vector:

$$z = \begin{bmatrix} x & y & \theta \end{bmatrix}^T,\qquad (7.31)$$

where x, y are the coordinates of the origin of the local coordinate frame RX_rY_r attached to the robot, located at its centre. The angle θ expresses orientation of the frame RX_rY_r with respect to $0X_0Y_0$. This orientation can be represented with use of the rotation matrix:

$$\mathbf{R}(\theta) = \begin{bmatrix} \cos\theta & \sin\theta & 0 \\ -\sin\theta & \cos\theta & 0 \\ 0 & 0 & 1 \end{bmatrix}.\qquad (7.32)$$

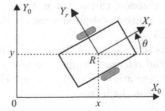

Fig. 7.3. Position and orientation of a mobile robot

The type (3,0) mobile robot, shown in Fig. 4(a), is equipped with three castor wheels, i.e. off-centered orientable wheels (see: Fig. 4(b)). Its geometric parameters

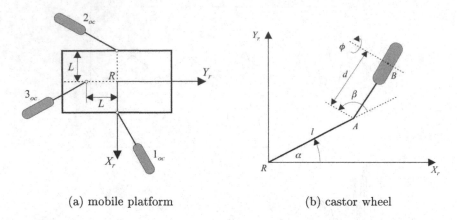

(a) mobile platform   (b) castor wheel

Fig. 7.4. Kinematic structure of the (3,0) robot

have been collected in Table 7.1 [1]. This structure can move in any direction at each time instant, without former re-orientation. It has been assumed that the particular parts of the robot (wheels, platform) cannot be deformed and the wheels roll on a plane without slippery effects.

Table 7.1. Parameters of the type (3,0) robot

Wheel	α	β	l
1_{oc}	0	var	L
2_{oc}	π	var	L
3_{oc}	$3\pi/2$	var	L

7.4.2 Control Scheme

The controller structure results from kinematics equations – it should be two-level for mobile robots. The master controller generates control signals vector $\boldsymbol{u}$ to minimize the position/orientation and velocity errors. The slave controller has to convert control signals, i.e. the coordinates of the vector $\boldsymbol{u}$, to the driving torques of motors. Taking this into account, the control system scheme for the trajectory tracking task may be as presented in Fig. 7.5.

The notation used in this scheme is as follows: v_r is the reference linear velocity of the robot's center, ω_r – the reference angular velocity of the robot, $\boldsymbol{z}_r$, $\boldsymbol{z}$ – the desired and actual position/orientation vectors, respectively, described by Eq. (7.31). The reference velocity vector has the form:

$$\dot{\boldsymbol{z}}_r = \begin{bmatrix} \dot{x}_r \\ \dot{y}_r \\ \dot{\theta}_r \end{bmatrix} = \begin{bmatrix} v_r \cos \theta_r \\ v_r \sin \theta_r \\ \omega_r \end{bmatrix}, \qquad (7.33)$$

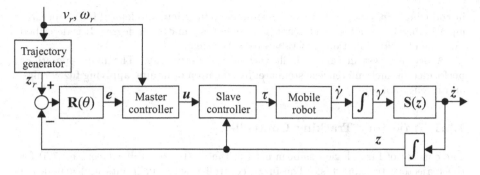

Fig. 7.5. Mobile robot control scheme for the trajectory tracking task

$\mathbf{R}(\theta)$ is the rotation matrix defined by Eq. (7.32), and e is the vector of position/orientation error:

$$e = \begin{bmatrix} e_x \\ e_y \\ e_\theta \end{bmatrix} = \mathbf{R}(\theta)(z_r - z) = \mathbf{R}(\theta) \begin{bmatrix} x_r - x \\ y_r - y \\ \theta_r - \theta \end{bmatrix} . \tag{7.34}$$

Next, $\gamma = \begin{bmatrix} v\ \omega \end{bmatrix}^T$ is a two-element velocity vector of the robot. It is related to the vector $\dot{z}$ of actual velocities by the following equation:

$$\dot{z} = \mathbf{S}(z)\gamma, \tag{7.35}$$

where $\mathbf{S}(z)$ is the kinematics matrix [1] of the mobile robot. The vector τ represents driving torques.

The slave controller makes use of the robot dynamics model and, generally, is independent from the choice of the master controller.

Performance of the trajectory tracking fuzzy control algorithm will be compared with the classical PD controller [4, 5], described by the following equation:

$$u = \begin{bmatrix} u_1 \\ u_2 \\ u_3 \end{bmatrix} = \begin{bmatrix} -k_{Px}e_x - k_{Dx}\dot{e}_x \\ -k_{Py}e_y - k_{Dy}\dot{e}_y \\ -k_{P\theta}e_\theta - k_{D\theta}\dot{e}_\theta \end{bmatrix} , \tag{7.36}$$

where

$$\dot{e}_x = v_r \cos\theta_r - v_x, \tag{7.37}$$

$$\dot{e}_y = v_r \sin\theta_r - v_y, \tag{7.38}$$

v_x, v_y are components of the current velocity in directions x and y. Moreover, k_{Px}, k_{Py}, $k_{P\theta}$ are gains of the proportional part of the controller, and k_{Dx}, k_{Dy}, $k_{D\theta}$ are gains of the differential part of the controller.

7.5 Implementation of the Fuzzy Controller

Fuzzy control is based on the idea of fuzzy logic, which, unlike classical two-value Boolean logic, operates on values from the range $\langle 0, 1 \rangle$. This approach allows introducing terms such as fuzzy set, membership function, and logical fuzzy operator [6]. Membership to a fuzzy set is not precise, but is a matter of degree. It is described by a membership function μ of values from the range $\langle 0, 1 \rangle$.

A decision system can be built with use of fuzzy logic. The main operations performed by such inference system are: fuzzification of inputs, applying fuzzy rules, and implication [6].

7.5.1 Trajectory Tracking Controller

The concept of fuzzy logic has been used in the master controller (see: Fig. 7.5) for the trajectory tracking task. The fuzzy controller with six inputs and generating three control signals has been proposed for the considered robot.

The input signals are as follows:

e_x, e_y − position errors of Cartesian coordinates (it has been assumed that $e_x, e_y \in \langle -1, 1 \rangle$ [m]), they are defined identically and for e_x may assume the following fuzzy values: ex_ud (negative large), ex_um (negative small), ex_z (zero), ex_dm (positive small), ex_dd (positive large) defined by membership functions μ in Fig. 6(a);

e_θ − orientation angle error (it has been assumed that $e_\theta \in \langle -\pi, \pi \rangle$), it may assume the following fuzzy values: eth_ud (negative large), eth_us (negative middle), eth_um (negative small), eth_z (zero), eth_dm (positive small), eth_ds (positive middle), eth_dd (positive large) defined by membership functions μ in Fig. 6(b);

e_{vx}, e_{vy} − linear velocity coordinates errors (assumed to be $e_{vx}, e_{vy} \in \langle -3, 3 \rangle$ [m/s]), they are defined identically and for e_{vx} may assume five fuzzy values: evx_ud, evx_um, evx_z, ev_dm, evx_dd defined by membership functions μ presented in Fig. 6(c);

ω − rotation, i.e. angular velocity, (assumed to be $\omega \in \langle -20, 20 \rangle$ [rad/s]) and may assume five fuzzy values: rot_ud, rot_um, rot_z, rot_dm, rot_dd; the membership functions are presented in Fig. 6(d).

Control signals are chosen as coordinates of the control vector $\boldsymbol{u} = \begin{bmatrix} u_x & u_y & u_\theta \end{bmatrix}$ and are as follows:

u_x, u_y − it has been assumed that $u_x, u_y \in \langle -40, 40 \rangle$, they are defined identically and for u_x may assume the following fuzzy values: ux_ud (negative large), ux_um (negative small), ux_z (zero), ux_dm (positive small), ux_dd (positive large) defined by membership functions μ in Fig. 6(e);

u_θ − it has been assumed that $u_\theta \in \langle -60, 60 \rangle$ and may assume five fuzzy values: uth_ud (negative large), uth_um (negative small), uth_z (zero), uth_dm (positive small), uth_dd (positive large); the membership functions are presented in Fig. 6(f).

Control rules are described independently for orientation and position of the robot.

Two inputs are used in case of orientation: angle error e_θ and angular velocity ω. Using them, there have been defined 19 rules describing output u_θ controlling the angular acceleration of the robot. The error e_θ is calculated from Eq. (7.34). The rules are formulated in order to minimize the orientation error with simultaneous

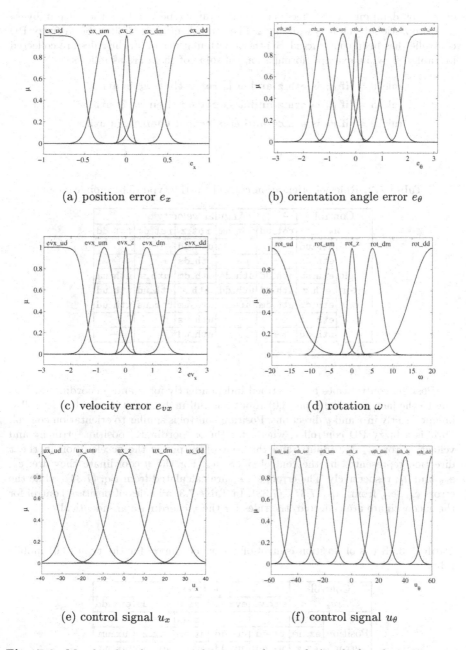

(a) position error e_x

(b) orientation angle error e_θ

(c) velocity error e_{vx}

(d) rotation ω

(e) control signal u_x

(f) control signal u_θ

Fig. 7.6. Membership functions of input and control signals for the trajectory tracking fuzzy controller

overshoot damping. This effect has been obtained be exerting the output by e_θ in a proportional manner, while by ω like a differential controller. So, a fuzzy PD controller has been constructed. Synthetic formulations of all the rules are collected in Table 7.2, while example formulations of some of them are as follows:

rule10 = 'if e_θ is eth_z and ω is rot_z then u_θ is uth_z',

rule15 = 'if e_θ is eth_z and ω is rot_dm then u_θ is uth_ud',

rule19 = 'if e_θ is eth_dm and ω is rot_dd then u_θ is uth_ud'.

Table 7.2. Rules of orientation control for the type (3,0) mobile robot

Control u_θ		Angular velocity ω				
		rot_ud	rot_um	rot_z	rot_dm	rot_dd
	eth_ud	–	–	uth_dd	–	–
	eth_us	–	–	uth_ds	–	–
Angle	eth_um	uth_dd	uth_ds	uth_dm	uth_z	uth_um
error	eth_z	uth_dd	uth_dd	uth_z	uth_ud	uth_ud
e_θ	eth_dm	uth_dm	uth_z	uth_um	uth_us	uth_ud
	eth_ds	–	–	uth_us	–	–
	eth_dd	–	–	uth_ud	–	–

Position control rules are described independently for x and y coordinates. It is due to the fact, that the type (3,0) robot has full mobility, so it may be controlled independently in x and y directions. Position control is similar to orientation control. So it is a fuzzy PD controller, where, for the x coordinate, position error e_x and velocity error e_{vx} are inputs, and the control output is the acceleration in the x direction represented by the control signal u_x. For the y coordinate they are: e_y, e_{vy} and u_y, respectively. The errors e_x, e_y are calculated from Eq. (7.34), while the errors e_{vx}, e_{vy} from Eqs. (7.37), (7.38). In Table 7.3, all rules of position control for the x coordinate are collected, the rules for the y coordinate are identical.

Table 7.3. Rules of position control of the x coordinate for the type (3,0) mobile robot

Control u_x		Velocity error e_{vx}				
		evx_ud	evx_um	evx_z	evx_dm	evx_dd
	ex_ud	–	–	ux_dd	–	–
Position	ex_um	ux_dd	ux_dd	ux_dm	ux_z	ux_um
error	ex_z	ux_dd	ux_dd	ux_z	ux_ud	ux_ud
e_x	ex_dm	ux_dm	ux_z	ux_um	ux_ud	ux_ud
	ex_dd	–	–	ux_ud	–	–

For several input combinations of e_θ and ω as well as e_x and e_{vx} the rules have not been implemented because, as tested in simulations, they do not improve control quality, but only increase computational costs.

7.5.2 Simulation Results

For comparison of trajectory tracking algorithms, a sinusoidal trajectory has been chosen. The robot moves with the reference velocity $v_r = 0,5$ m/s along the trajectory described by one period of the sinusoidal function and the movement lasted 10 s. The simulations have been carried out for the mobile robot of mass $m = 0.5$ kg, distance $L = 0.02$ m (see Fig. 4(a)) and all identical wheels of radius $r = 0.015$ m.

The simulation results of fuzzy control of the considered mobile robot are shown in Fig. 7.7. Movement path is shown in Fig. 7(a), and robot orientation θ in Fig. 7(b). The reference signals are marked with a dotted line, and the actual time transients with a solid one. Figures 7(c)-(e) present time transients of position errors e_x, e_y, orientation error e_θ and velocity error $e_v = \sqrt{e_{vx}^2 + e_{vy}^2}$, respectively, obtained in simulations. The time transients of control signals u_x, u_y, u_θ are presented in Fig. 7(f). The performance of the system with fuzzy controller has been compared to the classical PD controller described by Eq. (7.36) for the following values of parameters: $k_{Px} = k_{Py} = 40$, $k_{P\theta} = 60$, $k_{Dx} = k_{Dy} = k_{D\theta} = 5$.

The values of performance integral indices for fuzzy and PD control are collected in Table 7.4. Notice that fuzzy control is more efficient than classical control. Only in case of indices for the orientation error e_θ the PD controller is more efficient, moreover, its program implementation needs smaller computational costs.

Table 7.4. Control performance integral indices

	Classical PD	Fuzzy
$\int |e_x(t)|dt$	0.4207	0.0589
$\int |e_y(t)|dt$	0.1864	0.0461
$\int |e_\theta(t)|dt$	0.7435	1.2085
$\int |e_v(t)|dt$	0.4563	0.0673
$\int e_x^2(t)dt$	0.0256	0.0005
$\int e_y^2(t)dt$	0.0049	0.0005
$\int e_\theta^2(t)dt$	0.2486	0.3915
$\int e_v^2(t)dt$	0.4303	0.0035

7.6 Conclusions

In the paper, two example implementations of artificial intelligence methods have been presented. First, performance of two neural controllers has been tested in simulations for a double pendulum. With the exception of the initial phase of learning, the results are nearly the same, so usefulness of the optimal controller seems doubtful. The Riccati equation and the optimal solution based on it are

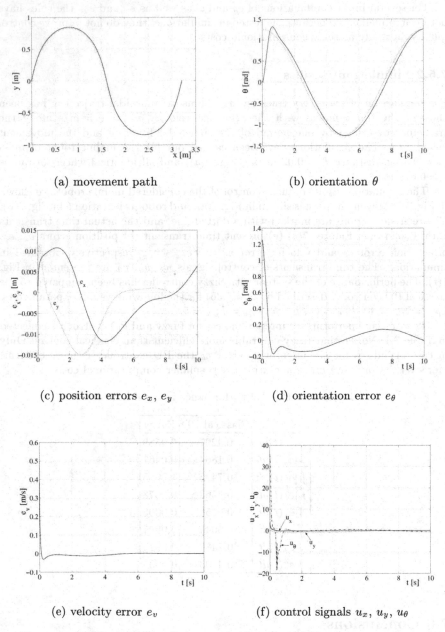

(a) movement path

(b) orientation θ

(c) position errors e_x, e_y

(d) orientation error e_θ

(e) velocity error e_v

(f) control signals u_x, u_y, u_θ

Fig. 7.7. Simulation results for the mobile robot

defined for linear systems, so the mathematical model of the manipulator has to be linearized many times during control. It is computationally time consuming and difficult. Contrary to this, relatively simple weight updates for the first controller lead to similar results.

Then, fuzzy inference methods have been implemented in construction of the fuzzy controller for a mobile robot. For the considered control tasks fuzzy algorithms lead to correct functioning of the system. However, the choice of membership functions and tuning their parameters has great influence on the control performance. Also extending the fuzzy rules set leads to larger computational costs.

References

1. Canudas de Wit C, Siciliano B, Bastin G (Eds.) (1996) Theory of robot control. Springer, Berlin Heidelberg New York
2. Ge S S, Lee T H, Harris C J (1998) Adaptive neural network control of robotic manipulators. World Scientific, London
3. Haykin S (1999) Neural networks: a comprehensive foundation, 2nd edition. Prentice-Hall, Upper Saddle River, NJ
4. Sciavicco L, Siciliano B (2000) Modeling and control of robot manipulators, 2nd edition. Springer, Berlin Heidelberg New York
5. Spong M, Vidyasagar M (1989) Robot dynamics and control. Wiley, New York
6. Yager R R, Filev D P (1994) Essentials of fuzzy modelling and control. Wiley, New York
7. Hu S, Ang Jr. M H , Krishan H (2000) Neural network controller for constrained robot manipulators. In: Proc. IEEE Int. Conf. on Robotics and Automation, San Francisco, CA 1906–1911.
8. Jungbeck M, Madrid M (2001) Optimal neural network output feedback control for robot manipulators. In: Proc. 2nd Int. Workshop on Robot Motion and Control, Bukowy Dworek, Poland 85–90
9. Kim Y, Lewis F (1999) Neural output feedback control of robot manipulators. IEEE Trans. on Robotics and Automation 15(2):301–309
10. Lewis F, Liu K, Yesildirek A (1995) Neural net robot controller with guaranteed tracking performance. IEEE Trans. on Neural Networks 6(3):703–715

8

Adaptive Control of Kinematically Redundant Manipulator along a Prescribed Geometric Path

Mirosław Galicki[1,2]

[1] Institute of Medical Statistics, Computer Science and Documentation
Friedrich Schiller University Jena, Jahnstrasse 3, D–07740 Jena, Germany
galicki@imsid.uni-jena.de
[2] Department of Mechanical Engineering, University of Zielona Góra,
Podgórna 50, 65-246 Zielona Góra, Poland

8.1 Introduction

Recently an interest has increased in applying redundant manipulators in useful practical tasks which are specified in terms of a geometric path to be followed by the end-effector. Redundant degrees of freedom make it possible to achieve some useful objectives such as collision avoidance in the work space with obstacles, joint limit avoidance and/or avoiding the singular configurations when the manipulator moves. Application of redundant manipulators to such tasks complicates their performance since these manipulators provide, in general, not unique solutions. Consequently, some objective criteria should be specified to solve the robot tasks uniquely. Minimization of the performance time is mostly considered in the literature. One may distinguish several approaches in this context. The task of time-optimal control of non-redundant manipulators has been solved both theoretically and practically by reducing the whole optimization problem to path co-ordinates [1]- [5]. However, these algorithms are not applicable in the collision avoidance tasks. Using the concept of a regular trajectory and the extended state space, numerical procedures have been proposed in [6]- [9] to find path-constrained time-optimal controls for kinematically redundant manipulators. Although all the aforementioned algorithms produce optimal solutions, they are not suitable to real-time computations due to their computational complexity. Therefore, it is natural to attempt other techniques in order to control the robot in real time. Using an integrable manifold concept and the inverse of the extended Jacobian matrix, an algorithm has been proposed in [10] to determine robot motions satisfying the end-effector path constraints and other useful objectives. A kinematic singular path tracking approach in an obstacle-free work space has been presented in [11], based on the null space-based method. Using on-line trajectory time scaling, a dynamic and computed torque laws, respectively, a nearly time-optimal path tracking control for non-redundant robot manipulators has been presented in works [12, 13]. Recently a technique which avoids solving an inverse of robot kinematic equations has been offered in [14, 15] for determining a collision-free trajectory of redundant manipulators.

K. Kozłowski (Ed.): Robot Motion and Control, LNCIS 335, pp. 129–139, 2006.
© Springer-Verlag London Limited 2006

In this work the problem of real-time controlling a kinematically redundant manipulator is considered, such that its end-effector follows the prescribed geometric path. Provided that a solution to the control problem of a redundant manipulator exists, the Lyapunov stability theory is used to derive the control scheme generating a corresponding trajectory whose equilibrium is asymptotically stable. The approach offered does not require any inverse of robot kinematic equations and the exact knowledge of robot dynamic equations. Instead, a generalized transpose Jacobian controller with adaptive component estimating the gravity term is introduced to generate robot controls. As is known, the classical transpose Jacobian control schemes with gravity term have been used, in fact, to a set point control problem (a regulation task) [16]- [23]. This study attempts to generalize the results obtained in the aforementioned works for a set point control problem to the path following task with an uncertain gravity term in the robot dynamic equations. Simultaneously, we also present an alternative approach to the inverse kinematic problem discussed in [24]- [26], which is purely kinematic. The paper is organized as follows. Section 2 formulates the robotic task to be accomplished in terms of a control problem. Section 3 describes how to employ the Lyapunov stability theory to determine controls (if they exist). Section 4 provides a computer example of generating the robot controls in a task space for a planar redundant manipulator comprising three revolute kinematic pairs. Finally, some conclusions are drawn.

8.2 Formulation of the Control Problem

The control scheme designed in the next section is applicable to holonomic mechanical systems comprising both non-redundant and redundant manipulators considered here which are described, in general, by the following dynamic equations, expressed in generalized co-ordinates (joint co-ordinates) $\mathbf{q} = (q_1, \ldots, q_n)^T \in \mathbb{R}^n$

$$\mathbf{M}(\mathbf{q})\ddot{\mathbf{q}} + \mathbf{C}(\mathbf{q}, \dot{\mathbf{q}})\dot{\mathbf{q}} + \mathbf{g}(\mathbf{q}) = \mathbf{u}, \qquad (8.1)$$

where $\mathbf{M}(\mathbf{q})$ denotes the $(n \times n)$ positive definite symmetric inertia matrix; $\mathbf{C}(\mathbf{q}, \dot{\mathbf{q}})\dot{\mathbf{q}}$ is the n-dimensional vector representing centrifugal and Coriolis forces; $\mathbf{g}(\mathbf{q}) = (g_1(\mathbf{q}), \ldots, g_n(\mathbf{q}))^T$ stands for the n-dimensional vector of gravity forces; $\mathbb{R}^n \ni \mathbf{u} = (u_1, \ldots, u_n)^T$ is the vector of controls (torques/forces); n denotes the number of kinematic pairs of the V-th class. In most applications of robotic manipulators, a desired path for the end-effector is specified in task space such as visual space or Cartesian space. The aim is to follow by the end-effector a prescribed geometric path (given in the m–dimensional task space, where $m \leq n$)) described by the following equations

$$\mathbf{p}(\mathbf{q}) - \boldsymbol{\theta}(s) = \mathbf{0}, \qquad (8.2)$$

where $\mathbf{p} : \mathbf{R}^n \longrightarrow \mathbf{R}^m$ denotes an m-dimensional, non-linear (with respect to vector $\mathbf{q}$) mapping constructed from the kinematic equations of the manipulator; $\mathbf{p}(\mathbf{q}) = (p_1(\mathbf{q}), \ldots, p_m(\mathbf{q}))^T$; $\boldsymbol{\theta}(s) = (\theta_1(s), \ldots, \theta_m(s))^T$ stands for the given geometric path; s is the current parameter of the path (e.g. its length); $s \in [0, s_{max}]$; s_{max} is the maximal path length. The mapping $\boldsymbol{\theta}$ is assumed not to be degenerated, i.e. $\|d\boldsymbol{\theta}/ds\| > 0$. The kinematic equations of a manipulator are independent (rank $(\partial \mathbf{p}/\partial \mathbf{q}) = m$). In general (i.e. when rank $(\partial \mathbf{p}/\partial \mathbf{q}) \leqslant m$), we should require that rank $\left(\frac{\partial \mathbf{p}}{\partial \mathbf{q}}\right) = $ rank $\left(\left[\frac{\partial \mathbf{p}}{\partial \mathbf{q}} \frac{d\boldsymbol{\theta}}{ds}\right]\right)$ in order to guarantee the consistency of the robotic task (8.2). For convenience of further considerations, the arguments of $\mathbf{M}$, $\mathbf{C}$ and $\mathbf{g}$ will be omitted. The problem is to determine control $\mathbf{u}$ which generates manipulator trajectory $\mathbf{q} = \mathbf{q}(t)$ and path parameterization $s = s(t)$ satisfying the equation (8.2) for each $t \in [0, T]$, where T denotes an (unknown) horizon of task performance. For simplicity, we consider such path following tasks that at the initial moment $t = 0$, for which $s(0) = 0$, a given (by definition) initial configuration $\mathbf{q}(0) = \mathbf{q}_0$ satisfies (8.2), i.e.

$$\mathbf{p}(\mathbf{q}_0) - \boldsymbol{\theta}(0) = \mathbf{0}. \tag{8.3}$$

Furthermore, at the initial and the final time moment the manipulator and path velocities equal zero, i.e.

$$\dot{\mathbf{q}}(0) = \dot{\mathbf{q}}(T) = \mathbf{0} \tag{8.4}$$

and

$$\dot{s}(0) = \dot{s}(T) = 0. \tag{8.5}$$

As is known, the task space velocity $\dot{\mathbf{p}}$ is related to joint space velocity $\dot{\mathbf{q}}$ as follows

$$\dot{\mathbf{p}} = \mathbf{J}(\mathbf{q})\dot{\mathbf{q}}, \tag{8.6}$$

where $\mathbf{J}(\mathbf{q}) = \partial\mathbf{p}/\partial\mathbf{q}$ is the $(m \times n)$ Jacobian matrix.

Without loss of generality, all the kinematic pairs of the robot are assumed to be revolute. Hence, the task space is bounded. Let us define errors $\mathbf{e}$, e_{m+1} of path following as $\mathbf{e} = (e_1, \dots, e_m)^T = \mathbf{p}(\mathbf{q}) - \boldsymbol{\theta}(s) = (p_1 - \theta_1, \dots, p_m - \theta_m)^T$ and $e_{m+1} = s - s_{\max}$. For simplicity of further considerations, $s_{\max}$ is assumed to be equal to 1, i.e. $s_{\max} = 1$. Several important properties of dynamic and kinematic equations (8.1), (8.2) are given, which will be utilized by designing the adaptive controller.

A. Matrix $\frac{1}{2}\dot{\mathbf{M}}(\mathbf{q}) - \mathbf{C}(\mathbf{q}, \dot{\mathbf{q}})$ is skew-symmetric so that

$$\forall \mathbf{q}, \dot{\mathbf{q}} \in \mathbb{R}^n \quad \left\langle \dot{\mathbf{q}}, \left(\frac{1}{2}\dot{\mathbf{M}}(\mathbf{q}) - \mathbf{C}(\mathbf{q}, \dot{\mathbf{q}})\right) \dot{\mathbf{q}} \right\rangle = 0, \tag{8.7}$$

where $\langle \, , \, \rangle$ is the scalar product of vectors.

B. The gravity term given in equation (8.1) is linear with respect to an ordered set of physical parameters $\boldsymbol{\Delta} = (\Delta_1, \dots, \Delta_d)^T$, i.e.

$$\mathbf{g}(\mathbf{q}) = \mathbf{D}(\mathbf{q})\boldsymbol{\Delta}, \tag{8.8}$$

where $\mathbf{D}(\mathbf{q})$ is called the $(n \times d)$ gravity regressor matrix; d stands for the number of the physical parameters.

C. The boundedness of the robot task space implies that there exists some constant $c_1 > 0$ such that

$$\|\mathbf{e}\| \leqslant c_1, \tag{8.9}$$

where $\| \cdot \|$ stands for the Euclidean norm.

D. Moreover, the following inequalities are also satisfied [27]:

$$\|\mathbf{M}\| \leqslant c_2, \qquad \|\mathbf{C}\| \leqslant c_3\|\dot{\mathbf{q}}\|, \tag{8.10}$$

where the matrix norm means the standard Euclidean (also called Frobenius) norm; c_2, c_3 denote some positive constants.

E. Due to the revolute kinematic pairs we also have

$$\|\mathbf{J}\| \leqslant c_4, \qquad \left\|\frac{\partial \mathbf{J}}{\partial \mathbf{q}}\right\| \leqslant c_5, \tag{8.11}$$

where c_4, c_5 are positive constants.

Expressions (8.1)-(8.6) formulate the robot task as a control problem. The fact that there exist state constraints makes the solution of this problem difficult. The next section will present an approach that renders it possible to solve the control problem (8.1)-(8.6) making use of the summarized properties (A)-(E) and the Lyapunov stability theory.

8.3 Path Control of the Manipulator

Our aim is to control the manipulator such that the end-effector fulfils (8.2)-(8.5). Based on (8.2) and (8.8), let us propose the following computationally simple adaptive control scheme

$$\mathbf{u} = -k_D \dot{\mathbf{q}} - k_P \mathbf{J}^T(\mathbf{q})\mathbf{e} + \mathbf{D}(\mathbf{q})\widehat{\boldsymbol{\Delta}}, \tag{8.12}$$

where $\widehat{\boldsymbol{\Delta}}$ stands for the estimated physical parameter vector defined below; k_D and k_P are positive scalars which could be replaced by diagonal matrices of positive constants without affecting the stability results obtained further on (this should lead to improved performance). The path parameterization $s = s(t)$ is computed by solving the following scalar differential equation

$$\ddot{s} = -k_D \dot{s} - (k_P + k_D)\left(\left\langle \mathbf{e}, -\frac{d\boldsymbol{\theta}(s)}{ds}\right\rangle + \gamma e_{m+1} + \frac{1}{2}\frac{d\gamma}{ds}e_{m+1}^2\right), \tag{8.13}$$

where γ is assumed to be a strictly positive function of s.

The estimated physical parameters $\widehat{\boldsymbol{\Delta}}$ of the dynamic equations are updated by the following law

$$\dot{\widehat{\boldsymbol{\Delta}}} = k_I \mathbf{D}^T(\mathbf{q})\left(-\mathbf{J}^T\mathbf{e} - \dot{\mathbf{q}}\right), \tag{8.14}$$

where k_I is a positive scalar. The closed loop error dynamics is obtained by substituting equations (8.12)- (8.14) into equation (8.1)

$$\dot{\mathbf{e}} = \mathbf{J}\dot{\mathbf{q}} - \frac{d\boldsymbol{\theta}}{ds}\dot{s},$$
$$\mathbf{M}\ddot{\mathbf{q}} + \mathbf{C}\dot{\mathbf{q}} + \mathbf{D}\widetilde{\boldsymbol{\Delta}} + k_D\dot{\mathbf{q}} + k_P\mathbf{J}^T\mathbf{e} = 0,, \tag{8.15}$$
$$\dot{\widetilde{\boldsymbol{\Delta}}} = k_I \mathbf{D}^T(\mathbf{q})\left(\mathbf{J}^T\mathbf{e} + \dot{\mathbf{q}}\right),$$

where $\widetilde{\boldsymbol{\Delta}} = \boldsymbol{\Delta} - \widehat{\boldsymbol{\Delta}}$. The choice of function γ is crucial for computational effectiveness of control scheme (8.12). One possibility is $\gamma = \gamma(s)$ with $d\gamma/ds > 0$. An alternative choice could be $\gamma = \gamma(\mathbf{e}^2)$, where γ attains its maximum for $\mathbf{e} = 0$ and smoothly decreases as $\|\mathbf{e}\|$ increases. In further analysis we take, for simplicity, the first form of γ. Moreover, throughout this paper, we assume that

$$\left|\frac{d\gamma}{ds}\right| \leqslant c_6 \tag{8.16}$$

where c_6 is a positive number.

Assumption 1 *Function γ is required not to satisfy differential equation* $\gamma + \frac{1}{2}\frac{d\gamma}{ds}e_{m+1} = 0$.

Assumption 1 results in asymptotic convergence of s to 1. Applying the Lyapunov stability theory, we derive the following result.

Theorem 1. *If $\mathbf{J}$ is the full rank matrix along the geometric path given by equation (8.2), inequality (8.16) and Assumption 1 are satisfied during the robot movement, and*

$$k_I > 0 \tag{8.17}$$

$$k_D > c_1c_3c_4 + c_1c_2c_5 + c_2(c_4)^2 \tag{8.18}$$

$$k_D\left(k_D - \left(c_1c_3c_4 + c_1c_2c_5 + c_2(c_4)^2\right)\right) - \frac{(c_2c_4c_6)^2}{4} > 0 \tag{8.19}$$

$$k_P + k_D > 2c_2(c_4)^2, \tag{8.20}$$

then the equilibrium $(\dot{\mathbf{q}}, \dot{s}, \mathbf{e}, e_{m+1}) = \mathbf{0}$ of systems (8.13), (8.15) is asymptotically stable.

Proof. Consider the following Lyapunov function candidate

$$V = \frac{1}{2}\left\langle \dot{\mathbf{q}} + \mathbf{J}^T\mathbf{e}, \mathbf{M}\left(\dot{\mathbf{q}} + \mathbf{J}^T\mathbf{e}\right)\right\rangle + \frac{1}{2k_I}\widetilde{\Delta}^T\widetilde{\Delta} + \frac{1}{2}\left(k_P + k_D\right)\gamma e_{m+1}^2 +$$
$$+ \frac{1}{2}\dot{s}^2 + \frac{k_D}{2}\mathbf{e}^T\mathbf{e} + \frac{k_P}{2}\mathbf{e}^T\mathbf{e} - \frac{1}{2}\left\langle\mathbf{J}^T\mathbf{e}, \mathbf{M}\mathbf{J}^T\mathbf{e}\right\rangle. \tag{8.21}$$

Using inequality (8.20), it is easy to see that the sum of the last three terms of function V is non-negative. Hence V is also non-negative. The time derivative of V along the trajectory of system (8.13), (8.15) can be written after some simplifications and using the skew-symmetric property of matrix $\frac{1}{2}\dot{\mathbf{M}} - \mathbf{C}$ as follows

$$\dot{V} = \left\langle\dot{\mathbf{q}}, \mathbf{C}\mathbf{J}^T\mathbf{e}\right\rangle - k_P\left\langle\mathbf{J}^T\mathbf{e}, \mathbf{J}^T\mathbf{e}\right\rangle - k_D\dot{\mathbf{q}}^T\dot{\mathbf{q}} + \left\langle\dot{\mathbf{q}}, \dot{\mathbf{M}}\mathbf{J}^T\mathbf{e}\right\rangle + \left\langle\dot{\mathbf{q}}, \mathbf{M}\mathbf{J}^T\dot{\mathbf{J}}\dot{\mathbf{q}}\right\rangle -$$
$$- \left\langle\dot{\mathbf{q}}, \mathbf{M}\mathbf{J}^T\frac{d\boldsymbol{\theta}(s)}{ds}\dot{s}\right\rangle - k_D\dot{s}^2 - \left\langle(k_P + k_D)\mathbf{e}, \frac{d\boldsymbol{\theta}(s)}{ds}\dot{s}\right\rangle.$$

On account of properties (8.2)-(8.11) and (8.16), we have

$$\dot{V} \leqslant - \left[\left(k_D - \left(c_1c_3c_4 + c_1c_2c_5 + c_2(c_4)^2\right)\right)\dot{\mathbf{q}}^T\dot{\mathbf{q}} + k_D\dot{s}^2 - c_2c_4c_6|\dot{s}|\|\dot{\mathbf{q}}\|\right] -$$
$$- k_P\left\langle\mathbf{J}\mathbf{J}^T\mathbf{e}, \mathbf{e}\right\rangle. \tag{8.22}$$

Under the condition, that controller gains k_P and k_D fulfil inequalities (8.18)-(8.20), $\dot{V}$ is negative for all $\dot{\mathbf{q}} \neq 0$, $\dot{s} \neq 0$ and $\mathbf{e} \neq 0$ (matrix $\mathbf{J}$ is by assumption non-singular). It equals zero only when $\dot{\mathbf{q}} = 0$, $\dot{s} = 0$ and $\mathbf{e} = 0$ which implies (using LaSalle-Yoshizawa theorem [28]) that $\dot{\mathbf{q}}$, $\dot{s}$, $\mathbf{e}$ tend to approach asymptotically zero, i.e. $\dot{\mathbf{q}}(T) \to 0$, $\mathbf{e}(T) \to 0$ and $\dot{s}(T) \to 0$, as $T \to \infty$. Hence $\ddot{\mathbf{q}}(T) \to 0$ and $\ddot{s}(T) \to 0$, as $T \to \infty$, too. Consequently, boundary conditions (8.3)-(8.5) are (asymptotically) fulfilled. The convergence of path velocity and acceleration yields the following equation (for $\gamma = \gamma(s)$)

$$\gamma e_{m+1}(\infty) + \frac{1}{2}\frac{d\gamma}{ds}e_{m+1}^2(\infty) = 0. \tag{8.23}$$

Hence $e_{m+1}(\infty) = 0$ or $\gamma + \frac{1}{2}\frac{d\gamma}{ds}e_{m+1}(\infty) = 0$. On account of Assumption 1, the second equality is not fulfilled. Thus $e_{m+1}(\infty) = 0$ (or equivalently $s(\infty) = 1$). Finally, stable equilibrium point $(\dot{\mathbf{q}}, \dot{s}, \mathbf{e}, e_{m+1})$ for differential equations (8.13) and (8.15) equals $(\mathbf{0}, 0, \mathbf{0}, 0)$. Let us emphasize that Lyapunov function (8.21) does not guarantee the convergence of $\widetilde{\boldsymbol{\Delta}}$ to zero (parameter estimations may differ from their nominal values).

On account of (8.3)-(8.5) we have $V_{t=0} = \frac{\widetilde{\boldsymbol{\Delta}}_{t=0}^T\widetilde{\boldsymbol{\Delta}}_{t=0}}{2\cdot k_I} + \frac{(k_P+k_D)\cdot\gamma_0}{2}$. For sufficiently large k_I, the first term in this equality may be omitted. Hence we obtain $V_{t=0} \simeq \frac{(k_P+k_D)\cdot\gamma_0}{2}$. Since $\dot{V}$ is not positive, function V fulfils the inequality $V \leqslant \frac{(k_P+k_D)\cdot\gamma_0}{2}$. Moreover, on account of (8.20) we have the following inequality $\frac{k_D}{4}\mathbf{e}^T\mathbf{e} + \frac{k_P}{4}\mathbf{e}^T\mathbf{e} - \frac{1}{2}\langle\mathbf{J}^T\mathbf{e}, \mathbf{MJ}^T\mathbf{e}\rangle > 0$. Consequently, the following bound on $\|\mathbf{e}\|$ may easily be obtained, based on (8.21) and the last dependence

$$\|\mathbf{e}\| = \|\mathbf{p}(\mathbf{q}) - \boldsymbol{\theta}(s)\| \leqslant \sqrt{2\gamma_0}. \tag{8.24}$$

An important remark may be derived from the proof. Namely, inequality (8.24) presents an upper bound (path independent) on the accuracy of path following by the end-effector according to the control law (8.12). Let us note that the upper bound on path following error (8.24) is very conservative. This does not require large values for k_I to achieve a good path following accuracy, as the numerical simulations (given in the next section) show. Moreover, several observations can be made regarding the control strategy (8.12). First note, that the proposed control law is very simple and requires, in fact, no information concerning the robot dynamic equations. Second, the choice of controller parameters k_P, k_D and k_I according to dependencies (8.17)-(8.20) guarantees asymptotic stability of the closed-loop error dynamics (8.15) during the manipulator movement. As is easy to see, for a special case $s = 1$ and the exact knowledge of gravity term, control scheme (8.12) is reduced to a classical transpose Jacobian controller with gravity compensation. Consequently, control strategy (8.12) generalizes those known from the literature. Moreover, the transpose of $\mathbf{J}$ (instead of a pseudoinverse) in control scheme (8.12) does not result in numerical instabilities due to (possible) kinematic singularities met on the robot trajectory. Nevertheless, (8.12) has been derived under the assumption of a full-rank Jacobian matrix along the path.

8.4 A Numerical Example

The aim of this section is to illustrate the performance of the proposed robot controller, given by (8.13). For this purpose, a planar manipulator schematically shown in Fig. 8.1 is considered, comprising three-revolute kinematic pairs of the V-th class ($n = 3$), and operating in two-dimensional task space ($m = 2$). In all numerical simulations, the SI units are used. The kinematic equations of the manipulator are

$$p_1(\mathbf{q}) = \sum_{i=1}^{3} l_i \sin(\sum_{k=1}^{i} q_k) \quad p_2(\mathbf{q}) = \sum_{i=1}^{3} l_i \cos(\sum_{k=1}^{i} q_k).$$

The components of the robot dynamic equations taken for the numerical simulations are as follows:

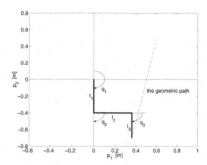

Fig. 8.1. A kinematic scheme of the manipulator and the task to be accomplished

- link lengths $l_1 = 0.4$, $l_2 = 0.36$, $l_3 = 0.3$;
- link masses $m_1 = 15.2$, $m_2 = 13.7$, $m_3 = 11.4$;
- the mass centers of the links represented for simplicity by the sections are located at their mass centers.

The gravity regressor matrix equals

$$\mathbf{D} = \begin{bmatrix} \sin(q_1) & \sin(q_1 + q_2) & \sin(q_1 + q_2 + q_3) \\ 0 & \sin(q_1 + q_2) & \sin(q_1 + q_2 + q_3) \\ 0 & 0 & \sin(q_1 + q_2 + q_3) \end{bmatrix}.$$

The nominal values of vector $\mathbf{\Delta}$ are $\Delta_1 = -13.07$, $\Delta_2 = -6.57$ and $\Delta_3 = -1.7$. The initial values for $\widehat{\mathbf{\Delta}}_i$ have been chosen as: $\widehat{\Delta}_1(0) = -15.0$, $\widehat{\Delta}_2(0) = -7.5$, and $\widehat{\Delta}_3(0) - -2.1$, respectively. The upper bound on the accuracy of the path following in all the computer simulations, is assumed to be equal to $\sqrt{2 \cdot \gamma_0} \simeq 10^{-1}$, where $\gamma(s) = 2.4 \cdot 10^{-3} + 2.4 \cdot 10^{-3} e^{2.7 \cdot s}$. The choice of control gains k_P, k_D and k_I based on coefficients c_i, $i = 1 : 6$ and relations (8.17)-(8.20) is conservative, which may result in large controls. Consequently, they were set experimentally as: $k_P = 2.5 \cdot 10^5$, $k_D = 17 \cdot (1 + 100 \cdot e^{-10 \cdot t}) + 385$, and $k_I = \mathrm{diag}(0.22,\ 0.02,\ 0.01)$. Let us note that small values for k_I result also in small values of controls. The task of the robot is to transfer the end-effector along the geometric path (the dotted line in Fig. 8.1), expressed by the following equations

$$\theta_1(s) = 0.36 + 0.24 \cdot s, \qquad \theta_2(s) = -0.7 + 1.2 \cdot s,$$

where $s \in [0, 1]$. The initial configuration $\mathbf{q}_0$ equals $\mathbf{q}_0 = (\pi,\ \frac{3\pi}{2},\ \pi/2)^T$. Let us introduce path following errors

$$e_1(t) = p_1(\mathbf{q}(t)) - \theta_1(s(t)), \qquad e_2(t) = p_2(\mathbf{q}(t)) - \theta_2(s(t)),$$

respectively, to evaluate the performance of the robot controller (8.12). The results of computer simulations are presented in Figs 8.2-8.6.

As one can observe from Figs 8.2-8.6, the time dependent damping function k_D eliminates errors and torques oscillations at the very beginning of time histories. As might be expected, the path following errors from Figs 8.2-8.3 are much smaller than those obtained from the conservative dependence (8.24). Figures 8.7–8.9 depict the time courses of $\widehat{\Delta}_1$, $\widehat{\Delta}_2$ and $\widehat{\Delta}_3$. As is seen from these figures, parameter estimations $\widehat{\Delta}_1$, $\widehat{\Delta}_2$ and $\widehat{\Delta}_3$ do not converge to their nominal values -13.07, -6.57 and -1.7.

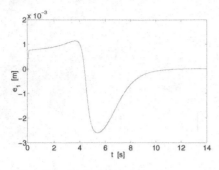

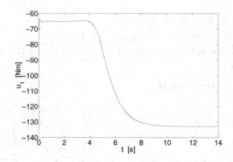

Fig. 8.2. Path following error e_1

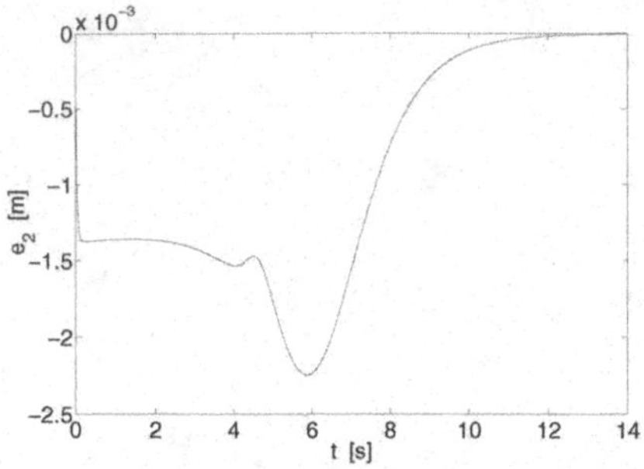

Fig. 8.3. Path following error e_2

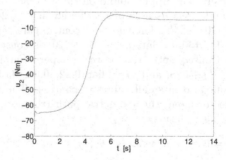

Fig. 8.4. Input torque u_1 in the path following task

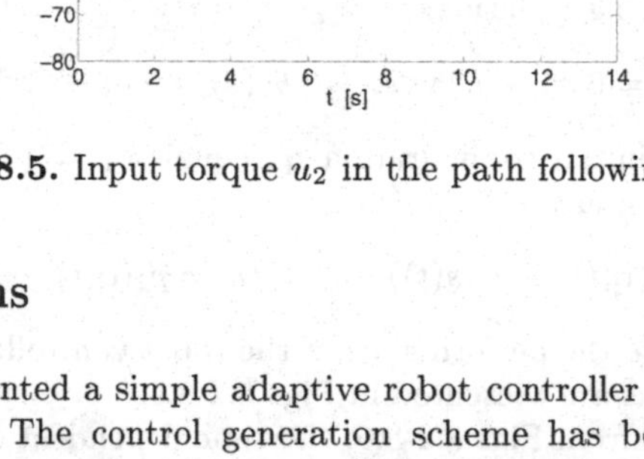

Fig. 8.5. Input torque u_2 in the path following task

8.5 Conclusions

This study has presented a simple adaptive robot controller for the path following by the end-effector. The control generation scheme has been derived using the Lyapunov stability theory. An advantage of the proposed control law (8.12) is that it requires, in fact, no information regarding the parameters of the robot dynamic equations (the adaptive component in controller (8.12) estimates the gravity term in the dynamic equations). The control strategy (8.12) is shown to be asymptotically stable (by fulfilment of practically reasonable assumptions). The proposed robot

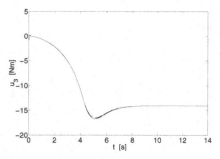

Fig. 8.6. Input torque u_3 in the path following task

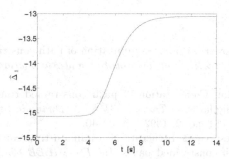

Fig. 8.7. Time course of adaptive estimate $\widehat{\Delta}_1$

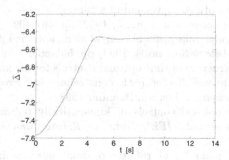

Fig. 8.8. Time course of adaptive estimate $\widehat{\Delta}_2$

controller has been applied to a planar redundant manipulator of three revolute kinematic pairs. Numerical simulations have shown that the results obtained are in accordance with the theoretical analysis. The novelty of the strategy proposed lies in its simplicity in design, program code and real-time implementation. The approach presented here may also be directly applicable to cooperating kinematically redundant manipulators.

Acknowledgement. This work was supported by the DFG Ga 652/1–1,2.

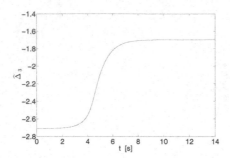

Fig. 8.9. Time course of adaptive estimate $\hat{\Delta}_3$

References

1. Z. Shiller, S. Dubowsky. Robust computation of path constrained time optimal motions. In: *Proc. IEEE Conf. on Robotics and Automation*, Cincinnati, USA 1990.
2. Z. Shiller, H.H. Lu. Computation of path constrained time optimal motions with dynamic singularities. *Trans. ASME: J. Dynamic Syst., Measurement, and Control*, vol. 114, no. 2, 1992, pp. 34-40.
3. Y. Chen, A.A. Desrochers. Structure of minimum-time control law for robotic manipulators with constrained paths. In: *Proc. IEEE Conf. on Robotics and Automation*, Scottsdale, USA 1989.
4. J.M. McCarthy, J.E. Bobrow. The number of saturated actuators and constraint forces during time-optimal movement of a general robotic system. *IEEE Trans. Robotics and Automation*. vol. 8, no. 3, 1992, pp. 407-409.
5. Z. Shiller. On singular time optimal control along specified paths. *IEEE Trans. on Robotics and Automation*, no. 4, 1994, pp. 561-566.
6. M. Galicki. The structure of time optimal controls for kinematically redundant manipulators with end-effector path constraints. In: *Proc. IEEE Conf. on Robotics and Automation*, Leuven, Belgium 1998.
7. M. Galicki. Time-optimal controls of kinematically redundant manipulators with geometric constraints. *IEEE Trans. on Robotics and Automation*, vol. 16, no. 1, 2000, pp. 89-93.
8. M. Galicki. The planning of robotic optimal motions in the presence of obstacles. *Int. J. Robotics Research*, vol. 17, no. 3, 1998, pp. 248-259.
9. M. Galicki, I. Pająk. Optimal motion of redundant manipulators with state equality constraints. In: *Proc. IEEE Int. Symp. on Assembly and Task Planning*, Porto, Portugal 1999.
10. I. Pająk, M. Galicki. The planning of suboptimal collision-free robotic motions. In: *Proc. First Workshop on Robot Motion and Control (RoMoCo)*, Kiekrz, Poland 1999.
11. K. Tchoń. A normal form appraisal of the null space-based singular path tracking. In: *Proc. First Workshop on Robot Motion and Control*, Kiekrz, Poland 1999.
12. O. Dahl. Path-constrained robot control with limited torques – Experimental evaluation. *IEEE Trans. on Robotics and Automation*, vol. 10, no. 5, 1994, pp. 658-669.

13. J. Kieffer, A.J. Cahill, M.R. James. Robust and accurate time-optimal path-tracking control for robot manipulators. *IEEE Trans. on Robotics and Automation*, vol. 13, no. 6, 1997, pp. 880-890.

14. M. Galicki. Real-time trajectory generation for redundant manipulators with path constraints. *Int. J. Robotics Research*, vol. 20, no. 7, 2001, pp. 673-690.

15. M. Galicki. Path following by the end-effector of a redundant manipulator operating in a dynamic environment. *IEEE Trans. on Robotics*, vol. 20, no. 6, 2004, pp. 1018-1025.

16. M. Takegaki, S. Arimoto. A new feedback method for dynamic control of manipulators. *Trans. ASME: J. Dynamic Syst., Measurement, and Control*, vol. 102, 1981, pp. 119-125.

17. S. Arimoto. *Control theory of non-linear mechanical systems*. Clarendon, Oxford, U.K. 1996.

18. C. Canudas de Wit, B. Siciliano, G. Bastin. *Theory of robot control*. Springer Verlag, New York 1996.

19. L. Sciavicco, B. Siciliano. *Modeling and control of robot manipulators*. McGraw-Hill, New York 1996.

20. S. Arimoto. Design of robot control systems. *Adv. Robot.*, vol. 4, no. 1, 1990, pp. 79-97.

21. F. Miyazaki, S. Arimoto. Sensory feedback for robot manipulators. *J. Robot. Syst.*, vol. 2, no. 1, 1985, pp. 53-71.

22. Y. Masutani, F. Miyazaki, S. Arimoto. Sensory feedback control for space manipulators. In: *Proc. IEEE Conf. on Robotics and Automation*, Scottsdale 1989.

23. R. Kelly. Regulation of manipulators in generic task space: an energy shaping plus damping injection approach. *IEEE Trans. on Robotics and Automation*, vol. 15, no. 2, 1999, pp. 377-381.

24. G. Antonelli, S. Chiaverini, G. Fusco. Kinematic control of redundant manipulators with on-line end-effector path tracking capability under velocity and acceleration constraints. In: *Prep. 6th IFAC Symposium on Robot Control (SYROCO)*, 2000.

25. G. Antonelli, S. Chiaverini, G. Fusco. An algorithm for on-line inverse kinematics with path tracking capability under velocity and acceleration constraints. In: *Proc. IEEE Conf. on Decision and Control*, 2000.

26. G. Antonelli, S. Chiaverini, G. Fusco. Real-time end-effector path following for robot manipulators subject to velocity, acceleration, and jerk joint limits. In: *Proc. IEEE/ASME International Conf. on Advanced Intelligent Mechatronics*, 2001.

27. M. Spong, M. Vidyasagar. *Robot dynamics and control*. Wiley, New York 1989.

28. M. Krstic, I. Kanellakopoulos, P. Kokotovic. *Nonlinear and adaptive control design*. Wiley, New York 1995.

9

Adaptive Visual Servo Control of Robot Manipulators via Composite Camera Inputs

Türker Şahin and Erkan Zergeroğlu

Department of Computer Engineering, Gebze Institute of Technology, PK. 141, 41400 Gebze/Kocaeli, Turkey   ezerger@bilmuh.gyte.edu.tr

9.1 Introduction

Non-contact sensing for control of robotic systems is essential for tracking and motion of manipulators in unstructured environments. A particularly important example to this is visual servoing, where information from vision sensors is used in the feedback loop to obtain the position information for the low level controllers. This work is based on a similar approach. Readers are referred to [1, 2] for more information on concurrent visual servoing.

Vision systems used for robotic applications are mostly classified according to the number of vision sensors they use. They are either monocular visual servoing, in which a single camera is utilized, or multi-camera vision systems, where multiple cameras located in the work-space are used to collect the task specific information. The monocular systems are cost effective, but in nearly all 3D applications the work space depth information is lost. In fact it is sometimes even not possible for some multi-camera configurations to capture the scene depth, and hence careful camera calibration is necessary. However, it is usually difficult to obtain the intrinsic camera parameters (i.e. the image center, magnification factors, and camera scaling factor) and the extrinsic camera parameters (the position and orientation of the camera within the work-space) exactly. Therefore a great deal of research has been applied to the camera calibration problem, but most solutions do not take the robot dynamics into account [3–5] and are limited to the kinematic level. It is crucial to incorporate the manipulator dynamics in the control loop for good performing visual servoing systems. However, so far such solutions have only been applied in monocular cases [6–8] and unfortunately most of related work constraints the problem to 2D case and ignores depth information. Recently, there has been presented in [9] an asymptotically stable position based visual feedback controller that can compensate the uncertainties of the homogenous transformation matrix between the task-space and the camera space. However, the controller assumes that the intrinsic camera parameters are perfectly calibrated.

In this paper, the result given in [7] have been extended to 3D position servoing case via the use of multiple cameras. The controller proposed uses a similar transformation matrix as in [7], which premultiplies the composite camera calibration matrix so that the resultant matrix is positive definite and can be used to

K. Kozłowski (Ed.): Robot Motion and Control, LNCIS 335, pp. 141–152, 2006.
© Springer-Verlag London Limited 2006

incorporate the camera parameters in the robot equation. A backstepping technique is then applied to achieve asymptotic end effector tracking. The controller requires the knowledge of position with respect to the base of the robotics manipulator for the cameras and still achieves asymptotic position tracking despite the uncertainties in the intrinsic camera calibration parameters and the robot dynamics.

The rest of the manuscript is organized in the following manner. In Section 2 the manipulator and the camera model used are presented. In Section 3 the control objective, controller development and stability analysis are presented, while the simulation results are shown in Section 4. Concluding remarks are summarized in Section 5.

9.2 Robot-camera Model

A schematic representation of the robot-camera system configuration considered in this work is given in Figure 9.1. We assume that the cameras are located at fixed points outside, but pointing towards the robot work-space such that: $i)$ camera1 and camera2's image planes are parallel to $X_r - Y_r$ and $X_r - Z_r$ planes of motion of the robot, respectively, and $ii)$ both cameras can capture images throughout the entire robot work-space.

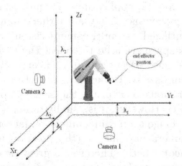

Fig. 9.1. Placement of the camera system with respect to robot coordinates

9.2.1 Robot Dynamics

The joint-space model for a three-link revolute direct-drive robot manipulator is assumed to be of the following form [10]

$$M(q)\ddot{q} + V_m(q,\dot{q})\dot{q} + G(q) + F(\dot{q}) = \tau, \tag{9.1}$$

where $q(t), \dot{q}(t), \ddot{q}(t) \in \Re^3$ denote the link position, velocity, and acceleration vectors, respectively, $M(q) \in \Re^{3 \times 3}$ represents the link inertia matrix, $V_m(q,\dot{q}) \in \Re^{3 \times 3}$ represents centripetal-Coriolis matrix, $G(q) \in \Re^3$ represents the gravity effects, $F(\dot{q}) \in \Re^3$ represents the friction effects, and $\tau(t) \in \Re^3$ represents the torque input vector.

Remark 1 *The above robot model has been developed for a non-redundant manipulator (i.e., we assume $n = 3$); and the formulation is based on only end effector position tracking problem for simplicity. However, the results presented in this paper can be extended to the redundant case with minor modifications (see [7] for details), and might serve as a stepping stone for the full order end effector position plus orientation tracking in Cartesian space (see [11] for details).*

9.2.2 Composite Camera Model Development

Using the standard pin-hole model for a camera system, a point $X_r = \begin{bmatrix} x_r & y_r & z_r \end{bmatrix}^T$ in a 3-D world frame can be represented in terms of the camera space coordinate frame as [12]

$$\begin{bmatrix} x_c \\ y_c \end{bmatrix} = \frac{f}{z_r} \begin{bmatrix} \beta_1 & 0 \\ 0 & \beta_2 \end{bmatrix} R(\theta) \left\{ \begin{bmatrix} x_r \\ y_r \end{bmatrix} - \begin{bmatrix} o_1 \\ o_2 \end{bmatrix} \right\} + \begin{bmatrix} c_1 \\ c_2 \end{bmatrix}, \qquad (9.2)$$

where $Y = \begin{bmatrix} x_c & y_c \end{bmatrix}^T$ denotes the corresponding position vector in camera space, f is the focal length of the lens used, β_1, β_2 are the magnification factors of the camera, $R(\theta) \in \Re^{2 \times 2}$ is the rotation matrix defined as

$$R(\theta) = \begin{bmatrix} \cos\theta & -\sin\theta \\ \sin\theta & \cos\theta \end{bmatrix} \qquad (9.3)$$

with θ being the rotation angle of the camera, $O = \begin{bmatrix} o_1 & o_2 \end{bmatrix}^T$ is the position of the optical center of the camera with respect to the world coordinate frame and $C = \begin{bmatrix} c_1 & c_2 \end{bmatrix}^T$ denotes the image center which is defined as the frame buffer coordinates of the intersection of the optical axis with the image plane. Using the camera transformation given in (9.2) the camera space variables for the camera 1 of the system shown in Figure 1 is obtained for as follows

$$\begin{bmatrix} x_{c1} \\ y_{c1} \end{bmatrix} = \frac{1}{z_r + \lambda_1} H_1 R(\theta_1) \begin{bmatrix} x_r - o_{11} \\ y_r - o_{12} \end{bmatrix} + \begin{bmatrix} p_{11} \\ p_{12} \end{bmatrix}, \qquad (9.4)$$

where $H_1 \in \Re^{2 \times 2}$ is defined as

$$H_1 = \begin{bmatrix} f_1\beta_{11} & 0 \\ 0 & f_1\beta_{12} \end{bmatrix}. \qquad (9.5)$$

Similarly camera 2 variables are

$$\begin{bmatrix} x_{c2} \\ z_{c2} \end{bmatrix} = \frac{1}{y_r + \lambda_2} H_2 R(\theta_2) \begin{bmatrix} x_r - o_{21} \\ z_r - o_{22} \end{bmatrix} + \begin{bmatrix} p_{21} \\ p_{22} \end{bmatrix}, \qquad (9.6)$$

where H_1, H_2, $R(\theta_1)$, $R(\theta_2) \in \Re^{2 \times 2}$ and $p_1 = \begin{bmatrix} p_{11} & p_{12} \end{bmatrix}^T$, $p_2 = \begin{bmatrix} p_{21} & p_{22} \end{bmatrix}^T \in \Re^{2 \times 1}$ are constant but unknown camera parameters, $\lambda_1, \lambda_2, o_{11}, o_{12}, o_{21}, o_{22}$ are positive constants representing the placement of the camera with respect to the origin of the robot world coordinate frame and are assumed to be known. Our first goal is to obtain a composite camera input from the two camera inputs placed in the work space that can lead us to obtain 3-dimensional position information (x_r, y_r, z_r) about the object in the work space as opposed to the normal 2-D projection using a standard camera. Using the properties of the rotation matrix and the fact that H_2 is a diagonal matrix, z_{c2} defined in (9.6) can be written in the following form

$$z_{c2} = \gamma_1 \frac{z_r - o_{22}}{y_r + \lambda_2} + \gamma_2 x_{c2} + \gamma_3, \tag{9.7}$$

where the constant parameters γ_1, γ_2, $\gamma_3 \in \Re$ are explicitly defined as

$$\gamma_1 = \frac{f_2 \beta_{22}}{\cos \theta_2}, \gamma_2 = \frac{\beta_{22} \sin \theta_2}{\beta_{21} \cos \theta_2}, \gamma_3 = p_{22} - p_{21} \frac{\beta_{22} \sin \theta_2}{\beta_{21} \cos \theta_2}. \tag{9.8}$$

Based on (9.4) and (9.8) we find the composite camera input representation as follows

$$\begin{bmatrix} x_{c1} \\ y_{c1} \\ z_{c2} \end{bmatrix} = \overbrace{\begin{bmatrix} H_1 R(\theta_1) & 0_{2\times1} \\ 0_{1\times2} & \gamma_1 \end{bmatrix}}^{\triangleq H \cdot R} \begin{bmatrix} \frac{x_r - o_{11}}{z_r + \lambda_1} \\ \frac{y_r - o_{12}}{z_r + \lambda_1} \\ \frac{z_r - o_{22}}{y_r + \lambda_2} \end{bmatrix} + \begin{bmatrix} p_{11} \\ p_{12} \\ \gamma_2 x_{c2} + \gamma_3 \end{bmatrix}. \tag{9.9}$$

For simplicity we defined the composite camera input vector $X_c = \begin{bmatrix} x_{c1} & y_{c1} & z_{c2} \end{bmatrix}^T$ and the off-setted Cartesian vector $X_{\overline{R}} \in \Re^3$ as

$$X_{\overline{R}} \triangleq \begin{bmatrix} \overline{x_r} \\ \overline{y_r} \\ \overline{z_r} \end{bmatrix} = \begin{bmatrix} x_r - o_{11} \\ y_r - o_{12} \\ z_r - o_{22} \end{bmatrix}. \tag{9.10}$$

Note that $X_{\overline{R}}$ can be calculated using the forward kinematics of the robot and the camera positioning values o_{11}, o_{12}, o_{22}. Taking the time derivative of (9.9) we obtain the following differential relationship between X_c and $X_{\overline{R}}$

$$\dot{X}_c = HRJ_c \dot{X}_{\overline{R}} + \begin{bmatrix} 0 \\ 0 \\ \gamma_2 \dot{x}_{c2} \end{bmatrix}, \tag{9.11}$$

where the composite camera image Jacobian, $J_c \in \Re^{3\times3}$, is defined explicitly as

$$J_c = \begin{bmatrix} \frac{1}{(\overline{z_r} + \lambda_1 + o_{22})} & 0 & -\frac{\overline{x_r}}{(\overline{z_r} + \lambda_1 + o_{22})^2} \\ 0 & \frac{1}{(\overline{z_r} + \lambda_1 + o_{22})} & -\frac{\overline{y_r}}{(\overline{z_r} + \lambda_1 + o_{22})^2} \\ 0 & -\frac{\overline{z_r}}{(\overline{y_r} + \lambda_2 + o_{12})^2} & \frac{1}{(\overline{y_r} + \lambda_2 + o_{12})} \end{bmatrix}. \tag{9.12}$$

In (9.12), for the composite image Jacobian matrix, J_c, to be invertible the analysis requires that $\det(J_c)$ exists and is positive. This is supplied when $\lambda_2 + o_{12}$, and $\lambda_1 + o_{22}$ are both selected to be positive. Note that we restrict the motion of the manipulator to one quadrant to provide positive $\overline{y_r}$, and $\overline{z_r}$ values.

9.3 Control Formulation and Design

We will assume that a smooth time-varying desired end effector trajectory generated in the camera space, denoted by $X_d(t) = \begin{bmatrix} x_d(t) & y_d(t) & z_d(t) \end{bmatrix}^T$, is constructed so that $X_d(t) \in C^2$. To provide a means of quantifying the position tracking control objective, we define the position tracking error signal in camera space $e(t) \in \Re^3$ as

$$e = X_d - X_c. \tag{9.13}$$

Taking the time derivative of (9.13) and multiplying the resultant equation by

$$A \triangleq \begin{bmatrix} A_1 & A_2 & 0 \\ A_3 & A_4 & 0 \\ 0 & 0 & A_5 \end{bmatrix} = (HR)^{-1} = \begin{bmatrix} \frac{\cos\theta_1}{f_1\beta_{11}} & \frac{\sin\theta_1}{f_1\beta_{12}} & 0 \\ -\frac{\sin\theta_1}{f_1\beta_{11}} & \frac{\cos\theta_1}{f_1\beta_{12}} & 0 \\ 0 & 0 & \frac{1}{\gamma_1} \end{bmatrix} \tag{9.14}$$

we obtain

$$A\dot{e} = A\left\{ \dot{X}_d - \begin{bmatrix} 0 \\ 0 \\ \gamma_2\dot{x}_{c2} \end{bmatrix} \right\} - J_c J \dot{q}. \tag{9.15}$$

Note that the forward kinematic relationship $\dot{X}_{\overline{R}} = J\dot{q}$ is used. Motivated by the subsequent stability analysis we pre-multiply both sides of (9.15) by the following transformation matrix

$$T = \begin{bmatrix} \frac{1}{A_1 A_4 - A_3 A_2} & \frac{A_3}{A_4} - \frac{A_2}{A_4(A_1 A_4 - A_3 A_2)} & 0 \\ 0 & 1 & 0 \\ 0 & 0 & 1 \end{bmatrix} \tag{9.16}$$

and obtain the following open loop dynamics for $e(t)$

$$Z\dot{e} = Z\left\{ \dot{X}_d - \begin{bmatrix} 0 \\ 0 \\ \gamma_2\dot{x}_{c2} \end{bmatrix} \right\} - TJ_c J \dot{q}, \tag{9.17}$$

where the constant matrix $Z \in \Re^{3\times3}$ is defined as

$$Z = \begin{bmatrix} \frac{1+A_3^2}{A_4} & A_3 & 0 \\ A_3 & A_4 & 0 \\ 0 & 0 & A_5 \end{bmatrix} \tag{9.18}$$

and is positive definite when the rotation angles of the cameras satisfy $-\pi/2 < \theta_1, \theta_2 < \pi/2$. Following a backstepping-like design procedure, we rewrite (9.17) in the following form;

$$Z\dot{e} = Z\left\{ \dot{X}_d - \begin{bmatrix} 0 \\ 0 \\ \gamma_2\dot{x}_{c2} \end{bmatrix} \right\} - Tv + TJ_I \eta, \tag{9.19}$$

where $J_I = J_c J$ is the image Jacobian and the auxiliary tracking-like signal $\eta(t) \in \Re^3$ is defined as

$$\eta = u - \dot{q} \tag{9.20}$$

with $u = J_I^{-1}v$. Using the definitions of (9.16) and (9.18) the open loop dynamics of (9.19) can be written in the following advantageous form

$$Z\dot{e} = \begin{bmatrix} \phi_4^{-1}(W_1\phi_1 - v_1) \\ W_2\phi_2 - v_2 \\ W_3\phi_3 - v_3 \end{bmatrix} + TJ_I \eta, \tag{9.21}$$

where $W_1(.)$, $W_2(.)$, $W_3(.)$ are known regression matrices defined explicitly as

$$W_1 = \begin{bmatrix} \dot{x}_d & \dot{y}_d & -v_2 \end{bmatrix}, W_2 = \begin{bmatrix} \dot{x}_d & \dot{y}_d \end{bmatrix}, W_3 = \begin{bmatrix} \dot{z}_d & \dot{x}_{c2} \end{bmatrix} \tag{9.22}$$

and ϕ_1, ϕ_2, ϕ_3, ϕ_4 represent unknown constant parameter vectors with proper dimensions that are defined as

$$\phi_1 = \left[\frac{1+A_3^2}{A_4}\phi_4 \ A_3\phi_4 \ \frac{A_3\phi_4 - A_2}{A_4} \right]^T, \qquad \phi_2 = \left[A_3 \ A_4 \right]^T,$$

$$\phi_3 = \left[\frac{1}{\gamma_1} \ -\frac{\gamma_2}{\gamma_1} \right]^T, \qquad \phi_4 = A_1 A_4 - A_2 A_3. \tag{9.23}$$

Based on the open-loop dynamics and the subsequent stability analysis the auxiliary internal control inputs are designed as follows

$$v_1 = W_1\hat{\phi}_1 + k_1 e_1, \qquad v_2 = W_2\hat{\phi}_2 + k_2 e_2, \qquad v_3 = W_3\hat{\phi}_3 + k_3 e_3, \tag{9.24}$$

where $e_i(t)$, $i = 1, 2, 3$ denote the elements of $e(t)$, k_1, k_2, k_3 are positive scalar control gains and $\hat{\phi}_1(t) \in \Re^3$, $\hat{\phi}_2(t) \in \Re^2$, $\hat{\phi}_3(t) \in \Re^2$ are dynamic parameter estimates that are updated according to

$$\dot{\hat{\phi}}_1 = \Gamma_1 W_1^T e_1, \qquad \dot{\hat{\phi}}_2 = \Gamma_2 W_2^T e_2, \qquad \dot{\hat{\phi}}_3 = \Gamma_3 W_3^T e_3, \tag{9.25}$$

where $\Gamma_1 \in \Re^{3\times3}$, $\Gamma_2 \in \Re^{2\times2}$ and $\Gamma_3 \in \Re^{2\times2}$ are diagonal positive-definite gain matrices. After substituting (9.24) into (9.21) we obtain the closed loop dynamics for $e(t)$ as

$$Z\dot{e} = \begin{bmatrix} \phi_4^{-1}(W_1\tilde{\phi}_1 - k_1 e_1) \\ W_2\tilde{\phi}_2 - k_2 e_2 \\ W_3\tilde{\phi}_3 - k_3 e_3 \end{bmatrix} + TJ_I\,\eta. \tag{9.26}$$

The backstepping type control design also requires the dynamics for the auxiliary signal $\eta(t)$. For deriving this, we take the time derivative of (9.20) and pre-multiply the resultant equation by the inertia matrix to obtain

$$M\dot{\eta} = -V_m\eta - J_I^T T^T e - \tau + Y\theta, \tag{9.27}$$

where $Y(.)$ is a regression matrix containing the known/measurable terms and θ is the vector containing unknown but constant system parameter with proper dimensions and defined as

$$Y\theta = M\dot{u} + V_m u + F(\dot{q}) + G + J_I^T T^T e. \tag{9.28}$$

Note that the term $J_I^T T^T e$ is injected to the dynamics to cancel the corresponding term in stability analysis.

Remark 2 *Based on the structure of (9.28) and standard adaptive controller design procedures, it is fairly easy to see that when the control torque input vector $\tau(t)$ with the parameter estimation vector $\hat{\theta}(t)$ are designed to have the following form*

$$\tau = Y\hat{\theta} + K_\eta\eta, \qquad \dot{\hat{\theta}} = \Gamma_\theta Y^T\eta, \tag{9.29}$$

where $K_\eta \in \Re^{n\times n}$ is a constant diagonal positive-definite gain matrix and Γ_θ is a constant diagonal positive-definite matrix with proper dimension, the tracking error can be driven to zero. However, as can be observed from (9.28) and (9.29), this approach requires calculation of the time derivative of the auxiliary control input, $u(t)$, which in turn necessitates on-line calculations of the derivatives of the input $v(t)$ and the image Jacobian J_I. Thus, the implementation would require massive computations and hinders the analysis, making the result unnecessarily complicated for practical purposes. In this work, instead of using the aforementioned standard adaptive controller approach, we utilize a high-gain controller that treats *the regression matrix multiplied by the uncertain parameter vector formulation as if it were a disturbance term and utilize a nonlinear damping argument to compensate for the unwanted effects of it. This method not only eases the implementation, but also preserves the asymptotic stability result, as is shown in the subsequent analysis.*

Using the fact that the system dynamics $Y\theta$ is bounded by a function of the form

$$\|Y\theta\| \leqslant \rho(\|x\|)\|x\| \tag{9.30}$$

with $x = \begin{bmatrix} e^T & \eta^T \end{bmatrix}$, the control torque input vector is designed to be:

$$\tau = k_n \rho(\|x\|)^2 \eta + K_\eta \eta, \tag{9.31}$$

where $K_\eta \in \Re^{n \times n}$ is a constant diagonal positive-definite gain matrix. After substituting (9.31) in (9.27) with the dynamics bounded by ω_R term, we obtain the closed loop dynamics for $\eta(t)$ as

$$M\dot{\eta} = -V_m \eta - J_I^T T^T e - K_\eta \eta - k_n \rho(\|x\|)^2 \eta + Y\theta. \tag{9.32}$$

We now state the following Theorem:

Theorem 1 *The adaptive control law proposed by (9.24) with the update laws (9.25) and the control law of (9.31) ensure the global asymptotic end effector tracking in the sense that*

$$\lim_{t \to \infty} e(t) = 0, \tag{9.33}$$

provided that the damping gain obeys the following inequality

$$k_n \geqslant \frac{1}{\min(\lambda_{\min}\{K_\eta\}, \min(\phi_4^{-1}k_1, k_2, k_3))} \tag{9.34}$$

and the camera orientations satisfy the following condition

$$-\pi/2 < \theta_i < \pi/2, \qquad i = 1, 2. \tag{9.35}$$

Proof 1 *We begin our proof by introducing the following non-negative scalar function*

$$V = \frac{1}{2}e^T Z e + \frac{1}{2}\eta^T M \eta + \frac{1}{2}\phi_4^{-1}\tilde{\phi}_1^T \Gamma_1^{-1}\tilde{\phi}_1 + \frac{1}{2}\tilde{\phi}_2^T \Gamma_2^{-1}\tilde{\phi}_2 + \frac{1}{2}\tilde{\phi}_3^T \Gamma_3^{-1}\tilde{\phi}_3. \tag{9.36}$$

Taking the time derivative of this expression, then inserting the closed loop dynamics from (9.26) and (9.32), and applying the parameter update terms in (9.25), we obtain the following relationship,

$$\dot{V} = e_1 \phi_4^{-1} W_1 \tilde{\phi}_1 - k_1 e_1^2 + e_2 W_2 \tilde{\phi}_2 - k_2 e_2^2 + e_3 W_3 \tilde{\phi}_3 - k_3 e_3^2 + e^T T J_I \eta$$

$$+ \frac{1}{2}\eta^T \dot{M} \eta + \eta^T (-V_m \eta - J_I^T T^T e - K_\eta \eta - k_n \rho(\|x\|)^2 \eta + Y\theta) \tag{9.37}$$

$$- \phi_4^{-1}\tilde{\phi}_1^T W_1^T e_1 - \tilde{\phi}_2^T W_2^T e_2 - \tilde{\phi}_3^T W_3^T e_3.$$

When we apply the well-known skew symmetric relationship between the inertia and centripetal-Coriolis matrices of (9.1) to (9.37) and eliminate the corresponding *parameter error and image jacobian (J_I) terms, followed by substitution of (9.30) for the system dynamics, we obtain*

$$\dot{V} \leqslant -\phi_4^{-1} k_1 e_1^2 - k_2 e_2^2 - k_3 e_3^2 - \eta^T K_\eta \eta \qquad (9.38)$$
$$-[\eta^T k_n \rho(\|x\|)^2 \eta) - \eta^T \rho(\|x\|) \|x\|].$$

By adding and subtracting $\frac{\|x\|^2}{4k_n}$ terms to the right-hand side of this equation, we can complete the squares of the nonlinear damping gain and the system dynamics bounding functions in the second line of (9.38), thus can further upperbound this result to have the following form:

$$\dot{V} \leqslant -\phi_4^{-1} k_1 e_1^2 - k_2 e_2^2 - k_3 e_3^2 - \eta^T K_\eta \eta \qquad (9.39)$$
$$- \left[\sqrt{k_n} \rho(\|x\|) \|\eta\| - \frac{\|x\|}{2\sqrt{k_n}} \right]^2 + \frac{\|x\|^2}{4k_n}.$$

As the completed squares term in (9.39) has a negative value for all times, its presence has no effect on the validity of this inequality, hence the rest of the term in (9.39) can be expressed to have the following form:

$$\dot{V} \leqslant -\beta \|x\|^2, \qquad (9.40)$$

where $\beta = \min\{\lambda_{\min}\{K_\eta\}, \min(\phi_4^{-1} k_1, k_2, k_3)\} - \frac{1}{4k_n}$ and the definition (9.30) has been applied for x. From (9.36) and (9.40), provided that (9.34) is satisfied, we can conclude that $V(t) \in \mathcal{L}_\infty$; hence, all the elements of $V(t)$, that is $e(t)$, $\eta(t)$, $\tilde{\phi}_1(t)$, $\tilde{\phi}_2(t)$, and $\tilde{\phi}_3(t) \in \mathcal{L}_\infty$. Furthermore, using the boundedness of the tracking error signal $e(t)$ and the estimation error terms $\tilde{\phi}_i(t)$, $i = 1, 2, 3$, the auxiliary control terms $v_i(t)$, $i = 1, 2, 3$ from (9.24) are also bounded. Due to the boundedness of $\eta(t)$ and $\rho(\|x\|)$, from equation (9.31), the control torque input $\tau(t)$ is also bounded.

Similarly from (9.25), the boundedness of $\dot{\hat{\phi}}_i(t)$, $i = 1, 2, 3$ is achieved. So we have proven that all the signals remain bounded during the closed loop operation. At this stage using the boundedness of $v_i(t)$ and $\eta(t)$ terms in equation (9.26) of closed loop error dynamics, we can also show that $\dot{e}(t)$ is bounded. Due to the boundedness of $\dot{e}(t)$, we can conclude that $e(t)$ is uniformly continuous. In addition, it is straightforward to use (9.40) to illustrate that $e(t) \in \mathcal{L}_2$. At this point from direct application of Barbalat's Lemma [13] we conclude (9.33). $\square$

9.4 Simulation Results

To illustrate the performance of the proposed controller we selected a 3 dof robot manipulator model with

$$M = \begin{bmatrix} M_{11} & 0 & 0 \\ 0 & M_{22} & M_{23} \\ 0 & M_{32} & M_{33} \end{bmatrix}, \qquad V_m = \begin{bmatrix} V_{11} & 0 & 0 \\ 0 & V_{22} & V_{23} \\ 0 & V_{32} & 0 \end{bmatrix}, \qquad G = \begin{bmatrix} 0 \\ G_2 \\ G_3 \end{bmatrix}, \qquad (9.41)$$

where the entries are formulated as

$$M_{11} = m_3(l_2 \cos(q_2) + l_{c3} \cos(q_2 + q_3))^2 + m_2 l_{c2}^2 \cos^2(q_2) + A_2 \sin^2(q_2)$$
$$\qquad + A_3 \sin^2(q_2 + q_3) + E_1 + E_3 \cos^2(q_2) + E_3 \cos^2(q_2 + q_3),$$
$$M_{22} = m_2 l_{c2}^2 \sin^2(q_2) + l_2 + l_3 + m_3(l_2^2 + l_{c3}^2 + 2l_2 l_{c3} \cos(q_3)),$$
$$M_{23} = M_{32} = m_3(l_{c3}^2 + l_2 l_{c3} \cos(q_3)) + l_3, \quad M_{33} = m_3 l_{c3} + l_3,$$

(9.42)

$$V_{11} = -m_3 l_2 l_{c3}(\dot{q}_2 + \dot{q}_3) \sin(q_2 + q_3) \cos(q_2) - m_3 l_2 l_{c3} \dot{q}_2 \sin(q_2) \cos(q_2 + q_3)$$
$$\qquad - (m_2 l_{c2}^2 + m_3 l_2^2 + E_3)\dot{q}_2 \sin(q_2) \cos(q_2)$$
$$\qquad - E_3(\dot{q}_2 + \dot{q}_3) \sin(q_2 + q_3) \cos(q_2 + q_3)$$
$$\qquad - m_3 l_{c3}^2(\dot{q}_2 + \dot{q}_3) \sin(q_2 + q_3) \cos(q_2 + q_3) + A_2 \dot{q}_2 \sin(q_2) \cos(q_2)$$
$$\qquad + A_3(\dot{q}_2 + \dot{q}_3) \sin(q_2 + q_3) \cos(q_2 + q_3),$$
$$V_{22} = -m_3 l_2 l_{c3} \dot{q}_3 \sin(q_3) + m_2 l_{c2}^2 \dot{q}_2 \sin(q_2) \cos(q_2),$$
$$V_{23} = V_{32} = -0.5 m_3 l_2 l_{c3} \dot{q}_3 \sin(q_3),$$

(9.43)

$$G_2 = 9.8(m_2 l_{c2} + m_3 l_2) \cos(q_2) + 9.8 m_3 l_{c3} \cos(q_2 + q_3),$$
$$G_3 = 9.8 m_3 l_{c3} \cos(q_2 + q_3).$$

(9.44)

In these matrices the coefficients of the entries are the lengths of the manipulator links $l_1 = 0.5$, $l_2 = 0.4$, $l_3 = 0.4$ in metres, the masses of the links $m_1 = 4$, $m_2 = 3$ and $m_2 = 3$ in kg, distances of the link joints $l_{c2} = 0.2$, $l_{c3} = 0.2$ in metres, and the cylindrical link radius $R = 0.05$. The torque limits are $\tau_i = \pm 50$ and the cylindrical link inertial parameters are obtained by $E_1 = m_1 R^2/2$, $E_i = m_i R^2/12$, $A_i = m_i R^2/2$ and $l_i = m_i R^2/12$, where $i = 2, 3$ for the robot links.

The selected internal controller gain matrix is

$$K \triangleq \text{diag}\{ k_1 \ k_2 \ k_3 \} = \text{diag}\{ 3 \ 1.5 \ 2 \} \tag{9.45}$$

while the applied adaptation term parameters are as below:

$$\Gamma_1 = \text{diag}\{ 0.5 \ 0.0001 \ 0.01 \}, \qquad \Gamma_2 = \text{diag}\{ 0.001 \ 0.005 \},$$
$$\Gamma_3 = \text{diag}\{ 0.00005 \ 0.00001 \}. \tag{9.46}$$

Similarly, the outer loop control has the control gain matrix and the damping gain coefficient selected as

$$K_\eta = \text{diag}\{ 50 \ 75 \ 25 \}, \quad k_n = 25. \tag{9.47}$$

For simulation results a sample desired position trajectory in the composite camera space is selected as

$$X_d(t) = \begin{bmatrix} 465 + 20 \sin(\frac{\pi}{6}t + \frac{\pi}{12}) \\ 490 + 25 \sin(\frac{\pi}{6}t + \frac{5\pi}{12}) \\ 480 + 20 \sin(\frac{\pi}{6}t + \frac{5\pi}{12}) \end{bmatrix} \tag{9.48}$$

with all components in pixels. Trajectories in similar ranges should be preferred as these do not force the singularities of the employed manipulator model.

The composite camera output for this trajectory is in Figure 9.2, from which the initial system transients can be observed to decay quickly. Similarly the Figure 9.3 depicts the quick convergence of the error terms to zero, verifying the asymptotically stable nature of the proposed system. The applied controller torque outputs are in Figure 9.4, while Figure 9.5 shows the parameter estimates for the inner loop, which also converge in short time periods. As an overall result, these figures depict the stable and efficient operation of the presented visual servo configuration and hence should verify the validity of the controller system developed in this paper.

9.5 Conclusion

We have presented a nonlinear adaptive 3-D end effector position tracking controller for a vision-based system composed of 2 fixed cameras. The proposed controller achieves asymptotic end effector tracking despite the presence of uncertainties in the intrinsic camera parameters of both vision sensors and the robot dynamics. This result was obtained by applying a novel approach to compose the camera inputs and form the 3-dimensional end effector position information in camera space and using a backstepping design scheme. Considering the similarities between the controller design procedures our controller might be considered as the 3-D extension of [7] with the use of multiple cameras. Unfortunately the theory presented is only supported by simulation results. In the current state, for an experimental verification, high-speed vision systems are required. Future work will concentrate on removing this drawback.

References

1. S. Hutchinson, G.D. Hager, P.I. Corke. A tutorial on visual servo control. In: *IEEE Trans. on Robotics and Automation*, vol. 12, no. 5, 1996, pp. 651-670.
2. B.J. Nelson, N.P. Papanikolopoulos, P.K. Khosla. Robotic visual servoing and robotic assembly tasks. In: *IEEE Robotics and Automation Mag.*, vol. 3, no. 2, 1996, pp. 23-31.
3. F. Miyazaki, Y. Masutani. Robustness of sensory feedback control based on imperfect Jacobian. In: Miura H. and Arimoto S. (Eds.) *Robotics Research: The Fifth Int. Symp.*, MIT Press, Cambridge, MA 1990, pp. 201-208.
4. G.D. Hager, W.C. Chang, A.S. Morse. Robot hand-eye coordination based on stereo vision. In: *IEEE Control Systems Mag.*, vol. 15, no. 1, 1995, pp. 30-39.
5. B.H. Yoshima, P.K. Allen. Active, uncalibrated visual servoing. In: *Proc. IEEE Int. Conf. on Robotics and Automation*, 1994, pp. 156-161.
6. R. Kelly. Robust asymptotically stable visual servoing of planar robots. In: *IEEE Trans. on Robotics and Automation*, vol. 12, no. 5, 1996, pp. 759-766.
7. E. Zergeroglu, D. Dawson, M.S. de Queiroz, A. Behal. Vision-based nonlinear tracking controllers with uncertain robot-camera parameters. In: *IEEE/ASME Trans. on Mechatronics*, vol. 6, no. 3, 2001, pp. 322-337.
8. E. Zergeroglu, D. Dawson, M.S. de Queiroz, P. Setlur. Robust visual-servo control of planar robot manipulators in the presence of uncertainty. In: *Journal of Robotic Systems*, vol. 20, no. 2, 2003, pp. 93-106.
9. Y. Shen, D. Sun, Y. Lui, K. Li. Asymptotic trajectory tracking of manipulators using uncalibrated visual feedback. In: *IEEE/ASME Trans. on Mechatronics*, vol.8 , no. 1, 2003, pp. 87-98.
10. M.W. Spong, M. Vidyasagar. *Robot dynamics and control*. Wiley, New York 1989.
11. B. Xian, M.S. de Queiroz, D. Dawson, I. Walker. Task-space tracking control of robot manipulators via quaternion feedback. In: *IEEE Trans. on Robotics and Automation*, vol. 20, no. 1, 2004, pp. 160-167.
12. R.K. Lenz, R.Y. Tsai. (1988) Techniques for calibration of the scale factor and image center for high accuracy 3-D machine vision metrology. In: *IEEE Trans. on Pattern Analysis and Machine Intelligence*, vol. 10, no. 5, 1988, pp. 713-720.
13. J.J. Slotine, W. Li. *Applied nonlinear control*. Prentice Hall, Englewood Cliff, NJ 1991.

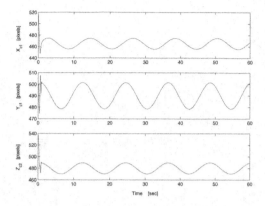

Fig. 9.2. The end effector trajectory as seen from the composite camera during simulation

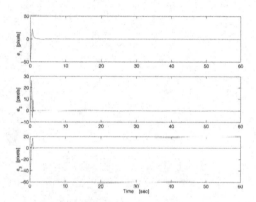

Fig. 9.3. The tracking error terms

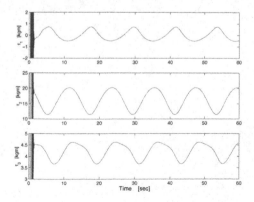

Fig. 9.4. Input control torques

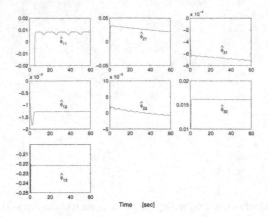

Fig. 9.5. Estimates for the uncertain parameters of the cameras

Flexible Robot Trajectory Tracking Control

Anthony Green and Jurek Z. Sasiadek

Department of Mechanical and Aerospace Engineering
Carleton University, Ottawa, Ontario, K1S 5B6, Canada
agreen2@connect.carleton.ca, jsas@ccs.carleton.ca

10.1 Introduction

Operational problems with space robots relate to several factors, one most importantly being structural flexibility and subsequently significant difficulties with position control. Elastic link vibrations with inherent nonminimum phase response coupled with large rotations create a complex nonlinear dynamic system for control. The ability of the flexible robot to follow a prescribe trajectory is an important issue. In case of more complex operation, good tracking control could potentially save much time and money. This paper compares the performance of control strategies with noise filtering, then combined with nonminimum phase and corrective control action modeled by time delays for a stationary spacecraft mounted flexible robot tracking a 12.6 m×12.6 m square trajectory. Finally a fuzzy logic system (FLS) vibration suppression strategy demonstrates its effectiveness in providing intelligent and autonomous precision control. The inclusion of this fuzzy logic approach to vibration suppression is motivated by the prospect of substituting for the control of space robotic systems typically teleoperated by astronauts. Tracking results are obtained initially without vibration suppression or noise filtering, then with EKF and FLAEKF state estimators to determine tracking accuracy for robot process and measurement noise. Time delays are included in the feedback and control action loops to simulate nonminimum phase response and corrective control. Finally, a FLS is used to adapt the control law and suppress residual vibrations. Good tracking results were previously achieved using an input shaping method to reduce residual vibrations coupled with an inverse kinematics control strategy and recursive order-n algorithm to model a two-link flexible robot tracking a square trajectory [1]. Studies were conducted on the control of a rigid two-link robot using linear quadratic Gaussian (LQG) control with Kalman filter, EKF and FLAEKF [2]. Fuzzy Kalman filter equations were derived in [3]. FLS adaptive control is used for vibration suppression of a flexible robot [4]. Flexible robot sensor location effects were investigated [5]. Figure 10.1 shows a flexible robot with planar motion and vibration modes. Gravity and friction effects are neglected. The robot parameters are taken from Banerjee and Singhose [1].

K. Kozłowski (Ed.): Robot Motion and Control, LNCIS 335, pp. 153–163, 2006.
© Springer-Verlag London Limited 2006

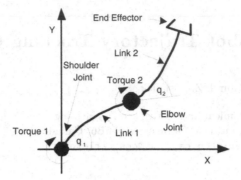

Fig. 10.1. Flexible robot

10.2 Flexible Dynamics

Assumed modes of vibration are used to model the flexible dynamics and time delays model nonminimum phase response. Assumed modes of vibration for an Euler-Bernoulli cantilever beam are combined with rigid-link dynamics to capture the nonlinear flexural multibody interactions of a two-link robot derived in Euler-Lagrange dynamics equation form. Assumed modes accommodate configuration changes during robot operation, whereas, natural modes must be continually recomputed. The Euler-Lagrange flexible dynamics matrix equations are given by [4–7]:

$$\tau = M(q)\ddot{q} + C(q,\dot{q})\dot{q} + Kq. \tag{10.1}$$

M contains rigid and flexible terms; C usually contains coupled rigid and elastic Coriolis and centrifugal effects but the elastic terms are neglected because of assumed mode orthogonality and small elastic deformation second-order terms; K contains link stiffness terms and q is a generalized coordinate vector of joint angles θ_i and link deformations δ_i, $i = 1, 2$. Full dynamics equations for a multi-degree-of-freedom (multi-dof) manipulator have been derived previously [8]. The complex mathematical model [8] was verified experimentally for multi-link flexible manipulator. The complex mathematical model was not suitable for real-time control and had to be simplified in control system design process. The simpler model used in real-time control was verified with experimental data and its performance compared with full model based on finite elements.

10.3 Control Strategies

10.3.1 LQG with EKF or FLAEKF Control

LQG control is shown in Fig. 10.4 with either EKF or FLAEKF noise filtering for which an intuitive Jacobian transpose control law is given by [2, 4]:

$$\tau_r = J^T(\theta) \left[K_p \begin{pmatrix} e_x \\ e_y \end{pmatrix} + K_d \begin{pmatrix} \dot{e}_x \\ \dot{e}_y \end{pmatrix} \right]. \tag{10.2}$$

The Jacobian $J(\theta)$ is dependent solely on rigid link angular motion only for the dominant vibration mode case. Using damping ratio $\zeta = 0.707$ and for $\omega_{c_1} = 12.28$ Hz, PD gains K_p and K_d are given by

$$K_p = \begin{bmatrix} \omega_{c_1} & 0 \\ 0 & \omega_{c_1} \end{bmatrix} = \begin{bmatrix} 150.79 & 0 \\ 0 & 150.79 \end{bmatrix}, \tag{10.3}$$

$$K_d = \begin{bmatrix} 2\zeta\omega_{c_1} & 0 \\ 0 & 2\zeta\omega_{c_1} \end{bmatrix} = \begin{bmatrix} 17.364 & 0 \\ 0 & 17.364 \end{bmatrix}. \tag{10.4}$$

10.3.2 Extended Kalman Filter

EKF state estimate, error covariance and Kalman gain equations are derived from linearized state-space equations for a rigid dynamics robot model [2,9]:

$$\dot{x} = Ax + Bu + Dw, \tag{10.5}$$

$$z = Fx + Gu + v, \tag{10.6}$$

where A, B, D, F and G are dynamic coefficient, control input coupling, process noise coupling, measurement sensitivity and output coupling matrices. State, state derivative, control and measurement, process and measurement noise vectors are given by $x, \dot{x}, u, z, w, v$ respectively. With appropriate transformation of equations (10.5) and (10.6) in discrete form may be given by [2,9]:

$$x_{k+1} = \Phi_k x_k + \Gamma_k u_k + \Delta_k w_k, \tag{10.7}$$

$$z_{k+1} = C_k x_k + v_k \tag{10.8}$$

for time increments $k \geqslant 0$ and $D = [0]$, where; $\Phi_k, \Gamma_k, \Delta_k$ and C_k are time-varying matrices and u_k a deterministic input vector. $w_k \approx N(\mu, Q_k)$ and $v_k \approx N(\mu, R_k)$ are zero-mean uncorrelated Gaussian process and measurement noise matrices. $Q_k = E[w_k w_k^T], R_k = E[v_k v_k^T]$ and $N_k = E[w_k v_k^T]$ are process and measurement noise covariance and cross-covariance matrices. The Kalman gain for the EKF with correlated process and measurement noise is given by [2,9]:

$$K_k = (P_k^- C_k^T + N_k)(C_k P_k^- C_k^T + R_k + C_k N_k + N_k^T C_k^T)^{-1}. \tag{10.9}$$

The *a posteriori* state estimate is given in terms of *total* measurements on a corrected trajectory rather than *incremental* measurements on a *nominal* trajectory typical of a Kalman filter:

$$\hat{x}_k = \hat{x}_k^- + K_k(z_k - \hat{z}_k^-), \tag{10.10}$$

where $\hat{z}_k^-$ and $z_k - \hat{z}_k^-$ are predictive measurements and measurement residuals. The *a posteriori* error covariance matrix is given as

$$P_k = (I - K_k C_k)P_k^- - K_k N_k^T. \tag{10.11}$$

Projected state estimate and error covariance matrices are

$$\hat{x}_{k+1}^- = \Phi_k \hat{x}_k, \tag{10.12}$$

$$P^-_{k+1} = \Phi_k P_k \Phi_k^T + Q_k. \tag{10.13}$$

The *a priori* state estimates and error covariance at time k = 0 are $\hat{x}_0^-$ and P_0^-. K_k, P^-_{k+1}, P_k^- and P_k are Kalman gain, projected error covariance, *a priori* error covariance, *a posteriori* error covariance matrices. $\hat{x}^-_{k+1}, \hat{x}_k^-$ and $\hat{x}_k$ are projected, *a priori* and *a posteriori* state estimates. Measured outputs q and noise v_k input to the state estimator. Process and measurement white noise inputs the EKF as a Gaussian distribution with Q_k and R_k noise co-variances represented by the distribution variance.

10.3.3 Fuzzy Logic Adaptive EKF

The weighted recursive equations for the FLAEKF were presented in previous work [3]. The fuzzy logic adaptive extended Kalman filter (FLAEKF) utilizes the EKF recursive equations but weighted with parameter α operating on Kalman gain and error covariance. Noise covariance and cross-covariance and error covariance matrices are given by [3]:

$$Q_\alpha = Q_k \alpha^{-k+1}, \tag{10.14}$$

$$R_\alpha = R_k \alpha^{-k+1}, \tag{10.15}$$

$$N_\alpha = N_k \alpha^{-k+1}, \tag{10.16}$$

$$P_k^{\alpha-} = P_k^- \alpha^{2k}, \qquad \alpha \geqslant 1. \tag{10.17}$$

The FLAEKF behaves as an EKF when $\alpha = 1$. For $\alpha \geqslant 1$, as k increases, Q_k and R_k decrease so the most recent measurement is weighted higher. The FLAEKF adapts the value of according to the magnitude of P_k and non-zero means such that optimality and a zero-mean white noise condition is maintained. Using (10.9), (10.15) and (10.16), the Kalman gain K_k for the FLAEKF with correlated process and measurement noise is given by [3]:

$$K_k = (\alpha^2 P_k^{\alpha-} + N_k)(\alpha^2 C_k P_k^{\alpha-} C_k^T + R_k + C_k N_k + N_k^T C_k^T)^{-1}. \tag{10.18}$$

The *a priori* state estimate is

$$\hat{x}_k^- = \Phi_k \hat{x}_k \tag{10.19}$$

and error covariance matrix is

$$P_{k+1}^{\alpha-} = \alpha^2 \Phi_k P_k^{\alpha-} \Phi_k^T + Q_k. \tag{10.20}$$

The *a posteriori* state estimate is

$$\hat{x}_k = \hat{x}_k^- + K_k(z_k - \hat{z}_k^-). \tag{10.21}$$

The *a posteriori* error covariance matrix is

$$P_k^\alpha = (I - K_k C_k)P_k^{\alpha-} - K_k\left\{\frac{N_k^T}{\alpha^2}\right\} \tag{10.22}$$

for initial condition: k = 0, $P_0^{\alpha-} = P_0$

Assuming Gaussian statistics and from a *conditional density* viewpoint the mean-square estimation error is

$$\hat{x}_k = E(x_k | z_k^+).$$ (10.23)

The mean state estimate is

$$\bar{x}_k = \hat{x}_k^- + M_k(z_k - \hat{z}_k^-),$$ (10.24)

where

$$M_k = P_k^- C_k^T (C_k P_k^- C_k^T) + R_k$$ (10.25)

and the error covariance of the residuals is

$$P_z = \left\{ (P_k^-)^{-1} + C_k^T R_k^{-1} C_k \right\}^{-1}.$$ (10.26)

Fuzzy system variables have triangular membership functions and all universes of discourse fall into the positive right half-plane to satisfy the condition $\alpha \geqslant 1$ [2,3,10]. Input variables are mean state estimate and error covariance of the residuals given by (10.24) and (10.26). Verbal descriptors for zero (ZERO), small (S), medium (M), and large (L) are used for membership functions shown in Fig. 10.2 to generate fuzzy rules of the form; IF Pz is S AND Mean is S THEN α is M.

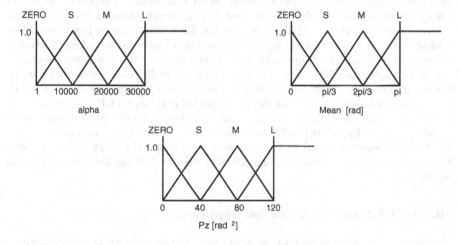

Fig. 10.2. Membership functions for α, mean and Pz

Each fuzzy variable has 4 membership functions resulting in 16 fuzzy rules derived in Table 10.1. Gaussian random number distributions represent white process and measurement noise for covariances Q_k and R_k. Cross covariance, $N_k = E[w_k v_k^T]$, models the cross-correlation of control inputs and measurement outputs and computed as the product of standard deviations of Q_k and R_k diagonal elements, i.e. $N_k = E[w_k v_k^T] \approx \sigma_{w_k} \sigma_{v_k}$. In this study $w_k = v_k = N(0,1)$ and $N_k = 1$. For both EKF and FLAEKF the LQG dynamic regulator control law is given by [3,9]:

$$\tau_k = -K\hat{x}_k,$$ (10.27)

where K is the linear quadratic (LQ) gain matrix for rigid-link robot dynamics. The LQG dynamic regulator control law is fed back to form a resultant control law given by [10]

Table 10.1. FLAEKF rule matrix

		Mean			
	α	Z	S	M	L
	Z	Z	S	M	L
P_z	S	S	M	L	L
	M	M	L	L	L
	L	L	L	L	L

$$\tau = \tau_r + \tau_k. \tag{10.28}$$

Within the fuzzy logic adaptive EKF system there are two groups of fuzzy logic controllers. The first group, with output α, detects filter divergence and adapts the EKF. Generally, when the covariance is becoming large, and mean value is moving away from zero, the Kalman filter becomes unstable. In this case, a large α is applied. When α is large, process noises are added and it can ensure all states in the model are sufficiently excited by the process noise. When the covariance is extremely large, correct measurements are problematic, and the filter cannot depend on them. Then a smaller α is applied. By selecting an appropriate α value the fuzzy logic controller adapts the Kalman filter optimally and strives to keep the innovation sequence acting as zero-mean white noise. The fuzzy logic controller uses 9 rules, such as: IF the covariance of residuals is large and the mean value is zero THEN α is zero. IF the covariance of residuals is zero and the mean value is large THEN α is small. The second group, in which output is scale, is used to detect the change of measurement noise covariance R_k, related to the covariance of residual. As the covariance of the residuals increases, measurement noise increases. When the fuzzy logic controller finds the covariance of residual is larger than expected, it applies a large scale to adjust α.

10.3.4 FLS Adaptive Vibration Suppression

Adaptive vibration suppression is achieved with the FLS shown in Fig. 10.5 where an output variable λ determined by elastic deformation inputs δ_1 and δ_2 from the flexible dynamics then multiplied by scaling gain K_s to operate on the control law in (10.2) to give [4]:

$$\tau = K_s \lambda \left\{ J^T(\theta) \left[K_p \begin{pmatrix} e_x \\ e_y \end{pmatrix} + K_d \begin{pmatrix} \dot{e}_x \\ \dot{e}_y \end{pmatrix} \right] \right\}. \tag{10.29}$$

Verbal descriptors; positive (P), positive maximum (PMAX), positive medium (PM), zero (ZERO) and negative (N) are used for the membership functions shown in Fig. 10.3 to generate 9 fuzzy rules of the form; IF δ_1 is N AND δ_2 is P THEN λ is PMAX [4]. The FLS is developed intuitively, such that, link deformation magnitudes vary positively or negatively to complement or counter deformation of the other link.

Then λ varies according to the resultant deformation and ranges from ZERO for zero deformation to PMAX at 1.0 thereby forming a symmetric fuzzy rule

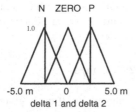

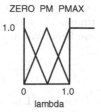

Fig. 10.3. Membership functions for δ_1, δ_2 and λ

Table 10.2. FLS Rule matrix

	λ	δ_2		
		N	ZERO	P
	N	PMAX	PM	PMAX
δ_1	ZERO	PM	ZERO	PM
	P	PMAX	PM	PMAX

matrix derived in Table 10.2. A scaling gain K_s operates on the λ fuzzy membership functions to modify their base widths and improve tracking accuracy as K_s increases [4, 10]. $K_s = 15$ is used for simulation.

10.4 Nonminimum Phase Response

Nonminimum phase (NMP) response is inherent in flexible robots and a problem for precision control. The phenomenon exists when joint actuation delays occur opposite in sense to that expected in reaction to a control input. Joint actuation induces flexing and momentary acceleration of the end effector in a direction opposite to that commanded. Analytical control theory describes this as transfer function poles or zeros occurring in the right-half s-plane and termed a phase shift or, transport lag, between the actuator and end effector. For continuous systems this phase shift creates a time delay between actual end effector position and corrective control action corresponding to the time required for mechanical waves to propagate through the link from joint to end-effector. Closely associated with nonminimum phase is noncollocation (distance between sensor and actuator) causing control action delay in response to sensed position errors. Collocated sensors provide joint angle data and typically mounted on rigid-link robot joints where nonminimum phase is not encountered. Flexible-link robots exhibit significant deformation at the end effector requiring accurate control by adjusting for the phase shift between actuator and end effector. An operational space control strategy, shown in Fig. 10.4, is suitable for noncollocated control as joint angles and rates are transformed through direct kinematics equations into end effector positions and velocities. Nonminimum phase response for a noncollocated sensor is modeled by a time delay in the feedback loop. Transport delay blocks (not shown) with a second-order Pad approximation are implemented in Matlab/Simulink^{TM} models. Nonminimum phase correction is given by a time delay in the forward control loop. Time delays are determined using the transverse beam vibration wave velocity c given by [11]:

$$c = \sqrt{\frac{E}{\rho}} = \sqrt{\frac{1745833}{21}} = 288.33 \text{ m/s.} \qquad (10.30)$$

Time delay from joint 1 to endpoint for two link lengths, i.e. 9 m, is given by:

$$t_{d_1} = \frac{9}{288.33} = 0.0312 \text{ s.} \qquad (10.31)$$

Time delay from joint 2 to endpoint for one link length of 4.5 m is given by:

$$t_{d_2} = \frac{4.5}{288.33} = 0.0156 \text{ s.} \qquad (10.32)$$

Average trajectory simulation time is 402 s for 16000 simulation steps (ss) at 0.001 step size, i.e. 0.0252 s per step.

Therefore, simulation time delay for joint 1 is:

$$d_1 = \frac{0.0312}{0.0252} = 1.238 \text{ ss} \qquad (10.33)$$

and simulation time delay for joint 2:

$$d_2 = \frac{0.0156}{0.0252} = 0.619 \text{ ss.} \qquad (10.34)$$

Time delay d_1 (worst case) is included in the control strategy position and velocity feedback loops.

10.5 Simulation Results

Figures 10.6(a) and (b) show clockwise tracking for the dominant vibration mode starting at bottom left. FLAEKF produces a slightly better result than EKF. Figures 10.6(c) and (e) show superposed tracking results for EKF, FLAEKF (collocated sensor), NMP, NMP correction and FLS vibration suppression. Figures 10.6(d) and (f) show zoomed views of transients at the first direction switch. Tracking times are 1 min 18 s for EKF with NMP correction, 1 min 21 s for FLAEKF with NMP correction and 1 min 50 s for FLS. Matlab/Simulink^{TM}, Control Systems and Fuzzy Logic Toolboxes were used for simulation [12].

10.6 Summary and Conclusions

Complex dynamics and inadequacies of classical and modern control systems are overcome by using a Fuzzy Logic System adaptive vibration suppression control strategy for a flexible robot manipulator. Results show the Fuzzy Logic System adaptive vibration suppression strategy achieves control accuracy superior to Linear Quadratic Gaussian method with Extended Kalman Filter or Fuzzy Logic Adaptive Extended Kalman Filter noise filtering algorithms derived from linear rigid dynamics equations and Non-Minimum-Phase time-delay corrective control action. The nonminimum phase response was also considered and briefly discussed. Nonminimum phase (NMP) response is inherent in flexible robot manipulators and a problem for precision control. To counter this problem a specific value of delay, determined experimentally, was introduced into the control strategy. The control strategy, based on Adaptive Kalman Filtering, includes an original method of Kalman filter gain adaptation that ensures stability within a broad range of conditions. The main contribution of this paper is the novel control strategy for a flexible manipulator. The dynamic model of a flexible manipulator was tested experimentally and against a more complex model in previous works.

References

1. Banerjee AK, Singhose W (1998) Command shaping in tracking control of a two-link flexible robot. J Guidance, Control and Dynamics, Engineering Note, 21:1012-1015
2. Green A, Sasiadek JZ (2001) Regular and fuzzy extended Kalman filtering for a two-link flexible robot manipulator. In: Proc AIAA Guidance, Navigation and Control Conf. AIAA, Reston, Virginia.
3. Sasiadek JZ, Wang Q (1999) Sensor fusion based on fuzzy Kalman filtering for autonomous robot vehicle. In: Proc IEEE Int Conf on Robotics and Automation, IEEE Press, Piscataway, New Jersey 2970-2975.
4. Green A, Sasiadek JZ (2005) J Guidance, Control and Dynamics, 28:36-42
5. Green A, Sasiadek J Z (2004) Sensor location effect for flexible robot control. In: Proc 4th Int Workshop on Robot Motion and Control (RoMoCo'04) Elsevier, Oxford 253-258.
6. De Luca A, Siciliano B (1991) Closed-form dynamic model of planar multilink lightweight robots. IEEE Trans. Systems, Man and Cybernetics, 21:826-839
7. Thomson WT (1981) Theory of vibration with applications, 2nd edn. Prentice-Hall, Upper Saddle River, New Jersey.
8. Beres W, Sasiadek JZ (1995) Finite element dynamic model of multilink flexible manipulators. App Math and Comp Sci, 5:231-262
9. Brown RG, Hwang PYC (1997) Introduction to random signals and applied Kalman filtering, John Wiley and Sons, New York.
10. Passino KM, Yurkovich S (1990) Fuzzy control. Addison-Wesley, Menlo Park, California.
11. Alexander HL (1988) Control of articulated and deformable space structures. In: Heer E, Lum H (eds) Machine Intelligence and Autonomy for Aerospace Systems. Progress in Astronautics and Aeronautics, AIAA, Washington, District of Columbia, pp 327-347
12. Matlab 6, Simulink 4 (2001) Control Systems and Fuzzy Logic Toolboxes, Release 12, The Mathworks Inc, Natick, Massachusetts.

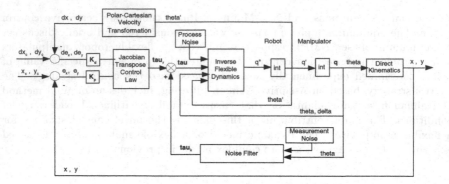

Fig. 10.4. LQG with EKF or FLAEKF control strategy

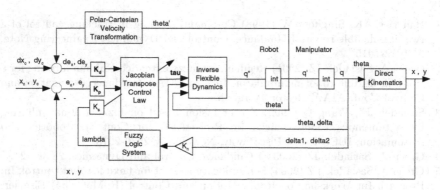

Fig. 10.5. FLS adaptive vibration suppression control strategy

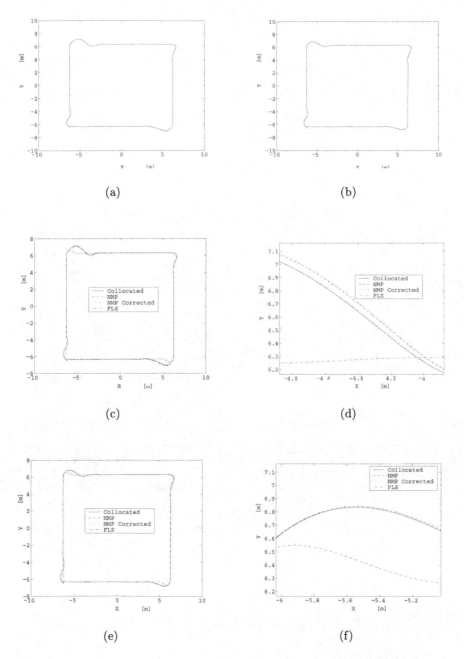

Fig. 10.6. Trajectories: (a) EKF, (b) FLAEKF, (c) EKF, NMP, NMP corrected and FLS, (d) EKF, NMP, NMP corrected and FLS zoomed at 1st direction switch, (e) FLAEKF, NMP, NMP corrected and FLS, (f) FLAEKF, NMP, NMP corrected and FLS zoomed at 1st direction switch

11

Modeling, Motion Planning and Control of the Drones with Revolving Aerofoils: an Outline of the XSF Project*

Lotfi Beji[1], Azgal Abichou[2], and Naoufel Azouz[1]

[1] LSC Laboratory, CNRS-FRE2494, Université d'Evry Val d'Essonne. 40, rue du Pelvoux, 91020, Evry Cedex, France {beji|azouz}@iup.univ-evry.fr
[2] LIM Laboratory, Ecole Polytechnique de Tunisie, BP 743, 2078 La Marsa, Tunisia azgal.abichou@ept.rnu.tn

11.1 Introduction

Aerial robotics has been known for several years as a considerable passion of private manufacturers as well as research laboratories. This interest is justified by the recent technological projections which make possible the design of powerful systems endowed with real capacities of autonomous navigation, at non-prohibitive cost. Today, the principal limitations which the researchers meet are related on one hand to the difficulty of controlling the apparatus in the presence of atmospheric turbulences, and on the other hand to the complexity of the problem of navigation requiring the perception of often constrained and evolutionary environment, in particular in the case of flights at low altitude. The applications are numerous. They initially relate to areas of safety (monitoring the airspace, urban and interurban traffic), natural risk management (monitoring the activity of volcanoes), environmental protection (measurement of the air pollution, monitoring the forests), intervention in hostile sites (radioactive media, mine clearance of the grounds without human intervention), management of large infrastructures (stoppings, high-voltage lines, pipelines), agriculture (detection and treatment of the cultures) and the catch of air sight in the production of films. All these missions require a powerful control of the apparatus and consequently precise information on its absolute and/or relative state with respect to its environment. Contrary to terrestrial mobile robots for which it is often possible to be limited to a kinematic model, control of aerial robots requires the knowledge of a dynamic model. The effects of gravity and the aerodynamic loads are the principal causes. These systems, for which the number of control inputs is lower than the number of degrees of freedom, are known as underactuated. The mechanism of control provides generally only one or two inputs for the translational dynamics and two or three inputs for the rotational dynamics. At the beginning of the nineties, the automation community

---

* This work is supported by the mini-flyer competition program organized by the DGA (Direction Générale des Armements) and the ONERA (Office Nationale d'Etude et de Recherche en Aérospatiale), France.

K. Kozłowski (Ed.): Robot Motion and Control, LNCIS 335, pp. 165–177, 2006.
© Springer-Verlag London Limited 2006

showed a renewal of interest in control of these systems. An outstanding example is a thorough study carried out on the dynamics of planes like VTOL (Vertical Take-Off and Landing) which made it possible to constitute an important source of knowledge and led to additional developments on the flat system theory and the techniques of linearization input-outputs [1, 2].

More recently, several researchers serve of their experiment on control mobile robots moving on the ground, in cooperation with control of flying machines.

Modelling and controlling aerial vehicles (blimps, mini rotorcraft) are the principal preoccupation of the *Lsc*, *Lim*-groups. Competition of the DGA-ONERA program won the XSF project which consists of a drone with revolving aerofoils. It is equipped with four rotors where two are directionals, what we call in the text *bidirectional* X4-flyer. In fact, the study of quad-rotor vehicles is not recent. However, combination of revolving aerofoils and directional rotors were attractive for the contest. In this topic, a mini-UAV was constructed by the *Lsc*-group taking into account industrial constraints. The aerial flying engine couldn't exceed 2 kg in mass, and 70 cm of scale with approximatively 30 min of flying-time. Compared to helicopters, named quad-rotor [3], the four-rotor rotorcraft has some advantages [4, 5]: given that two motors rotate counter clockwise while the other two rotate clockwise, gyroscopic effects and aerodynamic torques tend, in trimmed flight, to cancel. Vertical motion is controlled by collectively increasing or decreasing the power for all motors.

In a conventional X4-flyer, the longitudinal/lateral motion in x/y direction is achieved by differentially controlling the motors generating a pitching/rolling motion of the airframe that inclines the thrust (producing horizontal forces) and leads to lateral accelerations. In a bidirectional X4 flyer, without generating rotation of the engine, we can only rotate two rotors with respect to their local frame and have lateral motions. The motion planing and control of these configurations are new. With respect to the literature, we cite the VTOL stabilization by Hamel [5], where the dynamic motor effects are incorporated. The robust output regulation for autonomous vertical landing was studied by Marconi [8]. The stabilization results for the X4-flyer by the nested saturation algorithm were given by Castillo [6]. That work was illustrated by successful tests. In the case of the PVTOL (Planar), the stabilization was solved with the input/output linearization procedure [7] and the theory of flat systems [1, 2]. As we can see, the control of flying vehicles is limited to the stabilization problem. In this work, tracking results are presented.

This article is organized in the following way: Section 11.2 describes the worked out translational and rotational motions. Section 11.3 presents the developed ideas of control for the conventional X4-flyer and the bidirectional one. A strategy to solve the tracking problem through point to point steering is shown in Section 11.4. The simulation tests, results, comments and the prospects for evolution of research conclude the work.

11.2 Configuration Description and Modeling

The conventional and the bidirectional XSF are systems consisting of four individual electrical fans attached to a rigid cross frame. We consider a local reference airframe $\Re_G = \{G, E_1^g, E_2^g, E_3^g\}$ at G (mass center), while the inertial frame is $\Re_o = \{O, E_x, E_y, E_z\}$ such that the vertical direction E_z is upwards. Let the vector

$X = (x, y, z)$ denote the G position with respect to $\Re_o$. The rotation of the rigid body is defined by $R_{\phi,\theta,\psi} : \Re_o \rightarrow \Re_G$, where $R_{\phi,\theta,\psi} \in SO(3)$ is an orthogonal rotation matrix which is defined by the Euler angles, θ(pitch), ϕ(roll) and ψ(yaw), regrouped in $\eta = (\phi, \theta, \psi)$. A conceptual form is sketched in Fig. 11.1.

The XSF is a quadrotor of 68 cm × 68 cm of total size. It is designed in a cross form and made of carbon fiber. Each tip of the cross has a rotor including an electric brushless motor, a speed controller and a two-blade ducted propeller. In the middle one can find a central cylinder enclosing electronics, namely Inertial Measurement Unit, on-board processor, GPS, radio transmitter, cameras and ultrasound sensors, as well as the LI-POLY batteries. The operating principle of the XSF can be presented as follows: two rotors turn clockwise, and the other two rotors turn counter-clockwise to maintain the total equilibrium in yaw motion. With the equilibrium of the angular velocities of all the rotors maintained, the XSF is either in a fixed position, or moving vertically (changing altitude). A characteristic of the XSF compared to other quadrotors is the swiveling of the supports of the motors 1 and 3 around the pitching axis due to two small servomotors (Fig. 11.2). This permits a more stabilized horizontal flight and a suitable cornering.

Fig. 11.1. Conceptual form of the X4 Super-Flyer (XSF)

11.3 Aerodynamic Forces and Torques

In this part we will define the characteristics of the aerodynamic forces and torques resulting from the blade theory. The blade behaves as a rotating wing. Each element of the blade dr is in contact with the airflow with a speed V_R and according to an angle of attack α. One call pans the axisymmetric hooding of the hub, interdependent of the propeller in rotation. In the plan of the propeller, the pan is defined by the radius R_o.

Each elementary section of the blade of width dr creates a lift dL and a drag dD [9], such as

$$dL = \frac{1}{2}\rho C_L V_R^2 dS, \qquad dD = \frac{1}{2}\rho C_D V_R^2 dS, \qquad (11.1)$$

where ρ is the air density, C_L and C_D represent adimensional coefficients of lift and drag depending mainly on the angle of attack α ($\alpha = \gamma - \Phi$, see Fig. 11.3). According

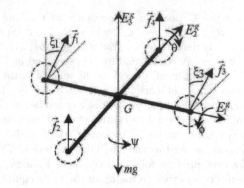

Fig. 11.2. Frames attached to the X4 bidirectional rotors rotorcraft

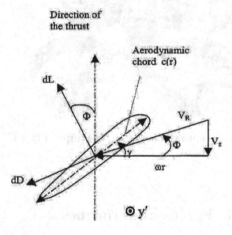

Fig. 11.3. Description of forces applied to the blade

to the fact that the XSF will hover or move at low speed, we made the assumption that the thrust f for a B blades propeller can be written as:

$$f = \frac{B}{2}\rho \int\limits_{R_0}^{R} V_R^2 c(r) C_L(r)\,dr = \frac{B}{2}\rho\omega^2 \int\limits_{R_0}^{R} r^2 c(r) C_L(r)\,dr. \tag{11.2}$$

We can eliminate this sentence which seems to be unclear or not necessary. Then the paragraph becomes:

$$f = K_T \omega^2. \tag{11.3}$$

The computation of the lift coefficient is often complex. It is thus essential to elaborate an experimental process, which permits to determine precisely the coefficient K_T as well as the limits of validity of the relation.

The drag torque is defined similarly as

$$M_D = \frac{B}{2}\rho \int_{R_0}^{R} V_R^2 c(r) C_D(r) r\, dr = \frac{B}{2}\rho\omega^2 \int_{R_0}^{R} c(r) C_D(r) r^3\, dr \qquad (11.4)$$

or simply

$$M_D = K_M \omega^2. \qquad (11.5)$$

The compensation of this torque in the center of gravity is established thanks to the use of contrarotating rotors 1-3 and 2-4. Recall that rotors 2 and 4 turn counterclockwise while rotors 1 and 3 turn clockwise.

11.4 Dynamics of Motion

We consider the translation motion of $\Re_G$ with respect to (wrt) $\Re_o$. The position of the mass center wrt $\Re_o$ is defined by $\overline{OG} = (x\ y\ z)^T$, its time derivative gives the velocity wrt to $\Re_o$ such that $\frac{dOG}{dt} = (\dot{x}\ \dot{y}\ \dot{z})^T$, while the second time derivative permits to get the acceleration: $\frac{d^2 OG}{dt^2} = (\ddot{x}\ \ddot{y}\ \ddot{z})^T$:

$$\begin{aligned}
m\ddot{x} &= S_\psi C_\theta u_2 - S_\theta u_3, \\
m\ddot{y} &= (S_\theta S_\psi S\phi + C_\psi C\phi)u_2 + C_\theta S_\phi u_3, \\
m\ddot{z} &= (S_\theta S_\psi C\phi - C_\psi S\phi)u_2 + C_\theta C_\phi u_3 - mg,
\end{aligned} \qquad (11.6)$$

where m is the total mass of the vehicle. The vector u_i, $i = 2,3$ combines the principal non conservative forces applied to the X4 bidirectional flyer airframe including forces generated by the motors and drag terms. Recall that the aerodynamic drag force is given by the classical relation : $F_D = \frac{1}{2}\rho_{air} V^2 S C_D$, where S is a reference area, C_D is the drag coefficient of the quadrotor in forward flight. We assume here that the apparent speed V of the XSF is small (< 4 m/s), thus we neglect force F_D in the dynamic model. Gyroscopic torques due to motors effects can be also neglected.

The lift (collective) force u_3 and the directional force u_2 are such that

$$\begin{pmatrix} 0 \\ u_2 \\ u_3 \end{pmatrix} = f_1 e_1' + f_3 e_3' + f_2 e_2 + f_4 e_4 \qquad (11.7)$$

with $f_i = K_T \omega_i^2$, $K_T > 0$ and constant (more details about K_T are given in Section 11.3), ω_i is the angular speed of motor i. In the local frame $\Re_G$ it is straightforward to verify that

$$e_1' = \begin{pmatrix} 0 \\ S_{\xi 1} \\ C_{\xi 1} \end{pmatrix}_{\Re_G}, \quad e_3' = \begin{pmatrix} 0 \\ S_{\xi 3} \\ C_{\xi 3} \end{pmatrix}_{\Re_G}, \quad e_2 = e_4 = \begin{pmatrix} 0 \\ 0 \\ 1 \end{pmatrix}_{\Re_G}.$$

Then we can write

$$u_2 = f_1 S_{\xi_1} + f_3 S_{\xi_3},$$
$$u_3 = f_1 C_{\xi_1} + f_3 C_{\xi_3} + f_2 + f_4, \tag{11.8}$$

where ξ_1 and ξ_3 are the two internal degrees of freedom of rotors 1 and 3, respectively. These bounded variables are controlled by DC servo-motors. e_2 and e_4 are the unit vectors along E_3^g which imply that rotors 2 and 4 are identical to that of the conventional X4-flyer. According to classical mechanics, the rotational dynamics is as follows:

$$\ddot{\theta} = \frac{1}{I_{xx}C_\phi}(\tau_\theta + I_{xx}S_\phi\dot{\phi}\dot{\theta}),$$
$$\ddot{\phi} = \frac{1}{I_{yy}C_\phi C_\theta}(\tau_\phi + I_{yy}S_\phi C_\theta\dot{\phi}^2 + I_{yy}S_\theta C_\phi\dot{\theta}\dot{\phi}), \tag{11.9}$$
$$\ddot{\psi} = \frac{1}{I_{zz}}\tau_\psi,$$

where the inertia matrix $I_G = \text{diag}(I_{xx}, I_{yy}, I_{zz})$, and the three inputs in torque

$$\tau_\phi = l(f_2 - f_4),$$
$$\tau_\theta = l(f_3 C_{\xi_3} - f_1 C_{\xi_1}), \tag{11.10}$$
$$\tau_\psi = l(f_3 S_{\xi_3} - f_1 S_{\xi_1}) + \frac{K_M}{K_T}(f_3 C_{\xi_3} - f_1 C_{\xi_1} + f_4 - f_2),$$

where l is the distance from G to the rotor i. Note that $\tau_\psi = \frac{K_M}{K_T}(f_3 - f_1 + f_4 - f_2)$, which is the case of the conventional machine ($\xi_1 = \xi_3 = 0$). Then with the bidirectional rotors, more manoeuver is obtained in yaw motion.

The equality from (11.9) is ensured, meaning that

$$\ddot{\eta} = \Pi_G(\eta)^{-1}(\tau - \dot{\Pi}_G(\eta)\dot{\eta}) \tag{11.11}$$

with $\tau = (\tau_\theta \; \tau_\phi \; \tau_\psi)^T$ as an auxiliary input, and

$$\Pi_G(\eta) = \begin{pmatrix} I_{xx}C_\phi & 0 & 0 \\ 0 & I_{yy}C_\phi C_\theta & 0 \\ 0 & 0 & I_{zz} \end{pmatrix}. \tag{11.12}$$

As a first step, the model given above can be input/output linearized by the following decoupling feedback laws

$$\tau_\theta = -I_{xx}S_\phi\dot{\phi}\dot{\theta} + I_{xx}C_\phi\tilde{\tau}_\theta,$$
$$\tau_\phi = -I_{yy}S_\phi C_\theta\dot{\phi}^2 - I_{yy}S_\theta C_\phi\dot{\theta}\dot{\phi} + I_{yy}C_\phi C_\theta\tilde{\tau}_\phi, \tag{11.13}$$
$$\tau_\psi = I_{zz}\tilde{\tau}_\psi$$

and the decoupled dynamic model of rotation can be written as

$$\ddot{\eta} = \tilde{\tau} \tag{11.14}$$

with $\tilde{\tau} = (\tilde{\tau}_\theta \; \tilde{\tau}_\phi \; \tilde{\tau}_\psi)^T$. Using the system of Equations (11.6) and (11.14), the dynamics of the system is defined by

$$m\ddot{x} = S_\psi C_\theta u_2 - S_\theta u_3,$$
$$m\ddot{y} = (S_\theta S_\psi S_\phi + C_\psi C_\phi)u_2 + C_\theta S_\phi u_3,$$
$$m\ddot{z} = (S_\theta S_\psi C_\phi - C_\psi S_\phi)u_2 + C_\theta C_\phi u_3 - mg, \tag{11.15}$$
$$\ddot{\theta} = \tilde{\tau}_\theta, \qquad \ddot{\phi} = \tilde{\tau}_\phi, \qquad \ddot{\psi} = \tilde{\tau}_\psi.$$

Remark 1. As shown in (11.15), the equivalent system of the control-inputs presents five inputs $U = (u_2, u_3, \tilde{\tau}_\theta, \tilde{\tau}_\phi, \tilde{\tau}_\psi)$, while the rotor force-inputs are of sixth order (see relations 11.7,11.10) $F = (f_1, f_2, f_3, f_4, \xi_1, \xi_3)$. Then the transformation $U \mapsto F$ is not a diffeomorphism.

11.4.1 Dynamic Motion of the Conventional X4 Flyer

One deduces this dynamic after substituting $\xi_1 = \xi_3 = 0$ in Eqs. (11.15) and (11.8):

$$m\ddot{x} = -S_\theta u_3,$$
$$m\ddot{y} = C_\theta S_\phi u_3,$$
$$m\ddot{z} = C_\theta C_\phi u_3 - mg, \tag{11.16}$$
$$\ddot{\theta} = \tilde{\tau}_\theta, \qquad \ddot{\phi} = \tilde{\tau}_\phi, \qquad \ddot{\psi} = \tilde{\tau}_\psi.$$

Remark 2. In the conventional X4-flyer (see Fig. 11.4, right) case, where we consider $\xi_1 = \xi_3 = 0$, the yaw motion is generated by $\tau_\psi = \frac{K_M}{K_T}(f_1 - f_2 + f_3 - f_4)$. The collective input in (11.16) is given by $u_3 = f_1 + f_3 + f_2 + f_4$.

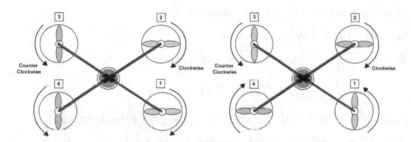

Fig. 11.4. Rotor rotations: bidirectional (left) and conventional (right)

11.5 Advanced Strategies of Control

Our aim in the XSF project is to test the capability of the drone while progressing in an area with approximatively ten-meter buildings. In this part, we will present briefly the point-to-point based tracking of the conventional X4-flyer. The reader can refer to Beji [10] for more details. More attention will be given to the bidirectional engine which was also studied by Beji [11] but not compared to the conventional one.

11.5.1 Conventional Aerial Vehicle

The tracking control problem was solved using the point-to-point steering. For the tracking problem, each point will be stabilized and tracked as a starting one. Flatness was proved with respect to the flat output (x, y, z) [10]. Recall that a system is flat if we can find a set of outputs (equal in number to the number of inputs) such that all states and inputs can be determined from these outputs without integration [2].

We proved that the state and the control vector are function of the flat output and their time derivatives. For a given smooth trajectory $\xi(t)=(x(t),y(t),z(t))$. We get $(u_3 > 0)$

$$u_3 = m\left(\ddot{x}^2 + \ddot{y}^2 + (\ddot{z}+g)^2\right)^{\frac{1}{2}},$$

$$\phi = \arctan\left(\frac{\ddot{y}}{\ddot{z}+g}\right) \quad \text{and} \quad \theta = -\arctan\left(\frac{c_\phi \ddot{x}}{\ddot{z}+g}\right). \tag{11.17}$$

Indeed, u_3, θ, ϕ, $\dot{u}_3$, $\dot{\theta}$, and $\dot{\phi}$ are functions of $\ddot{\xi}, \xi^{(3)}$. Thus it is straightforward to verify that states and inputs are functions of $(\xi, \dot{\xi}, \xi^{(2)}, \xi^{(3)})$. Moreover, we can derive $\theta(t)$, $\phi(t)$ and prove the ξ-dependency of $\tilde{\tau}_\theta$ and $\tilde{\tau}_\phi$ as function of $(\xi^{(2)}, \xi^{(3)}, \xi^{(4)})$.

From Eq. (11.16), the yaw motion can be naturally stabilized to a desired value (second order with constant coefficients) and the z altitude can be stabilized by u_3 (nonlinear decoupling feedback). The results are regrouped in the following theorem.

Theorem 1. *Consider the dynamics of the conventional aerial vehicle given by (11.16). The following control inputs stabilize any desired configuration $(x_d, y_d, z_d, \psi_d, \theta_d, \phi_d)$, where $\psi_d = n\pi$, $\theta_d \in [\theta_d^m, \theta_d^M]$ and $\phi_d \in [\phi_d^m, \phi_d^M]$:*

$$u_3 = mg + m\ddot{z}_d - mk_1^z(\dot{z}-\dot{z}_d) - mk_2^z(z-z_d),$$

$$\tilde{\tau}_\psi = \ddot{\psi}_d - k_1^\psi(\dot{\psi}-\dot{\psi}_d) - k_2^\psi(\psi-\psi_d),$$

$$\tilde{\tau}_\theta = -\frac{1}{g+f(z,z_d)}(\nu_x + \ddot{f}(z,z_d)\theta + 2\dot{f}(z,z_d)\dot{\theta}),$$

$$\tilde{\tau}_\phi = -\frac{1}{g+f(z,z_d)}(\nu_y + \ddot{f}(z,z_d)\phi + 2\dot{f}(z,z_d)\dot{\phi}), \tag{11.18}$$

$$\nu_x = x_d^{(4)} - k_1^x(x^{(3)} - x_d^{(3)}) - k_2^x(\ddot{x}-\ddot{x}_d) - k_3^x(\dot{x}-\dot{x}_d) - k_4^x(x-x_d),$$

$$\nu_y = y_d^{(4)} - k_1^y(y^{(3)} - y_d^{(3)}) - k_2^y(\ddot{y}-\ddot{y}_d) - k_3^y(\dot{y}-\dot{y}_d) - k_4^y(y-y_d).$$

Here, the function $f(z,z_d) = \ddot{z}_d - k_1^z(\dot{z}-\dot{z}_d) - k_2^z(z-z_d)$ is assumed to be regular wrt to their arguments. The k_i^z, k_i^ψ, k_i^x and k_i^y are positive and stable coefficients. 'd' denotes the desired trajectory.

Proof. Details of the proof are given in [10].

11.5.2 Bidirectional X4-flyer

Recall that the two rotors 1-3 can be directed together or differently leading to a longitudinal/lateral force. This force contributes to the displacement of the vehicle along the x/y axis, and is denoted by u_2. To bring the engine from one axis to another or both of them, one generates yaw motions first. As we can see in the following, the singularity occurs when $\psi = \pm\frac{\pi}{2}$. In order to avoid the singularity, u_2 should change structure. The static feedback linearization and feed-forward techniques are combined to ensure tracking.

Our first interest is the dynamics of (y,z), which can be decoupled by a static feedback law if the decoupling matrix is not singular. Then we have the following proposition.

Proposition 1. *Consider* $(\psi, \theta) \in]-\pi/2, \pi/2[$, *with the static feedback laws*

$$u_2 = m\nu_y C_\phi C_\psi^{-1} - m(\nu_z + g) S_\phi C_\psi^{-1},$$
$$u_3 = m\nu_y (S_\phi C_\theta^{-1} - C_\phi tg_\psi tg_\theta) + m(\nu_z + g)(C_\phi C_\theta^{-1} + S_\phi tg_\psi tg_\theta) \tag{11.19}$$

and

$$\nu_y = \ddot{y}_r - k_y^1(\dot{y} - \dot{y}_r) - k_y^2(y - y_r),$$
$$\nu_z = \ddot{z}_r - k_z^1(\dot{z} - \dot{z}_r) - k_z^2(z - z_r). \tag{11.20}$$

The dynamics of y and z are linearly decoupled and exponentially stable with the appropriate choice of the gain controller k_y^i and k_z^i.

Proof. From (11.15), one regroups the two dynamics as

$$\begin{pmatrix} \ddot{y} \\ \ddot{z} \end{pmatrix} = \frac{1}{m} \begin{pmatrix} S_\theta S_\psi S_\phi + C_\psi C_\phi & C_\theta S_\phi \\ S_\theta S_\psi C_\phi - C_\psi S_\phi & C_\theta C_\phi \end{pmatrix} \begin{pmatrix} u_2 \\ u_3 \end{pmatrix} - \begin{pmatrix} 0 \\ g \end{pmatrix}. \tag{11.21}$$

When incorporating inputs (11.19) in system (11.21), we get

$$\ddot{y} = \nu_y, \qquad \ddot{z} = \nu_z. \tag{11.22}$$

Taking θ close to zero, a feed-forward form of the x dynamic is given by

$$\ddot{x} = \frac{1}{m}(S_\psi u_2 - \theta u_3), \qquad \ddot{\theta} = \tilde{\tau}_\theta \tag{11.23}$$

while u_2 and u_3 were already defined in Proposition 1. Note that the dynamics of x seems to be a *zero dynamics*, but it is connected to θ, which is in its turn connected to $\tilde{\tau}_\theta$.

Proposition 2. *Suppose there exists a time T_f^1 such that $\forall t \in [T_0, T_f^1]$ $u_3(t) > 0$. Then the dynamics of x is decoupled under the following feedback controller*

$$\tilde{\tau}_\theta = \frac{1}{u_3}(-m\nu_x + \ddot{\psi} C_\psi u_2 - \dot{\psi}^2 S_\psi u_2 + 2\dot{\psi} C_\psi \dot{u}_2 + S_\psi \ddot{u}_2 - 2\dot{\theta}\dot{u}_3 - \theta\ddot{u}_3) \tag{11.24}$$

leading to $x^{(4)} = \nu_x$ with

$$\nu_x = x_d^{(4)} - k_x^1(x^{(3)} - x_d^{(3)}) - k_x^2(\ddot{x} - \ddot{x}_d) - k_x^3(\dot{x} - \dot{x}_d) - k_x^4(x - x_d) \tag{11.25}$$

which stabilizes the system exponentially $(x = x_d, \theta = 0)$. k_x^i are stable coefficients.

This last result is straightforward and can be shown after having derived twice the dynamics of x. Note that $u_3 > 0$ and $u_3 = mg$ $\forall t > T_f^1$, then the drone reaches the desired equilibrium $(x \to x_d, \theta \to \frac{-\ddot{x}}{g} \to 0)$. The drawback of the controller (11.24) is that it requires the first/second time derivative of u_3. Then a smooth reference trajectory is required. Moreover, and in order to reduce the calculation complexity of the controller, we can replace in (11.24) $\psi, \dot{\psi}, \ddot{\psi}$ by the references $\psi_d, \dot{\psi}_d, \ddot{\psi}_d$. In fact, ψ can be stabilized separately and reaches $\psi_d \in]-\pi/2, \pi/2[$.

Remark 3. The interval limit of $\psi_d \in]-\pi/2, \pi/2[$ is a handicap for the machine if one wants a displacement according to x with the force input u_2. In this case, the displacement along x requires an orientation of $\psi_d = \pi/2$, which is a singularity for u_2 (see (11.19)). We give a solution to this problem in the following.

Our second interest is the (x, z) dynamics, which can be also decoupled by a static feedback law as shown in the following proposition. Therefore, we keep the input u_3 for the altitude z stabilization and u_2 for the dynamics of x.

Proposition 3. *With the new inputs*

$$u_2 = (S_\psi C_\phi - S_\theta C_\psi S_\phi)^{-1}(m\nu_x C_\phi C_\theta + m(\nu_z + g)S_\theta),$$

$$u_3 = (S_\psi C_\phi - S_\theta C_\psi S_\phi)^{-1}(-m\nu_x(S_\psi S_\theta C_\phi - C_\psi S_\phi) + m(\nu_z + g)S_\psi C_\theta) \tag{11.26}$$

leading to

$$\ddot{x} = \nu_x, \qquad \ddot{z} = \nu_z, \tag{11.27}$$

the dynamics of x and z can be decoupled and stabilized through ν_x and ν_z as above.

Remark 4. The control inputs u_2 and u_3 in Proposition 3 ensure the tracking control in the $x - z$ plane. Further, one needs to stabilize the y motion. The problem can be formulated as in (11.23), where ϕ can be chosen as an input variable. The dynamics of y coupled to ϕ is in cascade form. Consequently $\tilde{\tau}_\phi$ can be deduced with the same procedure given in (11.24).

Remark 5. To switch between the two controllers (11.19) and (11.26) continuity is recommended which permits to avoid the peak phenomenon. This can be asserted if we impose constraints to the reference trajectories. In order to ensure this, we take $u_2(\psi = 0) = u_2(\psi = \pi/2) = 0$ with $\phi = \theta = 0$. For $\psi = 0$ one uses (11.19) and for $\psi = \pi/2$, one refers to expression (11.26). Then from (11.19), (11.26) we deduce that $\nu_y(T_f^2) = \nu_x(T_f^2)$ and $\ddot{y}_d(T_f^2) = \ddot{x}_d(T_f^2)$. Taking $\phi = \theta = 0$, we impose that $u_3(\psi = 0) = mg$ in Eq. (11.19) and $u_3(\psi = \pi/2) = mg$ in Eq. (11.26). T_f^2 is the necessary final time to reach $x_d(T_f^2)$.

11.6 Motion Planning and Simulation Results

In order to validate the motion planning algorithm with the proposed controller, the conventional and bidirectional X4-flyer are tested in simulation. The total mass of the drone is $m = 2$ kg. The technical characteristics of this flying vehicle were presented in [10]. We have considered the following reference trajectory

$$z_d(t) = h_d \frac{t^5}{t^5 + (T_f^1 - t)^5}, \tag{11.28}$$

where h_d is the desired altitude and T_f^1 is the necessary final time. The same trajectory is adopted for $x_d(t)$ and $y_d(t)$. The constraints to perform these trajectories and to ensure continuity in the proposed control schemes are such that $z_d(0) = x_d(T_f^2) = y_d(T_f^2) = 0$, $z_d(T_f^1) = h_d$, $\dot{z}_d(0) = \dot{x}_d(T_f^1) = \dot{y}_d(T_f^2) = 0$, $\dot{z}_d(T_f^1) = \dot{x}_d(T_f^2) = \dot{y}_d(T_f^3) = 0$, $\ddot{z}_d(T_f^1) = \ddot{x}_d(T_f^2) = \ddot{y}_d(T_f^3) = 0$, $\ddot{z}_d(0) = \ddot{x}_d(T_f^1) = \ddot{y}_d(T_f^2) = 0$. T_f^2 and T_f^3 are the final times to execute motion along x and y, respectively. Minimizing the time of displacement here implies that the drone accelerates at the beginning and decelerates at the arrival.

We connect the reference $z - x$ as in Fig. 11.7, considering

$$x_d(t) = -\rho\cos(\alpha(t)) + \rho \qquad \text{and} \qquad z_d(t) = \rho\sin(\alpha(t)) + h_d \tag{11.29}$$

and for the reference transition $x - y$, we take

$$x_d(t) = \rho \sin(\alpha(t)) + x_d(T_f^2) \qquad \text{and} \qquad y_d(t) = -\rho \cos(\alpha(t)) + \rho. \qquad (11.30)$$

To ensure continuity between $x_d(t)$ and $y_d(t)$ derivatives, we take

$$\alpha(t) = \frac{\alpha_d(t - t_f)^5}{(t - t_f)^5 + (t_a - t)^5}, \qquad (11.31)$$

where $\rho = 2$ m, $\alpha_d = k_\alpha \pi/2$ $(k_\alpha = 1, 2)$, $t_a = 6$ s, $t_f = 8$ s, $x_d(T_f^2) = y_d(T_f^3) = h_d + \rho = 10$ m.

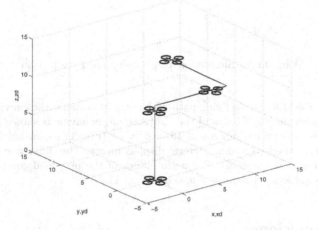

Fig. 11.5. xyz-flying path: conventional X4-flyer

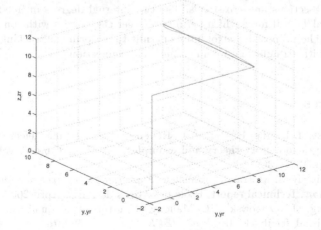

Fig. 11.6. xyz-flying path: bidirectional X4 flyer

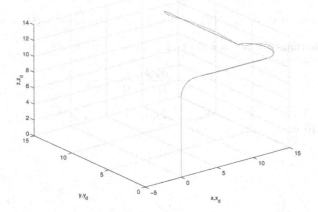

Fig. 11.7. Bidirectional xyz-flying path realization

Figures 11.5-11.6 show a flying path with straight corners: the conventional case result is sketched in Fig. 11.5, and the bidirectional behavior is shown in Fig. 11.6. The engine is maintained horizontal ($\theta = \phi = 0$) (Fig. 11.5), meaning that u_2 is used to perform the longitudinal/lateral displacements. The difference between the two realizations is not notable. The performance of the proposed control laws with the bidirectional flying machine is shown in Fig. 11.7.

11.7 Conclusions

We have considered in this work the modeling and stabilizing/tracking control problem for the three decoupled displacements of a conventional and bidirectional X4-flyer machine. The objectives were to test the capability of engines to fly with rounded intersections and crossroads. The two internal degrees of freedom lead to a longitudinal/lateral force which permits to steer the system with rounded corner flying path. The proposed control inputs permit to perform the tracking objectives: flying path with straight and round corners like connection.

References

1. M. Fliess, J. Levine, P. Martin, P. Rouchon. Flatness and defect of nonlinear systems: introductory theory and examples. *Int. J. of Control*, vol. 61, no. 6, 1995, pp. 1327-1361.
2. P. Martin, R.M. Murray, P. Rouchon. *Flat systems, equivalence and trajectory generation*. Technical report, Ecole des Mines de Paris, April 2003.
3. E. Altug, J. Ostrowski, R. Mahony. Control of a quadrotor helicopter using visual feedback. In: *Proc. IEEE Conf. on Robotics and Automation*, Washington, DC 2002, pp. 72-77.

4. P. Pound, R. Mahony, P. Hynes, J. Roberts. Design of a four rotor aerial robot. *Proc. of the Australasian Conf. on Robotics and Automation*, Auckland 2002.

5. T. Hamel, R. Mahony, R. Lozano, J. Ostrowski. Dynamic modelling and configuration stabilization for an X4-flyer. In: *Proc. IFAC 15th World Congress on Automatic Control*, Barcelona 2002.

6. P. Castillo, A. Dzul, R. Lozano. Real-time stabilization and tracking of a four rotor mini-rotorcraft. *IEEE Trans. on Control Systems Technology*, vol. 12, no. 4, 2004, pp. 510-516.

7. J. Hauser, S. Sastry, G. Meyer. Nonlinear control design for slightly non minimum phase systems: application to V/STOL aircraft. *Automatica*, vol. 28, no. 4, 1992, pp. 665-679.

8. L. Marconi, A. Isodori. Robust output regulation for autonomous vertical landing. In: *Proc. 39th IEEE Conf. on Decision and Control*, Sydney 2000.

9. B.W. McCormick, Jr. *Aerodynamics of V/STOL flight*. Academic Press, New York 1967.

10. L. Beji, A. Abichou. Streamlined rotors mini rotorcraft: trajectory generation and tracking. *Int. J. of Control, Automation, and Systems*, vol. 3, no. 1, 2005, pp. 87-99.

11. L. Beji, A. Abichou, M.K. Zemalache. Smooth control of an X4 bidirectional rotors flying robot. In: *Proc. IEEE Int. Workshop on Robot Motion and Control (RoMoCo)*, Dymaczewo, Poland 2005.

Part III

Climbing and Walking Robots

12

Absolute Orientation Estimation
for Observer-based Control of a Five-link
Walking Biped Robot

Vincent Lebastard, Yannick Aoustin, and Franck Plestan

IRCCyN, UMR CNRS 6597, Ecole Centrale de Nantes, Université de Nantes
1 rue de la Noë, BP 92101, 44321 Nantes Cedex 3, France
{Vincent.Lebastard|Yannick.Aoustin|Franck.Plestan}@irccyn.ec-nantes.fr

12.1 Introduction

The design of control laws and reference trajectories for biped robots is difficult, is still a challenging problem and will be not properly solved as long as the dynamics of the robots under interest is not thoroughly understood (see for example some recent papers such as [1–10]). Furthermore, the improvement of desired performance and the increasing sophistication of such mechanical systems induce that the complexity of the control, based on the nonlinear model of the biped robots and using all state variables, increases. Accurate sensors are necessary to measure the joint variables, the absolute orientation of the robot and the ground reactions. The determination of the absolute orientation of a biped robot with feet is easy if the sole of the foot has a flat contact with the ground. However, fluid walking gaits induce partial contact of the soles of the feet with the ground: in this case, the biped is underactuated and an accurate measurement of its absolute orientation (with an unstable posture) is, from a technical point-of-view, difficult. Note that it is easier to obtain an accurate measurement of a joint variable equipped with an encoder sensor, especially if this joint is actuated with a motor equipped with a gearbox reducer. Then, there is a real interest to develop observers in order to estimate absolute angular positions and velocities from only the knowledge of the relative angular variables. In order to define the strategy of observation, the biped under interest has point feet and is underactuated in single support: then, its walking gait is dynamically stable. To our best knowledge, little work has been done for the design of such observers. Previous work on observer design concerned especially the estimation of velocities (for noiseless differentiation) by supposing that all the angular variables are measured [11]. A first attempt at the design of observer/controller using only the measurement of joint link angular variables (relative angles) for a three-link biped with no actuator in the ankles, in both cases of its stabilization in a vertical position and its walking, has been made by the authors and is based on high-gain observers [12]. This class of observers [13] is based on the concept of *uniform observability* [14] of nonlinear systems and their principle consists in reducing the effects of the nonlinear terms in the estimation error equation with a high-gain correction. This approach gives an asymptotic convergence observer. Due

K. Kozłowski (Ed.): Robot Motion and Control, LNCIS 335, pp. 181–199, 2006.
© Springer-Verlag London Limited 2006

to the latter property, no controller-observer superposition stability proof has been proposed: this proof is very difficult to establish for nonlinear systems for which the used observers are not finite-time convergent. An estimation of the absolute orientation of a two-links biped without foot, with a Kalman filter is proposed in [15]. The latter observer [16] is designed as an extended stochastic filter by linearizing the nonlinear system around the current state estimate. However, as stability proof of observer-based control is clearly easier to establish with finite-time convergence observers, such a class of observers is interesting.

The finite-time convergence property is one of the main characteristics of sliding mode observers (with robustness of estimation/observation versus uncertainties) [17] whose dynamics depend on discontinuous output terms. However, this class of observers presents the main drawback of sliding mode, the *chattering effect*, as the estimation error dynamics directly depend on a discontinuous function. This phenomenon could generate high-frequency oscillations on observation, which could be negative for the control and the system. In the control context, in order to limit this problem a solution has been given in the 90's using the higher order sliding mode approach [18], which removes the sliding mode restrictions (chattering) while preserving its features (robustness, finite-time convergence) and improving the accuracy: in fact, the system dynamics does not directly depend on a discontinuous function. As standard sliding modes, higher order sliding mode approach has been used to design observers. A second order sliding mode observer, based on the twisting algorithm [18], has been designed for an electrical motor [19]: its robustness and its finite-time convergence have been proved. A step-by-step higher order sliding mode observer [20] ensures, step-by-step, the finite-time convergence of the estimation error to zero. These solutions have been successfully evaluated on 3-links walking biped robot [21, 22] and allow to propose a formal proof of walking stability.

The biped studied in this paper is a five link biped without feet. Then this biped is underactuated in single support. The purpose of this paper is to prove, using simulation work, the feasibility of estimating in a closed loop the absolute orientation of the biped and its velocity for a walking gait, supposing that only the joint variables are available. The walking gait is composed of single support phases and impulsive impacts. Reference trajectories of the actuated joint variable are polynomials based on the absolute orientation of the virtual stance leg (see [8]) designed with a parameterized optimization algorithm. The control law is based on the exact linearization of the joint variables. Then the great difficulty of the problem which is considered in this paper is that the estimated orientation is absolutely necessary to feed, at each sampling time, the control law and the reference trajectories. The physical parameters of *Rabbit* [23] are considered. The current chapter proposes an extension of both previous references through two observer-based controllers. The chapter is organized as follows: the model of the robot is presented in Section 12.2. The control law is presented in Section 12.3. Section 12.4 is devoted to the observation problem. The definition of observability is given in Subsection 12.4.1. High-gain and step-by-step observers are detailed in Sections 12.4.2-12.4.3. Simulation results are shown in Section 12.5. Our conclusion and perspectives are presented in Section 12.6.

12.2 Model of a Planar Five-link Biped Robot

12.2.1 General and Reduced Dynamic Models

A planar *five-link biped* is considered and composed of a torso, hips, two identical legs with knees, but without ankles and feet (see Figure 12.1). The general dynamic model can be determined from Lagrange's equations and is given by

$$\mathbf{D_e\ddot{q}_e + H_e\dot{q}_e + G_e = B_e\Gamma + D_R R} \tag{12.1}$$

with[1] $\mathbf{q_e} = [\mathbf{q}^T\ x_t\ z_t]^T$. Vector $\mathbf{q}$ is composed of the joint variables and the absolute orientation of the torso, $\mathbf{q} = [q_{31}\ q_{41}\ q_{32}\ q_{42}\ q_1]^T$; (x_t, z_t) are the Cartesian coordinates of the center of mass of the torso (see Figure 12.1). The angles are defined positive for counter-clockwise motion. $\mathbf{D_e(q)}(7 \times 7)$ is the symmetric positive inertia matrix. Matrix $\mathbf{H_e(q, \dot{q})}(7 \times 7)$ is the Coriolis and centrifugal effects matrix and $\mathbf{G_e(q_e)}(7 \times 1)$ is the gravity effects vector. $\mathbf{B_e}$ is a constant 7×4-matrix composed of 1 and 0 and $\mathbf{D_R(q_e)}$ is the 7×4-Jacobian matrix converting the external forces into the corresponding joint torques. $\Gamma = [\Gamma_1\ \Gamma_2\ \Gamma_3\ \Gamma_4]^T$ is the actuators torques vector. $\mathbf{R} = [R_{N1}\ R_{T1}\ R_{N2}\ R_{T2}]^T$ represents the ground reaction acting on the swing / stance leg tips. Assume that

$\mathbf{H_1}$ During the swing phase of the motion, the stance leg is acting as a pivot; the contact of the swing leg with the ground results in no rebound and no slipping of the swing leg.

Then, equation (12.1) can be reduced to

$$\mathbf{D\ddot{q} + H\dot{q} + G = B\Gamma}. \tag{12.2}$$

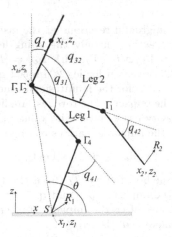

Fig. 12.1. Biped robot in the sagittal plane

---

[1] Notation ^{T} means transposition.

As the kinetic energy of the biped is invariant under a rotation of the world frame [24], and viewed that q_5 defines the orientation of the biped, the 5×5-symmetric positive inertia matrix is independent of this variable, *i.e.* $\mathbf{D} = \mathbf{D}(q_{31}\ q_{41}\ q_{32}\ q_{42})$; $\mathbf{H}(\mathbf{q}, \dot{\mathbf{q}})(5 \times 5)$ is the Coriolis and centrifugal effects matrix, and $\mathbf{G}(\mathbf{q})(5 \times 1)$ is the gravity effects vector. $\mathbf{B}(5 \times 4)$ is a constant matrix composed of 1 and 0. Equation (12.2) can be written as

$$\dot{\mathbf{x}} = \begin{bmatrix} \dot{\mathbf{q}} \\ \mathbf{D}^{-1}(-\mathbf{H}\dot{\mathbf{q}} - \mathbf{G} + \mathbf{B}\mathbf{\Gamma}) \end{bmatrix} \tag{12.3}$$
$$= \mathbf{f}(\mathbf{x}) + \mathbf{g}(\mathbf{q}_{rel}) \cdot \mathbf{\Gamma}$$

with $\mathbf{x} = [\mathbf{q}^T\ \dot{\mathbf{q}}^T]^T$ and $\mathbf{q}_{rel} = [q_{31},\ q_{32},\ q_{41},\ q_{42}]^T$, the joint angles vector. The state space is taken such that $\mathbf{x} \in \mathcal{X} \subset I\!\!R^{10} = \{\mathbf{x} = [\mathbf{q}^T\ \dot{\mathbf{q}}^T]^T \mid \dot{\mathbf{q}} \in \mathcal{N},\ \mathbf{q} \in \mathcal{M}\}$, where $\mathcal{N} = \{\dot{\mathbf{q}} \in I\!\!R^5 \mid |\dot{\mathbf{q}}| < \dot{\mathbf{q}}_M < \infty\}$ and $\mathcal{M} = (-\pi, \pi)^5$. From these definitions, note that all the state variables are bounded.

12.2.2 Passive Impact Model

The impact occurs at the end of a single-support phase, when the swing leg tip touches the ground. Let T_I denote impact time. State the subscripts 2 for the swing leg and 1 for the stance leg during the single-support phase. An impact occurs when the swing leg tip touches the ground, *i.e.* $\mathbf{x} \in \mathcal{S} = \{\mathbf{x} \in \mathcal{X} \mid z_2(\mathbf{q}) = 0\}$ with $z_2(\mathbf{q})$ the altitude of the swing leg tip. Assume that

$\mathbf{H_2}$ The impact is passive and absolutely inelastic.

$\mathbf{H_3}$ The swing leg touching the ground does not slip and the previous stance leg takes off the ground.

$\mathbf{H_4}$ At the impact, the angular positions are continuous and the angular velocities are discontinuous.

Given these hypotheses, the ground reactions at the instant of the impact can be considered impulsive forces acting on only the swing leg (leg 2) and defined by Dirac delta-functions $\mathbf{R_2} = \mathbf{I_{R_2}}\Delta(t - T)$, with $\mathbf{I_{R_2}} = [I_{R_{2N}}\ I_{R_{2T}}]^T$ the vector of magnitudes of impulsive [25] reaction for leg 2. Impact equations can be obtained through the integration of (12.1) for the infinitesimal time from T_I^- (just before the impact) to T_I^+ (just after the impact). The torques supplied by the actuators at the joints and Coriolis and gravity forces have finite values: thus, they do not influence the impact. Consequently, the impact equations can be written as

$$\mathbf{D_e}(\dot{\mathbf{q}}_e{}^+ - \dot{\mathbf{q}}_e{}^-) = \mathbf{D_R}\mathbf{I_{R_2}}, \tag{12.4}$$

$\mathbf{q_e}$ is the configuration of the biped at $t = T_I$ (from $\mathbf{H_4}$, this configuration does not change at the instant of the impact), $\dot{\mathbf{q}}_e{}^-$ and $\dot{\mathbf{q}}_e{}^+$ are, respectively, the angular velocities just before and just after the impact. Furthermore, the velocity of the stance leg tip before the impact equals zero

$$\begin{bmatrix} \dot{x}_1(\mathbf{q}_e{}^-, \dot{\mathbf{q}}_e{}^-) & \dot{z}_1(\mathbf{q}_e{}^-, \dot{\mathbf{q}}_e{}^-) \end{bmatrix}^T = \mathbf{0}_{2 \times 1} \tag{12.5}$$

with (x_1, z_1) the Cartesian coordinates of the stance leg tip. The swing leg after the impact becomes the supporting leg. Therefore, its tip velocity becomes zero after the impact

$$\left[\dot{x}_2(\mathbf{q_e}^+, \dot{\mathbf{q}}_e{}^+) \quad \dot{z}_2(\mathbf{q_e}^+, \dot{\mathbf{q}}_e{}^+) \right]^T = \mathbf{0} =_{2 \times 1} \qquad (12.6)$$

with $(x_2 \ z_2)$ the Cartesian coordinates of the swing leg tip. The final result is an expression for $\mathbf{x}^+ = [\mathbf{q}^{+T} \ \dot{\mathbf{q}}^{+T}]^T$ (state just before the impact) in terms of $\mathbf{x}^- = [\mathbf{q}^{-T} \ \dot{\mathbf{q}}^{-T}]^T$ (state just after the impact), which can be written as [26] $\mathbf{x}^+ = \Delta(\mathbf{x}^-)$.

12.2.3 Nonlinear Model All over the Step

The overall biped model can be expressed as a system with impulse effects as

$$\begin{aligned} \dot{\mathbf{x}} &= \mathbf{f}(\mathbf{x}) + \mathbf{g}(\mathbf{q}_{rel})\boldsymbol{\Gamma}, \quad \text{for } \mathbf{x}^- \notin \mathcal{S}, \\ \mathbf{x}^+ &= \Delta(\mathbf{x}^-), \qquad\qquad \text{for } \mathbf{x}^- \in \mathcal{S}, \end{aligned} \qquad (12.7)$$

where $\mathcal{S} = \{\mathbf{x} \in \mathcal{X} \mid z_2(\mathbf{q}) = 0 \}$.

12.3 Design of the Controller

12.3.1 Strategy

The control for the walking gait [26] consists in tracking the four joints reference angles q_{31d}, q_{41d}, q_{32d} and q_{42d} of the biped. During the single-support phase, the degree of under-actuation equals one, as only four outputs can be driven. Then, the robot gets a walking motion if the controller drives to zero the output vector $\mathbf{h}(\mathbf{x})$ defined as

$$\mathbf{h}(\mathbf{x}) = [q_{31} - q_{31d} \quad q_{32} - q_{32d} \quad q_{41} - q_{41d} \quad q_{42} - q_{42d}]^T .$$

As the relative degree of each output component equals 2, and using standard Lie derivative notation [27], one gets $\ddot{\mathbf{h}} = \mathbf{L}_f^2\mathbf{h}(\mathbf{x}) + \mathbf{L_g}\mathbf{L_f}\mathbf{h}(\mathbf{x})\boldsymbol{\Gamma}$. The control strategy consists in decoupling the system and in forcing the system to evolve by arbitrarily stated dynamics. Knowing that, for $\mathbf{x} \in \mathcal{X}$, the decoupling matrix $\mathbf{L_g}\mathbf{L_f}\mathbf{h}$ never equals zero, the control law $\boldsymbol{\Gamma}$ reads as

$$\boldsymbol{\Gamma} = [\mathbf{L_g}\mathbf{L_f}\mathbf{h}]^{-1}[-\mathbf{L}_f^2\mathbf{h} + \mathbf{v}]. \qquad (12.8)$$

One gets a linear input-output behavior of the output vector $\ddot{\mathbf{h}} = \mathbf{v}$. In the current work, the control law $\mathbf{v}$ is chosen to be *finite time convergent*. The feedback function $\mathbf{v}$ comes from [28] (with $\epsilon > 0$)

$$\mathbf{v} = \boldsymbol{\Upsilon}(\mathbf{h}, \dot{\mathbf{h}}) = \frac{1}{\epsilon} \cdot \begin{bmatrix} \Upsilon_1(h_1, \epsilon \cdot \dot{h}_1) \\ \Upsilon_2(h_2, \epsilon \cdot \dot{h}_2) \\ \Upsilon_3(h_3, \epsilon \cdot \dot{h}_3) \\ \Upsilon_4(h_4, \epsilon \cdot \dot{h}_4) \end{bmatrix} . \qquad (12.9)$$

Each function $\Upsilon_i(h_i, \epsilon \cdot \dot{h}_i)$ is defined as

$$\Upsilon_i = -\text{sign}(\vartheta_i(h_i, \epsilon \cdot \dot{h}_i)) \cdot |\vartheta_i(h_i, \epsilon \cdot \dot{h}_i)|^{\frac{\alpha}{2-\alpha}} - \text{sign}(\epsilon \cdot \dot{h}_i) \cdot |\epsilon \cdot \dot{h}_i|^\alpha \qquad (12.10)$$

with $\vartheta_i(\cdot) = h_i + \frac{1}{2-\alpha}\text{sign}(\epsilon \cdot \dot{h}_i) \cdot |\epsilon \cdot \dot{h}_i|^{2-\alpha}$ and $0 < \alpha < 1$. The real parameter $\epsilon > 0$ allows the settling time of the controllers to be adjusted.

12.3.2 Reference Motion for the Swing Phase

The objective is to describe a cyclic gait with single supports and impacts. Let the reference trajectories of actuated joints q_i ($i \in \{31, 32, 41, 42\}$) be defined as polynomial functions of 4th order in θ, which is the angle between the virtual stance leg (between the hips and the stance leg tip) and the ground (Figure 12.1),

$$q_i(\theta) = a_{i0} + a_{i1}\theta + a_{i2}\theta^2 + a_{i3}\theta^3 + a_{i4}\theta^4. \qquad (12.11)$$

The 4th order polynomials have been chosen in order to prescribe for one step, the initial position $\mathbf{q}(\theta_{ini})$ and the final positions $\mathbf{q}(\theta_{final})$, the initial and final velocities and an intermediate configuration $\mathbf{q}_{int} = \mathbf{q}_{\theta_{int}}$ for the biped. Then, if $q_i(\theta)$'s are exactly tracked, it is ensured to reach the final configuration for the final value of θ. Since the biped is under-actuated, the evolution of θ results from (12.3) with the feedback control law (12.8). The polynomial functions (12.11) are defined with 20 coefficients. By taking into account the algebraic equation of impact and that the motion is periodic and continuous between each single support phase, it is possible to reduce the number of parameters to only 12 [29]. Then vector $\mathbf{p}$ of parameters is finally

$$\begin{aligned}
\mathbf{p} =& [d \; q_{1f} \; x_{hf} \; y_{hf} \; (\dot{q}_{31}/\dot{\theta})_f \; (\dot{q}_{41}/\dot{\theta})_f (\dot{q}_{32}/\dot{\theta})_f \; \cdots \\
& (\dot{q}_{42}/\dot{\theta})_f \; q_{31}(\theta_{int}) \; q_{41}(\theta_{int}) \; q_{32}(\theta_{int}) \; q_{42}(\theta_{int})].
\end{aligned}$$

Remarks:

- Only the four independent variables: the final position of the hips x_{hf}, z_{hf}, the final orientation of the trunk q_{1f}, and the final length step d are necessary to define the initial and final configurations which are related by simple permutation of the legs.
- using impact equation (12.4) and parameters $(\dot{q}_{31}/\dot{\theta})_f$, $(\dot{q}_{41}/\dot{\theta})_f$, $(\dot{q}_{32}/\dot{\theta})_f$, $(\dot{q}_{42}/\dot{\theta})_f$ it is possible to calculate $(\dot{q}_{31}/\dot{\theta})_i$, $(\dot{q}_{41}/\dot{\theta})_i$, $(\dot{q}_{32}/\dot{\theta})_i$, $(\dot{q}_{42}/\dot{\theta})_i$ and the ratio $\alpha = \dot{\theta}_f/\dot{\theta}_i$ which is necessary to find the cyclic gait.

As a conclusion, an optimal trajectory can be defined by using only 12 parameters. To find the cyclic gait a vector p is chosen from an optimization process minimizing energy criterion, under physical constraints such as no slipping and no take off of the stance leg tip, torque limit, minimization of energy (see for more details [30–32]).

12.4 Observer Design

12.4.1 Analysis of Observability

Consider the dynamical system (12.3), with y the vector composed of the measured variables $\mathbf{y} = [y_1 \; y_2 \; y_3 \; y_4]^T = [q_{31} \; q_{32} \; q_{41} \; q_{42}]^T = \mathbf{q}_{\mathbf{rel}}$

$$\dot{\mathbf{x}} = \mathbf{f}(\mathbf{x}) + \mathbf{g}(\mathbf{y})\boldsymbol{\Gamma},$$

$$\mathbf{y} = \underbrace{\begin{bmatrix} 1 & 0 & 0 & 0 & 0 & 0 & 0 & 0 & 0 & 0 \\ 0 & 1 & 0 & 0 & 0 & 0 & 0 & 0 & 0 & 0 \\ 0 & 0 & 1 & 0 & 0 & 0 & 0 & 0 & 0 & 0 \\ 0 & 0 & 0 & 1 & 0 & 0 & 0 & 0 & 0 & 0 \end{bmatrix}}_{\mathbf{C}} \mathbf{x} \qquad (12.12)$$

with $\mathbf{x} \in \mathcal{X}$, $\Gamma \in \mathbb{R}^4$ and $\mathbf{y} \in \mathbb{R}^4$. In the biped context, this model describes the swing motion and is studied over one step, i.e. for $t \in [T_I^i, T_I^{i+1}[$, with T_I^i (resp. T_I^{i+1}) the initial (resp. final) impact time of the step i. As $\mathbf{g}(\mathbf{y})\Gamma$, the *input-output injection* term of (12.12), is fully known, an observer for (12.12) can be designed by the following way. With abuse of notation, consider the next nonlinear system, which is the part of (12.12) without the input-output injection term $\mathbf{g}(\mathbf{y})\Gamma$,

$$\begin{aligned} \dot{\mathbf{x}} &= \mathbf{f}(\mathbf{x}), \\ \mathbf{y} &= \mathbf{C}\mathbf{x}. \end{aligned} \qquad (12.13)$$

The observers presented in Subsections 12.4.2 and 12.4.3 are designed from this model (12.13). Let $\mathcal{O}$ denote the generic observability space defined by

$$\mathcal{O} = \tilde{\mathcal{X}} \cap \tilde{\mathcal{Y}}, \qquad (12.14)$$

with[2] $\tilde{\mathcal{X}} = \text{Span}_{\mathcal{K}}\{d\mathbf{x}\}$ and $\tilde{\mathcal{Y}} = \text{Span}_{\mathcal{K}}\{d\mathbf{y}^{(j)}, j \geqslant 0\}$. Function $\mathbf{y}^{(l)}$ denotes the lth time derivative of $\mathbf{y}$.

Definition 1 *System (12.13) is generically observable if* $\dim \mathcal{O} = 10$. ∎

This condition is called *Rank condition of generic observability*. In fact, this definition has to be detailed, because the observability property of (12.13) depends on $\mathbf{x}$. As a matter of fact, the dimension of $\mathcal{O}$ can fail in $\mathcal{X}$. Let $\mathcal{T}$ denote an open set of $\mathcal{X}$ such that the condition of the following definition is fulfilled.

Definition 2 *System (12.13) is observable if there exist* $\mathcal{T} \subset \mathcal{X}$ *and 4 integers* $\{k_1, k_2, k_3, k_4\}$, *called observability indices, such that*

- $\displaystyle\sum_{i=1}^{4} k_i = 10$,

- *The transformation*

$$\left[y_1 \;\cdots\; y_1^{(k_1-1)} \;\cdots\; y_4 \;\cdots\; y_4^{(k_4-1)} \right]^T = \Phi(\mathbf{x})$$

is a diffeomorphism for $\mathbf{x} \in \mathcal{T}$, *which is equivalent to*

$$\text{Det}\left[\tfrac{d\Phi(\mathbf{x})}{d\mathbf{x}} \right] \neq 0 \text{ for } \mathbf{x} \in \mathcal{T}. \qquad (12.15)$$

∎

Proposition 1. *There exist* $\mathcal{T} \subset \mathcal{X}$ *and observability indices* $[k_1 \; k_2 \; k_3 \; k_4]^T$ *such that system (12.13) is observable for* $\mathbf{x} \in \mathcal{T}$. ∎

Sketch of proof During the swing phase, and along the desired trajectories, for $[k_1 \; k_2 \; k_3 \; k_4]^T = [3\ 3\ 2\ 2]^T$, the determinant of $\frac{d\Phi(\mathbf{x})}{d\mathbf{x}}$ crosses zero (see Figure 12.2). At this singular point, system (12.13) is not observable ($\mathbf{x} \in (\mathcal{X}/\mathcal{T})$, the union of $\mathcal{T}$ and $\mathcal{X}$ without their intersection); elsewhere, system (12.13) is observable ($\mathbf{x} \in \mathcal{T} \subset \mathcal{X}$).

---

[2] $\text{Span}_{\mathcal{K}}$ is a space spanned over field $\mathcal{K}$ of meromorphic functions of $\mathbf{x}$.

Remark 3 *The previous observability indices choice is not unique which induces the existence of several subsets $\mathcal{T}$. Furthermore, loss of observability (see proof of Proposition 1) induces necessity to propose solutions in order to estimate the state variable over $\mathcal{X}$, which will be done in Section 12.4.4.* ∎

Suppose that system (12.13) is observable with $[k_1 \ k_2 \ k_3 \ k_4]^T$ its observability indices and the associated state coordinate transformation,

$$\mathbf{z} = \mathbf{\Phi}(\mathbf{x}) = \left[y_1 \ \cdots \ y_1^{(k_1-1)} \ \cdots \ y_4 \ \cdots \ y_4^{(k_4-1)} \right]^T. \tag{12.16}$$

From (12.16), system (12.13) reads as the canonical form

$$\begin{aligned} \dot{\mathbf{z}} &= \mathbf{A}\mathbf{z} + \varphi(\mathbf{z}), \\ \mathbf{y} &= \mathbf{C}\mathbf{z} \end{aligned} \tag{12.17}$$

with $\mathbf{A} = \mathrm{diag}\,[\mathbf{A}_1 \ \cdots \ \mathbf{A}_4]_{10\times 10}$, $\mathbf{C} = [\mathbf{C}_1 \ \cdots \ \mathbf{C}_4]^T_{4\times 10}$, $\varphi(\mathbf{z}) = [\varphi_1^T \ \cdots \ \varphi_4^T]^T$, $\mathbf{A}_{k_i}$, $\mathbf{C}_i$ and φ_i defined, respectively, as

$$\mathbf{A}_{k_i} = \begin{bmatrix} 0 & 1 & 0 & \cdots & 0 \\ 0 & 0 & 1 & \ddots & \vdots \\ \vdots & \vdots & \ddots & \ddots & \vdots \\ 0 & 0 & \cdots & 0 & 1 \\ 0 & 0 & \cdots & 0 & 0 \end{bmatrix}_{k_i \times k_i}, \quad \mathbf{C}_i = \begin{bmatrix} 1 & 0 & \cdots & 0 \end{bmatrix}_{1\times k_i}, \quad \varphi_i = \begin{bmatrix} 0 \\ \vdots \\ 0 \\ \varphi_{ik_i}(z) \end{bmatrix}_{k_i \times 1}.$$

Function $\mathbf{z} = \mathbf{\Phi}(\mathbf{x})$ is a diffeomorphism from $\mathcal{T}$ onto $\mathcal{Z} = \mathbf{\Phi}(\mathcal{T}) \subset I\!\!R^{10}$.

12.4.2 High-gain Observer

This section displays the design of a high-gain observer [13]. Let $\mathbf{K}$ denote a matrix of appropriate dimensions. Then, the system

$$\dot{\hat{\mathbf{z}}} = \mathbf{A}\hat{\mathbf{z}} + \varphi(\hat{\mathbf{z}}) + \mathbf{\Lambda}^{-1}\mathbf{K}(\mathbf{y} - \mathbf{C}\hat{\mathbf{z}}) \tag{12.18}$$

with $\hat{\mathbf{z}} \in \mathcal{Z}$, is an asymptotic observer for (12.17) if

- **HG1.** $\varphi(\hat{\mathbf{z}})$ is Lipschitzian with respect to $\hat{\mathbf{z}} \in \mathcal{Z}$. This assumption means that the nonlinear term $\varphi(\hat{\mathbf{z}})$ and its derivative are bounded. Then, it is always possible to define a correction term to limit their effects.
- **HG2.** Matrix $(\mathbf{A} - \mathbf{KC})$ has its eigenvalues in the left half-plane, the choice of $\mathbf{K}$ acting on the dynamics of the estimation.
- **HG3.** $\mathbf{\Lambda}(\mathbf{T}) = \mathrm{diag}[\Lambda_1 \ \Lambda_2 \ \cdots \ \Lambda_p]^T$ with $\Lambda_i = \mathrm{diag}[T_i \ T_i^2 \ \cdots \ T_i^{k_i-1}]$, with $0 < T_i < 0$.

Furthermore, the dynamics of this observer can be made arbitrarily fast. Then, an observer for (12.3) is designed as

$$\dot{\hat{\mathbf{x}}} = \mathbf{f}(\hat{\mathbf{x}}) + \mathbf{g}(\hat{\mathbf{x}})\mathbf{\Gamma} + \left[\frac{\mathrm{d}\mathbf{\Phi}(\hat{\mathbf{x}})}{\mathrm{d}\hat{\mathbf{x}}} \right]^{-1} \mathbf{\Lambda}^{-1}\mathbf{K}(\mathbf{y} - \mathbf{C}\hat{\mathbf{x}}). \tag{12.19}$$

12.4.3 Step-by-step Observer

In this section, the observer is based on triangular form one [20] and its main property is the finite time convergence to zero of the estimation error. System (12.17) is still on (particular) triangular form. Such an observer of (12.17) can be written as [20]

$$\dot{\hat{z}} = \mathbf{A}\hat{z} + \varphi(\hat{z}) + \mathbf{E}(t)\chi(\cdot) \tag{12.20}$$

with $\hat{z} \in \mathcal{Z}$ the estimated vector of $\mathbf{z}$. Knowing that the principle of this class of observers consists in forcing, each in turn, estimated state variables to corresponding real ones, in finite time, this latter property is based on an adequate choice of $\mathbf{E}(t)$ and $\chi(\cdot)$, *i.e.* $\mathbf{E}(t)$ and χ are defined such that the estimation error $\mathbf{e} = \hat{z} - \mathbf{z}$ converges to zero in finite time.

Finite Time Convergence Observer

For the sake of clarity, and without loss of generality, only the observer design for a third-order system is fully displayed in the sequel, *i.e.* $\mathbf{z} \in \mathcal{Z} \in I\!\!R^3$. Then, in this case, system (12.17) reads as

$$\dot{z}_1 = z_2, \qquad \dot{z}_2 = z_3, \qquad \dot{z}_3 = f_3(z), \qquad y = z_1 \tag{12.21}$$

with $\mathbf{z} = [z_1 \; z_2 \; z_3]^T$. Then, an observer for (12.21) reads as

$$\dot{\hat{z}}_1 = \hat{z}_2 + E_1(t)\chi_1(\cdot), \quad \dot{\hat{z}}_2 = \hat{z}_3 + E_2(t)\chi_2(\cdot), \quad \dot{\hat{z}}_3 = f_3(\hat{z}) + E_3(t)\chi_3(\cdot) \tag{12.22}$$

with $E_i(t)$ and χ_i $(1 \leqslant i \leqslant 3)$ defined such that each estimation error $e_i = \hat{z}_i - z_i$ converges to zero in finite time. To ensure a finite time convergence, the function $\chi_1(\cdot)$ is based on the *twisting algorithm* [18] and reads as

$$\dot{\chi}_1 = -\Lambda_1 \, \text{sign}(\hat{z}_1 - z_1) \tag{12.23}$$

with

$$\Lambda_1 = \begin{cases} \lambda_m & \text{if } e_1 \dot{e}_1 \leqslant 0, \\ \lambda_M & \text{if } e_1 \dot{e}_1 > 0, \end{cases} \tag{12.24}$$

$$\lambda_m > \text{Max}(|e_2|), \quad \lambda_M > 3\lambda_m,$$

the operator $\text{Max}(\cdot)$ denoting the boundary value of the current variable. Functions $\chi_2(\cdot)$ and $\chi_3(\cdot)$ use the standard sliding mode approach and read as

$$\begin{aligned} \chi_2 &= -\lambda_2 \, \text{sign}(\tilde{z}_2 - \hat{z}_2), \\ \chi_3 &= -\lambda_3 \, \text{sign}(\tilde{z}_3 - \hat{z}_3) \end{aligned} \tag{12.25}$$

with $\lambda_2 > 0$, $\lambda_3 > 0$,

$$\tilde{z}_j = \hat{z}_j + E_{j-1}(t)\chi_{j-1} \tag{12.26}$$

for $j \in \{2, 3\}$.

Determination of E_i and Proof of Finite Time Convergence

From (12.21)-(12.22), the estimation error dynamics reads as

$$\dot{e}_1 = e_2 + E_1(t)\chi_1(\cdot),$$
$$\dot{e}_2 = e_3 + E_2(t)\chi_2(\cdot),$$
$$\dot{e}_3 = f_3(\hat{z}) - f_3(z) + E_3(t)\chi_3(\cdot)$$

(12.27)

with $e_i = \hat{z}_i - z_i$.

- **Step 1.** Suppose that $e_1(0) \neq 0$, and observer (12.22) is initialized such that $E_1 = 1$ and $E_2 = E_3 = 0$. The error dynamics reads as

$$\dot{e}_1 = e_2 + \chi_1,$$
$$\dot{e}_2 = e_3,$$
$$\dot{e}_3 = f_3(\hat{z}) - f_3(z).$$

(12.28)

As χ_1 is based on the twisting algorithm with appropriate tuning of Λ_1, e_1 reaches zero in finite time at $t = t_1$. Then, $\forall\, t \geqslant t_1$, $e_1(t) = \dot{e}_1(t) = 0$, i.e.

$$e_1 = 0, \quad \dot{e}_1 = e_2 + \chi_1 = \hat{z}_2 - z_2 + \chi_1 = 0.$$

(12.29)

From (12.29), one gets $\hat{z}_2 + \chi_1 = z_2$ and from (12.26), $z_2 = \tilde{z}_2$.

- **Step 2.** For $t \geqslant t_1$, one states $E_1 = E_2 = 1$ and $E_3 = 0$. From (12.25)-(12.29), dynamics of e_2 and e_3 read as

$$\dot{e}_2 = e_3 - \lambda_2 \,\mathrm{sign}(e_2),$$
$$\dot{e}_3 = f_3(\hat{z}) - f_3(z).$$

(12.30)

By tuning $\lambda_2 > \mathrm{Max}(|e_3|)$, a finite time convergence to $e_2 = 0$ is ensured at $t = t_2$. Then, $\forall\, t \geqslant t_2$, $e_2(t) = \dot{e}_2(t) = 0$, i.e.

$$e_1 = 0, \quad \dot{e}_1 = 0,$$
$$e_2 = 0, \quad \dot{e}_2 = e_3 + \chi_2 = 0.$$

(12.31)

From the second line of (12.31), one gets $\hat{z}_3 + \chi_2 = z_3$ and from (12.26), $z_3 = \tilde{z}_3$.

- **Step 3.** For $t \geqslant t_2$, one states $E_1 = E_2 = E_3 = 1$. From (12.25)-(12.31), dynamics of e_3 reads as

$$\dot{e}_3 = f_3(\hat{z}) - f_3(z) - \lambda_3 \,\mathrm{sign}(\hat{z}_3 - z_3).$$

(12.32)

By tuning $\lambda_3 > \mathrm{Max}(|f_3(\hat{z}) - f_3(z)|)$, a finite time convergence to $e_3 = 0$ is ensured, which implies a finite time convergence of the observer.

12.4.4 Loss of Observability and Observation Algorithm

As mentioned previously, during the swing phase, along the *nominal* trajectories, for all observability indices possibilities, there is loss of observability (see Figure 12.2 for $[k_1 \; k_2 \; k_3 \; k_4]^T = [3 \; 3 \; 2 \; 2]^T$: determinant of $\frac{d\Phi(x)}{dx}$ is crossing zero). Of course, this induces a problem for the observer design. A solution consists in designing observers for several combinations of observability indices, for which the observability singularity appears for different articular positions. As a matter of fact,

$[k_1\ k_2\ k_3\ k_4]^T = [3\ 2\ 2\ 3]^T$ is also eligible: the variation of determinant of $\frac{d\Phi(\mathbf{x})}{d\mathbf{x}}$ is displayed in Figure 12.3 (dotted line). Then, there exists $\mathcal{T}' \subset \mathcal{X}$ such that, $\forall \mathbf{x} \in \mathcal{T}'$, the function

$$\mathbf{z} = \Phi(\mathbf{x}) = \begin{bmatrix} y_1 & \dot{y}_1 & \ddot{y}_1 & y_2 & \dot{y}_2 & y_3 & \dot{y}_3 & y_4 & \dot{y}_4 & \ddot{y}_4 \end{bmatrix}^T$$

is a state transformation.

Observation Algorithm

Let $\Phi_1(\mathbf{x}) = \begin{bmatrix} y_1 & \dot{y}_1 & \ddot{y}_1 & y_2 & \dot{y}_2 & \ddot{y}_2 & y_3 & \dot{y}_3 & y_4 & \dot{y}_4 \end{bmatrix}^T$ ($\mathbf{x} \in \mathcal{T}$ and $[k_1\ k_2\ k_3\ k_4]^T =$ $[3\ 3\ 2\ 2]^T$), and $\Phi_2(\mathbf{x}) = \begin{bmatrix} y_1 & \dot{y}_1 & \ddot{y}_1 & y_2 & \dot{y}_2 & y_3 & \dot{y}_3 & y_4 & \dot{y}_4 & \ddot{y}_4 \end{bmatrix}^T$ ($\mathbf{x} \in \mathcal{T}'$ and $[k_1\ k_2\ k_3\ k_4]^T = [3\ 2\ 2\ 3]^T$). Let $T_{SW} = \text{Min}(t)$ such that $t \in [T_I^i, T_I^{i+1}[$ and $\text{Det}\left[\frac{d\Phi_1(\mathbf{x})}{d\mathbf{x}}\right](t) = 0$ (see Figure 12.3). Then, one has

- For $t \in [T_I^i, T_{SW}[$, the observer is designed with $[k_1\ k_2\ k_3\ k_4]^T = [3\ 3\ 2\ 2]^T$,
- From $t \in [T_{SW}, T_I^{i+1}[$, the observer is designed with $[k_1\ k_2\ k_3\ k_4]^T = [3\ 2\ 2\ 3]^T$.

∎

Proposition 2. *An observer for system (12.12) reads as*

$$\dot{\hat{\mathbf{x}}} = \mathbf{f}(\hat{\mathbf{x}}) + \mathbf{g}(\mathbf{y})\mathbf{\Gamma} + \mathbf{M}\mathbf{F}(\hat{\mathbf{x}}, \mathbf{y}) \tag{12.33}$$

$$with \quad \mathbf{M} = \begin{cases} \left[\frac{d\Phi_1(\hat{\mathbf{x}})}{d\hat{\mathbf{x}}}\right]^{-1} & for\ t \in [T_I^i, T_{SW}[\\ \left[\frac{d\Phi_2(\hat{\mathbf{x}})}{d\hat{\mathbf{x}}}\right]^{-1} & for\ t \in [T_{SW}, T_I^{i+1}[\end{cases}$$

$$and \quad \mathbf{F}(\hat{\mathbf{x}}, \mathbf{y}) = \begin{cases} \mathbf{\Lambda}^{-1}\mathbf{K}(\tilde{\mathbf{y}} - \mathbf{C}\hat{\mathbf{x}}) & for\ high\text{-}gain\ observer \\ \mathbf{E}(t)\chi(\cdot) & for\ step\text{-}by\text{-}step\ observer \end{cases}$$

with $\mathbf{\Lambda}$, $\mathbf{K}$, $\mathbf{E}(t)$ *and* $\chi(\cdot)$ *appropriate-dimensional matrices respectively defined through (12.18)-HG1-HG2-HG3-(12.22).* ∎

Practical Point-of-view

- The commutation from the first observer structure to the second one is made through the condition that $T_{SW} = \text{Min}(t)$ such that $t \in [T_I^i, T_I^{i+1}[$ and $\left|\text{Det}\left[\frac{d\Phi_1(\hat{\mathbf{x}})}{d\hat{\mathbf{x}}}\right]\right|(t) = D_{SW}$, where $D_{SW} > 0$ a real fixed by the user (see Figure 12.3). The choice of D_{SW} is made in order that the condition number with respect to inversion of $\left[\frac{d\Phi_2(\hat{\mathbf{x}})}{d\hat{\mathbf{x}}}\right]$ is not "too large".
- The finite-time convergence observer previously exposed ensures that the estimation errors *exactly* converge to zero. In practice, this property is ensured for a neighborhood of zero [20] which the estimation errors are forced to reach and to stay in.

Remark 4 *Observer (12.33) is designed for the swing phase. But, a step is composed of both swing and impact phases. At the impact event, the impact model is applied to the estimated state [33]. Then, over one step, an observer of (12.7) reads as*

$$\begin{aligned} \dot{\hat{\mathbf{x}}} &= \mathbf{f}(\hat{\mathbf{x}}) + \mathbf{g}(\mathbf{y})\mathbf{\Gamma} + \mathbf{MF}(\hat{\mathbf{x}}, \mathbf{y}) & \hat{\mathbf{x}}^-(t) \notin \hat{S} \\ \hat{\mathbf{x}}^+ &= \Delta(\hat{\mathbf{x}}^-) & \hat{\mathbf{x}}^- \in \hat{S} \end{aligned} \qquad (12.34)$$

with $\hat{S} = \{\hat{\mathbf{x}} \in \hat{\mathcal{X}} \mid \hat{z}_2(\hat{\mathbf{q}}) = 0\}$.

12.5 Simulations

The parameters of the five-link biped prototype *Rabbit* [23] are used to design the five-links biped parameters. The masses and lengths of the links (Indices 31, 41, 32, 42, 1: swing leg (femur, tibia), stance leg (femur, tibia), torso, resp.) are $m_{31} = m_{32} = 3.2$ kg, $m_{41} = m_{42} = 6.8$ kg, $m_1 = 17.0528$ kg, $l_{31} = l_{32} = l_{41} = l_{42} = 0.4$ m, $l_1 = 0.625$ m. The distances between the joint and the mass center of each link are $s_{31} = s_{32} = 0.127$ m, $s_{41} = s_{42} = 0.163$ m, $s_1 = 0.1434$ m. The inertia moments around the mass center of each link are $I_{31} = I_{32} = 0.0484$ kg·m^2, $I_{41} = I_{42} = 0.0693$ kg·m^2, $I_1 = 1.8694$ kg·m^2. The inertia of the rotor for each DC motor is $I = 3.32\,10^{-4}$ kg·m^2. The ratio N of the gearbox reducers equals 50. The value U of the applied torques equals 150 N·m.

The reference trajectories are obtained by an optimization algorithm under constraint (for example, the swing leg tip must be above a parabola), this optimization problem being founded on the model, the physical parameters and the static/dynamic performances of *Rabbit* [23]. Note that the trunk motion obtained from this optimization process has small amplitudes. It can be explained by the fact that the four actuators of *Rabbit* are located in the base of the trunk. The control law described in Section 12.3 is applied with parameters $\alpha = 0.9$ and $\epsilon = 30$ and is using the estimated state variables from high-gain and step-by-step observers (12.33). The initial real and estimated state variables have been, respectively, stated as

$$\begin{aligned} \mathbf{q}^T(0) &= [-2.4508 \quad -3.1325 \quad -0.3552 \quad -0.1000 \quad -0.3665], \\ \dot{\mathbf{q}}^T(0) &= [-1.5149 \quad 0.3263 \quad 0.9674 \quad -2.5863 \quad -0.4701], \\ \hat{\mathbf{q}}^T(0) &= [-2.4508 \quad -3.1325 \quad -0.3552 \quad -0.1000 \quad -0.2618], \\ \dot{\hat{\mathbf{q}}}^T(0) &= [-1.4149 \quad 0.4263 \quad 1.0674 \quad -2.4863 \quad -0.3701]. \end{aligned}$$

Parameter D_{min} has been stated to 4. The high-gain and step-by-step observer parameters are tuned such that

- The high-gain observer parameters are fixed such that the eigenvalues of $(\mathbf{A} - \mathbf{KC})$ are located at -10, and $T = 0.1$.
- The sliding mode observer parameters: $\lambda_m = 5$, $\lambda_M = 25$, $\lambda_2 = 5$, $\lambda_3 = 75$.

The choice of observer and control law parameters has been made with respect to closed-loop dynamics and admissible maximum value for input (saturation).

Figures 12.4-12.5 show the absolute position and the velocity estimation errors. Each estimation error converges to zero before each impact, asymptotically for high-gain observer (see Figure 12.4), and in finite time for the sliding mode observer (see Figure 12.5). The control law has been tuned such that the control output vector $\mathbf{h}(\hat{\mathbf{x}})$ equals zero before the impact (see Figure 12.6 for the state estimated by the high-gain observer and Figure 12.7 for the state estimated by the sliding mode observer). Then, the biped robot is reaching a stable periodic cycle after several steps (Figure 12.8). The estimation error presents a higher transient with the high-gain observer with respect to the step-by-step one. However, the main drawback of the sliding mode, high frequency oscillation of the estimation error, clearly appears in Figures 12.5-12.7, The estimated variables with the high-gain observer-based control are smoother.

12.6 Conclusion

In future, in order to economize energy or to obtain fast motions, the walking gaits of biped robots will be composed of underactuated phases with partial contact between the ground and the sole of the stance foot. The knowledge of all the components of the state vector will be necessary to define efficient control laws. However, from a technological point-of-view, a highly accurate measurement of the absolute orientation of a legged robot in non-equilibrium postures is a hard task. In this paper the design of orientation observers has been made for a biped with point feet. Numerical simulations, based on the *Rabbit* prototype, show that the estimation of the absolute orientation for the cyclic gait of a five-link biped robot without feet is possible. Both designed observers are nonlinear and based on the canonical observability form. The first observer is based on a high-gain approach (which ensures asymptotic convergence), and its behaviour is relatively smooth, which is a crucial point for future real applications. The second observer is based on sliding modes. As the chattering effect is the main drawback of sliding mode solutions, a second-order sliding mode approach is used, which induces lower chattering, robustness with respect to parameter variations and disturbances, and finite time convergence. Associated with an efficient control law, both observers lead to a cyclic motion close to the nominal motion. The perspectives are the experimental evaluation of these techniques on the prototype *Rabbit*.

References

1. A.A. Grishin, A.M. Formal'sky, A.V. Lensky, and S.V. Zhitomirsky. Dynamic walking of a vehicle with two telescopic legs controlled by two drives. *Int. J. of Robotics Research*, 13(2):137–147, 1994.
2. J. Pratt, C.M. Chew, A. Torres, P. Dilworth, and G. Pratt. Virtual model control: an intuitive approach for bipedal locomotion. *Int. J. of Robotics Research*, 20(2):129–143, 2001.
3. G. Capi, S. Kaneko, K. Mitobe, L. Barolli, and Y. Nasu. Optimal trajectory generation for a primatic joint biped robot using genetic algorithms. *Robotics and autonomous systems*, 38(2):119–128, 2002.
4. F. Zonfrilli, M. Oriolo, and T. Nardi. A biped locomotion strategy for the quadruped robot sony ers-210. In *Proc. IEEE Int. Conf. on Robotics and Automation ICRA*, pages 2768–2774, Washington, D.C., USA, 2002.
5. A. Albert and W. Gerth. New path planning algorithms for higher gait stability of a bipedal robot. In *Proc. Int. Conf. on Climbing and Walking Robots CLAWAR*, pages 521–528, Paris, France, 2001.

6. G. Cabodevilla, N. Chaillet, and G. Abba. Energy-minimized gait for a biped robot. In *Autonome mobile systems*, pages 90–99, Springer-Verlag, Berlin, Germany, 1995.

7. Y. Aoustin and A.M. Formal'sky. Control design for a biped: reference trajectory based on driven angles as functions of the undriven angle. *Int. J. of Computer and Systems Sciences*, 42(4):159–176, 2003.

8. F. Plestan, J.W. Grizzle, E.R. Westervelt, and G. Abba. Stable walking of a 7-dof biped robot. *IEEE Transactions on Robotics and Automation*, 19(4):653–668, 2003.

9. K Kaneko, F. Kanehiro, S. Kajita, H. Hirukawa, T. Kawasaki, M. Hirata, K.Akachi, and T. Isozumi. Humanoid robot hrp-2. In *Proc. IEEE Int. Conf. on Robotics and Automation ICRA*, pages 1083–1090, New-Orleans, Louisiana, USA, 2004.

10. M. Vukobratovic and B. Borovac. Zero-moment point-thirty five years of its life. *Int. J. of Humanoid Robotics*, 1(1):157–173, 2004.

11. P. Micheau, M.A. Roux, and P. Bourassa. Self-tuned trajectory control of a biped walking robot. In *Proc. Int. Conf. on Climbing and Walking Robot CLAWAR*, pages 527–534, Catania, Italy, 2003.

12. V. Lebastard, Y. Aoustin, and F. Plestan. Observer-based control of a biped robot. In *Proc. Int. Workshop on Robot Motion and Control ROMOCO*, pages 67–72, Puszczykowo, Poland, 2004.

13. G. Bornard and H. Hammouri. A high gain observer for a class of uniformly observable systems. In *Proc. IEEE Conf. on Decision and Control CDC*, pages 1494–1496, Brighton, England, 1991.

14. J. P. Gauthier and G. Bornard. Observablity for any $u(t)$ of a class of nonlinear systems. *IEEE Transactions on Automatic Control*, 26(4):922–926, 1981.

15. Y. Aoustin, G. Garcia, and A. Janot. Estimation of the absolute orientation of a two-link biped using discrete observers. In *Proc. Mechatronics and Robotics Conf. MECHROB*, pages 1315–1320, Aachen, Germany, 2004.

16. R.E. Kalman, P.L. Falb, and M.A. Arbib. *Topics in mathematical system theory*. McGraw-Hill, New York, USA, 1969.

17. J.J.E. Slotine and W. Li. *Applied nonlinear control*. Prentice Hall, New-York, USA, 1991.

18. A. Levant. Sliding order and sliding accuracy in sliding mode control. *Int. J. of Control*, 58(6):1247–1263, 1993.

19. T. Floquet, J.P. Barbot, and W. Perruquetti. A finite time observer for flux estimation in the induction machine. In *Proc. IEEE Conf. on Control Applications CCA*, pages 1303–1308, Glasgow, Scotland, 2002.

20. T. Boukhobza and J.P. Barbot. High order sliding modes observer. In *Proc. IEEE Conf. on Decision and Control CDC*, pages 1912–1917, Tampa, Florida, USA, 1998.

21. V. Lebastard, Y. Aoustin, and F. Plestan. Second order sliding mode observer for stable control of a walking biped robot. In *Proc. IFAC World Congress*, Praha, Czech Republic, 2005.

22. V. Lebastard, Y. Aoustin, and F. Plestan. Step-by-step sliding mode observer for control of a walking biped robot by using only actuated variables measurement. In *Proc. IEEE Int. Conf. on Intelligent Robots and Systems IROS*, pages 3102–3107, Edmonton, Canada, 2005.

23. C. Chevallereau, G. Abba, Y. Aoustin, F. Plestan, E.R. Westervelt, C. Canudas de Wit, and J.W. Grizzle. Rabbit: a testbed for advanced control theory. *IEEE Control Systems Magazine*, 23(5):57–79, 2003.

24. M.W. Spong and M. Vidyasagar. *Robot dynamics and control*. John Wiley, New-York, USA, 1991.

25. A.M. Formal'sky. *Locomotion of Anthropomorphic Mechanisms*. [In Russian], Nauka, Moscow, Russia, 1982.

26. J.W. Grizzle, G. Abba, and F. Plestan. Asymptotically stable walking for biped robots : analysis via systems with impulse effects. *IEEE Transactions on Automatic Control*, 46(1):51–64, 2001.

27. A. Isidori. *Nonlinear Control Systems*. Springer-Verlag, London, England, 1995.

28. S.P. Bhat and D.S. Bernstein. Continuous finite-time stabilization of the translational and rotationnal double integrator. *IEEE Transaction on Automatic Control*, 43(5):678–682, 1998.

29. S. Miossec and Y. Aoustin. Walking gait composed of single and double supports for a planar biped without feet. In *Proc. Int. Conf. on Climbing and Walking Robots CLAWAR*, pages 767–774, Paris, France, 2002.

30. S. Miossec and Y. Aoustin. Mouvement de marche composé de simple et double supports pour un robot bipède planaire sans pieds. In *Proc. Conf. Int. Francophone d'Automatique CIFA*, [in French], Nantes, France, 2003.

31. C. Chevallereau and Y. Aoustin. Optimal reference trajectories for walking and running of a biped. *Robotica*, 19(5):557–569, 2001.

32. D. Djoudi, C. Chevallereau, and Y. Aoustin. Optimal reference motions for walking of a biped robot. In *Proc. IEEE Int. Conf. on Robotics and Automation ICRA*, pages 2014–2019, Barcelona, Spain, 2005.

33. Menini L. and Tornambè A. Velocity observers for linear mechanical systems subject to single non-smooth impacts. *Systems and Control Letters*, 43:193–202, 2001.

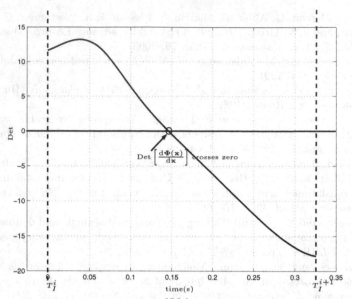

Fig. 12.2. Determinant of $\frac{d\Phi(x)}{dx}$ vs. time (s) along one step

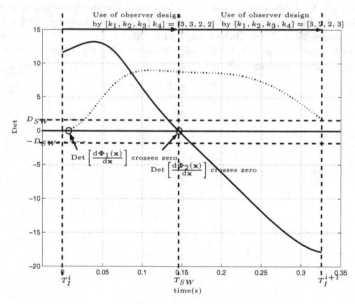

Fig. 12.3. Determinants of $\frac{d\Phi_1(\hat{x})}{d\hat{x}}$ (bold) and $\frac{d\Phi_2(\hat{x})}{d\hat{x}}$ (dotted) vs. time (s) over one step

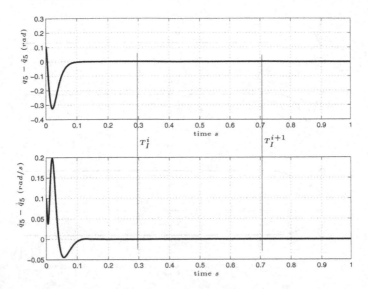

Fig. 12.4. High-gain observer. Orientation $q_5 - \hat{q}_5$ (top) and velocity $\dot{q}_5 - \dot{\hat{q}}_5$ (bottom) estimation errors versus time (s). The vertical solid lines display the contact events

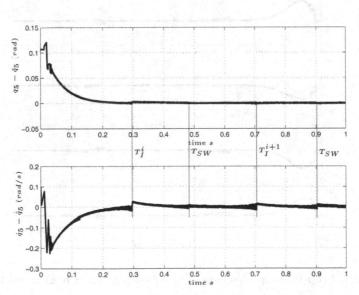

Fig. 12.5. Step-by-step observer. Orientation $q_5 - \hat{q}_5$ (top) and velocity $\dot{q}_5 - \dot{\hat{q}}_5$ (bottom) estimation errors versus time (s). The vertical solid lines display the contact events (T_I^i and T_I^{i+1}) and the commutation time (T_{SW})

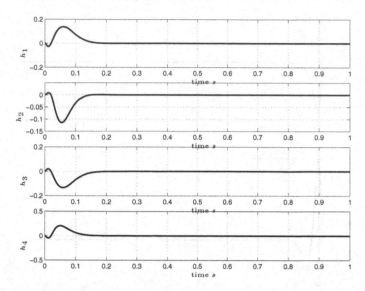

Fig. 12.6. High-gain observer. Control output vector $\mathbf{h}(\hat{\mathbf{x}})$ versus time (s)

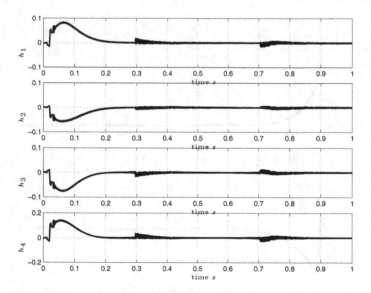

Fig. 12.7. Step-by-step observer. Control output vector $\mathbf{h}(\hat{\mathbf{x}})$ versus time (s)

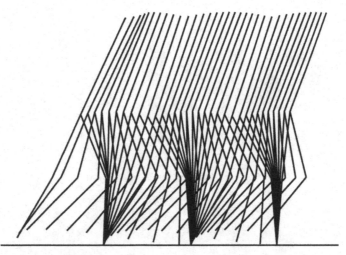

Fig. 12.8. Plot of walking as a sequence of stick figures

13

Biologically Inspired Motion Planning in Robotics⋆

Teresa Zielinska[1] and Chee-Meng Chew[2]

[1]Warsaw University of Technology, Institute of Aeronautics and Applied
Mechanics (WUT–IAAM), ul. Nowowiejska 24, 00-665 Warsaw, Poland
teresaz@meil.pw.edu.pl
[2]National University of Singapore 119260, Department of Mechanical Engineering,
(NUS–MPE) mpeccm@nus.edu.sg

13.1 Introduction

Walking machines are special types of mobile robots realizing *discrete locomotion*, where the motion path is not continuous but is separated to the footprints. Most walking machines in their structure take some biological systems as templates, with two, four or six legs. The level of autonomy of machine actions (*intelligence*) is determined by the properties of mechanical structure and abilities of the control system. In animals the complexity of the nervous system is related to the complexity of the body build – the more complex is the body in the biological sense the more advanced control it needs. But this does not mean that the increase in autonomy of walking robots can be obtained only by increasing the complexity of mechanics and control. In the animal world the body structure matches the living conditions. Simple animals, with primitive bodies and *control centers* can survive well due to the proper spontaneous reactions (arising directly from an impulse, without reasoning). Transferring this observation into technical world it means that the mechanical structure of walking devices must be properly chosen to the assumed working conditions and the control system with sensors and software must be dedicated to the task fulfilled by the device.

Recent research focuses on biology inspired robots. The designers of new prototypes (biomimetic robots) take advantages of new materials, technologies, actuators, sensors, control, and motion planning methods. Research requires close cooperation between biologists, neuro-biologists and engineers. Biomimetic walking machines are not very different than traditional devices, the design target is to build them agile, relatively cheap, and able to deal well with real environment. The motion principles use neurological backgrounds (e.g. analogies to neural control of insect legs). The design of these robots requires understanding of the biomechanics and neurology of biological systems on which they are based [46].

In this publication we discuss biologically inspired methods of gait generation. The methods were elaborated for multi-legged machines and for bipeds.

⋆ This work was supported by IAAM statutory funds and EERSS NUS program

K. Kozłowski (Ed.): Robot Motion and Control, LNCIS 335, pp. 201–219, 2006.
© Springer-Verlag London Limited 2006

13.2 Adaptive Motion Planning for a Multi-legged Robot

The brain is such a complex system that science is very far from full knowledge and understanding of the essence of its functioning. Nevertheless there exists a hypothetical description of its role in planning limb motion [20, 28]. According to it, neurons in the sensing and motion regions execute senso-motoric associations, i.e. associate signals from receptors with potential elementary motions, a senso-motoric mapping is formed. The premotoric brain stem network relates receptor signals to the motion programs generated by the cerebellum. The definition of senso-motoric mapping is equated, by the researchers, to the creation of vector of preferred motion directions. According to the above hypothesis cerebellum defines the motion trajectory (and so the velocity components also). Motion planning is performed in two subsequent stages. It was observed that the brain has a *where* subsystem – evaluating motion direction and sense – independent of muscles, and a *when* subsystem dependent on muscles – taking into account their efficiency and evaluating motion dynamics. During motion planning first is the action of *where* and then a *where* subsystem.

13.2.1 Basic Relations

In our research the above introduced hypothesis assuming two-stage motion planning by the brain was utilized. There is no difference in motion planning for a free gait with no-periodic sequence of leg transfers and for rhythmic gait with periodic sequence of leg transfers. The planning is performed on-line taking into account the information obtained from diverse sensors.

The following variables were defined:

p^i – *realized state of the walking machine* in the instant t^i ($p^i = p(t_i)$); this state is defined as a set of leg-end coordinates (in the instant t_i) expressed in coordinate frame (or frames) affixed to the machine body; e.g. this state can be described by the set of leg-end coordinates expressed in Cartesian coordinate frames affixed to the hips, i.e (x^i_j, y^i_j, z^i_j) – state of the j-th leg expressed in the j-th coordinate frame in the i-th instant; $p^i = ((x^i_1, y^i_1, z^i_1), \ldots, (x^i_k, y^i_k, z^i_k))$, for a k-legged walking machine,

s_a – *current sensor state*; state s_a describes the state of the so-called virtual sensor, that is a sensor created artificially by the control system by adequately aggregating the information gathered by real sensors [43]. The state of such sensors is defined as the information about the environment or the machine obtained from a set of real sensors (e.g. distance from obstacles),

st_{sens} – *real* static (longitudinal) or dynamic *stability margin* computed on the basis of sensor readings (state); $st_{sens} = st_{sens}(s_a)$,

o_{sens} – *motion limits* determined on the basis of sensor state in the current instant of time; $o_{sens} = o_{sens}(s_a)$,

o – *known motion limits*, e.g. the maximum possible forward stretch of the leg,

mt – *motion comfort criterion*, e.g. keeping the maximum acceleration below a certain limit or keeping the velocity of motion constant, etc.

g_f – *global motion goal*, e.g. motion duration, distance.

The term of motion comfort was introduced using the reference to the human gait features know in biomechanics. During a human motion several conditions are fulfilled assuring walking comfort. Between others human walks reducing accelerations of his mass center. A good discussion of human gait features can be found i.e. in [4,36]. For explanation we only mention that those indices or conditions are not absolute and always valid perfectly optimized criterions.

13.2.2 Motion Planning

The terms of *current motion goal* and *motion intention* are introduced. The current motion goal g^i depends on: the global goal, the realized machine state, the motion limits (o_{sens}, o), the stability margin st_{sens} and the *realized* current motion goal for the previous time instant $r(g^{i-1})$:

$$g^i = g^i (g_f , \boldsymbol{p}^{i-1}, o_{sens} , o, st_{sens} , r(g^{i-1})). \qquad (13.1)$$

The possible limb displacements are classified into the groups – sets named preferred motion directions. The trajectory shapes during each motion belonging to the specific set can differ. The sets are distinguished considering what is the purpose of a movement (for what it is performed).

The determination of the *current motion goal* in our method is equivalent to the choice to what set of preferred motion directions belongs the currently planned movement. It must be underscored that in this first phase of planning only the name of the set is evaluated but not the range and exact direction of the planned motion. The selection of the current goal is based on the analysis of an adequate decision scheme (reflecting rel.(13.1)). Each scheme describes the selection of a new g^i for the currently realized motion goal $r(g^{i-1})$. The second stage of motion planning – determination of *motion intention* in^i is equivalent to the evaluation of the velocity components for the planned movement belonging to the just chosen set. For each set different relations and conditions used for velocity calculation are formulated. It is expected that the motion with the evaluated velocity components will be performed until the next decision is taken concerning the current motion goal and the motion intention. This goes with a given time period imitating the human decision latency time. The latency time is not kept in extreme situations – as obstacle collision, reaching the motion limits, the new decision will be taken immediately. In general, with the time and velocity the nominal trajectory that the machine would reproduce in the absence of obstacles is specified. Obstacles cause modification of this trajectory, hence it might not be executed as intended.

$$in^i = in^i (mt , st_{pr}(in^i), o_{pr}(in^i), o , \boldsymbol{p}^{i-1}, g^i), \qquad (13.2)$$

where:

o_{pr} – forecasted motion limits for the planned motion intention in^i; the notation $o_{pr} = o_{pr}(in^i)$ describes the forecast for the intention in^i,

st_{pr} – forecasted value of the static/dynamic stability margin; notation $st_{pr} = st_{pr}(in^i)$ describes the forecast for the intention in^i.

13.2.3 Example

The simple example summarizes gait generation with the introduced method. Due to a limited size of this publication not every detail is commented; a wider description can be found in the book [39], and a summary of the method is submitted in [46]. A quadruped walking machine is considered. Legs move in one plane (it is not side motion). The machine moves with statically stable gait (the geometry of the machine is chosen in such way that stability is kept when three or four legs remain on the ground) [45]. The height of the obstacles is not greater than the maximum lift of the leg, and their length is not greater than half the maximum step length. Only straight line motion on flat terrain is considered. The above assumptions are not due to the limitations of presented method but for sake of simplicity and clarity of presentation. Motion planning is performed on-line during walking. Planning consist of two short phases which are repeated one after another one during the motion (as possible human is doing). The current motion goal is first determined, and for it the motion intention is evaluated.

We evaluated that leg-end can move upward (when leg is risen over the ground), forward (when leg-end moves forward in relation to the hip joint), downward (when leg-end moves toward the ground) and backward (when leg supports the body and pushes it forward). In the considered example with flat terrain (and only obstacles protruding out of it) during backward and forward motion the leg-end does not change its vertical distance to the hip joint, and during downward motion this distance is increasing with opposite for upward motion. Referring to it, the names of the direction sets (i.e. possible current motion goals) for each leg are the following:

$$g^i = \{BACKWARD,\ FORWARD,\ UPWARD,\ DOWNWARD\}. \qquad (13.3)$$

BACKWARD means that the leg-end must remain in the support phase when, due to pushing the body forwards, the leg-end moves backward relative to the trunk, UPWARD means that the leg should be moved above the ground and the distance between the ground and the leg-end will increase, FORWARD means transferring the leg-end towards the front of the body and DOWNWARD – the leg-end motion towards the ground (Fig. 13.1). The time and velocity components (distance) during this motion with chosen motion goal depend on limits dictated by device kinematics, on posture stability requirements and on obstacles presence.

Figure 13.2 presents the considered motion limits referred to a leg-end. The following constraints are considered:

o_{xmin} – largest possible backward displacement of a leg supported by the ground (BACKWARD motion limit),

o_{xp} – largest possible forward displacement of an upraised leg (FORWARD, UPWARD motion limit),

o_{xmax} – largest possible forward displacement of a leg supported by the ground (FORWARD, UPWARD, DOWNWARD motion limit),

o_{yp} – leg coordinate at the moment of a contact with the ground (DOWNWARD motion limit),

o_{ymax} – largest possible leg–end elevation over the ground (UPWARD motion limit),

o_{odl} – distance of the leg–end from an obstacle measured along the direction of motion (it can be the constraint of FORWARD, UPWARD, DOWNWARD motion).

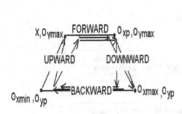

Fig. 13.1. Sets of leg-end motion goals

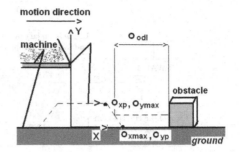

Fig. 13.2. Leg-end trajectory; illustration of motion limits

The current goal choice is based on the analysis of an adequate *decision scheme* reflecting relationship (13.1). Each scheme describes the selection of g^i for the currently realized motion goal $r(g^{i-1})$.

The variables used in the relationship (13.1) have the following meaning:

1. *Global motion goal:*

$$g_f = \{traverse\ the\ path\ segment,\ avoid\ obstacles,$$
$$move\ with\ velocity\ v_{xk},\ statically\ stable\ motion\}.$$

2. *Realized state of walking machine:* (coordinates expressed in the frames affixed to the centers of hip joints)

$$\boldsymbol{p}^{i-1} = (\ x_1^{i-1},\ y_1^{i-1},\ ...\ ,\ x_4^{i-1},\ y_4^{i-1}\ ).$$

3. *Static stability margin:*

$$st_{sens} = \{\ three\ leg-ends\ on\ the\ ground\}.$$

4. *Motion limits:*

$$o\ ,\ o_{sens}\ (s_a\ ) = \{o_{xmin}, o_{xp}, o_{xmax},\ o_{ymax},\ o_{odl}\}.$$

The determination of *motion intention* in_i (rel.(13.2)) is equivalent to the definition of velocity components of each leg-end. In this case rel.(13.2) describes the method of evaluating velocity components of leg motion for the selected current motion goal. The following notation is introduced:

a) $x_{akt},\ y_{akt}$ - current leg coordinates,

b) $v_x^{nameg^i},\ v_y^{nameg^i}$ - components of velocity:
$nameg^i = \{BACKWARD,\ FORWARD,\ UPWARD,\ DOWNWARD\}$

c) $x_{BACKWARD}^{min_1},\ x_{BACKWARD}^{min_2}$ - coordinates of two legs nearest to o_{xmin} standing on the ground.

Two situations are distinguished:

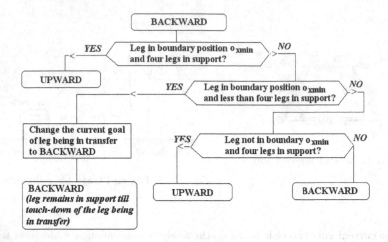

Fig. 13.3. Block diagram for a choice of current motion goal for realized goal – BACKWARD

- two legs supported by the ground, with coordinates $x^{min_1}_{BACKWARD}$ and $x^{min_2}_{BACKWARD}$ not differing much. This results in both legs approaching the o_{xmin} limit nearly simultaneously, and so both would have to be lifted. Such a situation is called a *blockade* – dead-lock (due to the stability requirement the motion of both legs would be blocked). When a blockade is detected, the o_{xmax}, o_{xp} are modified,
- there is *no blockade*.

The variables used in relationship (13.2) have the following meaning:

1. *Motion comfort criterion:*

$$mt = \{move\ with\ constant\ velocity,\ body\ at\ constant\ height\ \}.$$

2. *Forecast static stability* (for the prognosis of statically stable position at least three legs must support the body in the planned motion):

$$st_{pr} = \{static\ stability\ maintained:\ three\ legs\ in\ support\}.$$

3. *Forecasted motion limits;* modified, as discussed above:

$$o_{pr} = \{modified:\ o_{xmin},\ o_{xp},\ o_{yp}, o_{ymax},\ o_{odl},\ blockade(yes/no)\}.$$

The velocity components of the legs remaining on the ground result from the assumed velocity of the machine. The motion intention is determined in each control step, hence in each step the leg-end velocity components are calculated. The method of computing the velocity of the transferred leg is different in the case when the leg is blocked and when it is free. Below only the method of computing the velocity of a not blocked leg will be summarized. If $nameg^i = \{FORWARD,\ UPWARD,\ DOWNWARD\}$ then the time which elapses until the next leg is raised is computed:

$$\delta t = \frac{x_{BACKWARD}^{min_1} - o_{xmin}}{v_{xk}}. \tag{13.4}$$

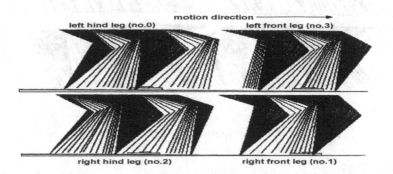

motion direction ────────►

left hind leg (no.0)   left front leg (no.3)

right hind leg (no.2)   right front leg (no.1)

Fig. 13.4. Rhythmic gait; leg transfer sequence is not predefined

If $nameg^i = \{FORWARD, \ UPWARD\}$, then the time to the instant in which the lowering of the leg commences is determined (the instant in which the threshold point o_{xp} is attained):

$$\delta t_p = \frac{o_{xp} - x_{akt}}{v_x^{nameg^i}}. \tag{13.5}$$

The leg–end velocity components are the following:

$$v_x^{nameg^i} = \frac{o_{xmax} - x_{akt}}{\delta t}, \tag{13.6}$$

$$v_y^{nameg^i} = \frac{y_{akt} - o_{ymax}}{\delta t_p}. \tag{13.7}$$

To validate the method a simulation program controlling the motion of a quadruped walking machine was coded. The legs of the machine had to avoid small obstacles. It had been assumed that each leg was equipped with a distance sensor. The machine moves on a horizontal surface, out of the ground are protruding obstacles detectable by sensors. The machine can pass the legs over, but can not step on the obstacle.

The length of trunk was $L = 0.35m$ (units are given only for better reference), width – $W = 0.25m$, leg length (sum of thigh and shank) $= 0.3m$. The body height is kept constant and is equal to $H = 0.24m$, for this height coordinate of support is $o_{yp} = 0.0m$. The leg-end forward shift in relation to the hip (x-coordinate) was limited to $o_{xmax} = 0.066m$, the backward shift was symmetric ($o_{xmin} = -0.066m$. The maximum upward shift was $o_{ymax} = 0.02m$. The maximum forward shift of an upraised leg was $o_{xp} = 0.05m$. With those limits each leg was able to move over a rectangular obstacle of height smaller than $0.02m$ and length smaller than $0.10m$. With this size proportions for the stable walk the body must be supported in every moment by not less than three legs. Despite this, the simulation software checked the static stability condition and the minimal acceptable static stability margin was equal to $W/12$ (the distance between the vertical projection of machines center of

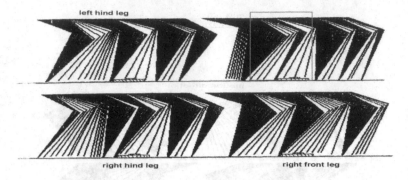

Fig. 13.5. Non-rhythmic gait; obstacles avoiding

mass to the ground and boundary of support polygon measured along the motion direction).

Figure 13.3 shows the applied decision scheme for selection of g^i for $r(g^{i-1})$ being BACKWARD. The change of the current motion goal to BACKWARD for the leg being above the ground (not for the leg for which this scheme will be analyzed) means that during evaluation of the motion intention for this leg the velocity components will be calculated for the shortest possible leg-end support. Figure 13.4 illustrates the history of leg transfers when the obstacles are not present, and Fig. 13.5 illustrates walking in the presence of obstacles. For better comparison the obstacles considered in Fig. 13.5 are marked in Fig. 13.4 (but are ,,transparent" to the machine). From those stick diagrams is possible to notice that without obstacles the step lengths are equal for all the legs. The legs were transferred in a cyclic sequence: 0, 2, 3, 1. This sequence of leg transfers was not specified but resulted only from the initial position, and no external factor (obstacles) changed this sequence. As we can notice the support points are placed where later the obstacles will be located. In the vicinity of obstacles the velocities of the raised legs change considerably, and the legs are transferred in such a way that obstacles are avoided. In the presence of obstacles the leg transfer sequence was not cyclic, e.g. 0, 2, 3, 0, 2, 3, 1, 2, 0, 3, 1, 2, 1, 0,... (non-rhythmic gait) and not predefined. The transfer time depended on obstacle distribution, motion kinematic limits and stability conditions as it was discussed above. The step lengths are different (Fig. 13.5), the legs are placed just before obstacles which is more ,,convenient" for obstacle avoiding (this prevents from going out of motion limits o_{xmax}, o_{xmin} due to the obstacle). The short steps before and after obstacles result from the coordination of the relative positions for all the legs.

The diagrams in Fig. 13.6 are for left front leg shown on right upper part of stick diagrams in Fig. 13.4 and Fig. 13.5. Discussed below fragment of this leg motion including almost two steps with one support phase was put in frame in Fig. 13.5. Figure 13.6a) illustrates this leg-end's velocity components expressed in relation to the control step used in simulation. The control steps are marking flow of simulation time, they can also be directly related to the decision latency and control steps considered in real-time control. The first part of the diagram illustrates the situation when the obstacle was not detected, the leg first moves upward, v_y is positive ($-v_y$ is negative), from the step no.7 the leg starts to move down, till

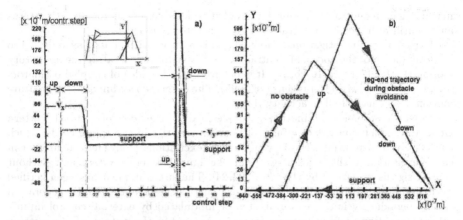

Fig. 13.6. Leg-end velocity v_x, v_y – a), leg-end trajectories with and without avoiding obstacles – b)

step 22 when the leg is supported by the ground (v_y is equal to zero) and leg-end supporting the body moves backward in relation to the hip (v_x is negative). From the step 76-77 starts the leg „reaction" to the detected obstacle, the leg is moved very fast upward (only during 3 steps) next is moves fast downward, to be placed on the ground for step 81. Figure 13.6b) shows fragments of the leg-end trajectory with and without avoiding obstacles. The first trajectory was without obstacle avoiding and the leg-end velocity components along this trajectory were recorded for control steps 0-74 in Figure 13.6b), the velocity components for avoiding obstacle are from step 74. During the motion with obstacles, the leg-end velocity components in transfer phase increase greatly at a certain instants. That is the result of motion synchronization for all the legs. The shape of the leg-end transfer trajectory for the consecutive steps is not similar, it is not pre-planned and results from the obstacle distribution.

13.3 Biped Gait Pattern Generator

The kinematic structures of biped walking machines usually imitate the structure of a human skeleton. Both the gait of biped machines and the human gait are dynamically stable [31, 35]. The scientists elaborating methods of controlling the motion of biped walking machines use diverse models and follow different ways of solving dynamics problems (an interesting overview is presented in [32]. Generally speaking, the solution to the dynamics problem supplies the information about the force and torque transients, which should assure the proper posture (i.e. vertical) and the forward motion of the machine. The desired changes of leg coordinate trajectories result from geometrical considerations, e.g. determination of joint angle changes based on an assumed motion of the device's centre of gravity [11] or the assumption that the motion of leg links resembles the oscillations of an inverted pendulum [21,31], so the machine motion (gait pattern) differs from the human gait pattern. The above mentioned situation induced the author to investigate biped machine gait generators. The author decided to look closely at biological patterns and to utilize the idea of Central Pattern Generator (CPG).

Experimental investigations of living organism locomotion control systems led to the formulation of the notion of Central Pattern Generator generating the oscillatory motion rhythm [7,13,14,16,26,37]. It is believed that the idea of coupled oscillators originated with Christian Huygens (1629-95). The observed rhythm of living creature motion is an effect of CPG activity [13].

Research of biological motion principles led to the idea of Central Pattern Generator, which consists of groups of neurons (usually in the spinal cord), which collectively realize sequences of cyclic muscle excitations [13]. These neurons can cause excitation without feedback from the musco-skeletal system and without control signals generated by the brain [23,24]. The last statement has been verified by experiments with animals having their cortical stem cut. Pattern generation can be periodically initiated, terminated or modulated by external control inputs. Information derived from sensory inputs may modify the output of the pattern generator so as to adapt locomotion to the environment [2]. It is not confirmed whether the leg movements in all animals are driven by coupled centralized biological oscillators. Current models of vertebrate locomotion are predominantly based on the concept of distributed self-organized generators [17, 18, 25]. The investigations of CPG encompass:

a) modeling of possible biological structures of CPG: e.g. [15,37],
b) mathematical models describing the specific operation of CPG (coupling between oscillators, existence of stable states, etc.), [15,18,30],
c) research into properties of real neural structures, e.g. [7],
d) construction of oscillator models generating the gait rhythm and modeling changes in that rhythm: [7,14,29,33].
    Researchers mainly look for different analytical models, which would generate only sequences of leg transfers (pace markers) [7]. Most of the investigations deal with the generation of insect gaits: [7,14]. Quadruped [16,25] and human gait [19,33,41] are also generated. The majority of investigations limit themselves to the generation of rhythm for leg-end motions.

The presented method of gait generation produces a gait pattern similar to the human gait. It uses the idea of a gait generator that generates an output similar to the output of a biological generator, but does not imitate the internal structure of the biological generator.

13.3.1 Model of Coupled Oscillators

Out of the group of the most widespread oscillator models, van der Pol coupled oscillators are investigated as the human locomotion rhythm generators [8,19,41,46].

The equations describing the dynamical properties of oscillators have the following general form:

$$\ddot{x} - \mu \cdot (p^2 - x^2) \cdot \dot{x} + g^2 \cdot x = q. \tag{13.8}$$

The variables μ, p^2, g^2, q influence the properties of oscillators.

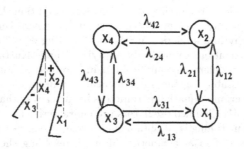

Fig. 13.7. Denotation of angles used by the gait generator and coupling factors between them

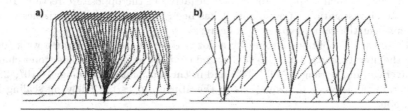

Fig. 13.8. Stick diagram of human gait - a), gait generated by coupled oscillators - b)

Cyclic solutions of those oscillator equations can be interpreted as the values marking the change of thigh or shank angles during walking. The changes in angles during motion are described by the following equations:

$$\ddot{x}_1 - \mu_1 \cdot (p_1^2 - x_a^2) \cdot \dot{x}_1 + g_1^2 \cdot x_a = q_1, \qquad (13.9)$$
$$\ddot{x}_2 - \mu_2 \cdot (p_2^2 - x_b^2) \cdot \dot{x}_2 + g_2^2 \cdot x_b = q_2,$$
$$\ddot{x}_3 - \mu_3 \cdot (p_3^2 - x_c^2) \cdot \dot{x}_3 + g_3^2 \cdot x_c = q_3,$$
$$\ddot{x}_4 - \mu_4 \cdot (p_4^2 - x_d^2) \cdot \dot{x}_4 + g_4^2 \cdot x_d = q_4,$$

where
$$x_a = x_1 - \lambda_{21} \cdot x_2 - \lambda_{31} \cdot x_3, \qquad x_b = x_2 - \lambda_{12} \cdot x_1 - \lambda_{42} \cdot x_4,$$
$$x_c = x_3 - \lambda_{13} \cdot x_1 - \lambda_{43} \cdot x_4, \qquad x_d = x_4 - \lambda_{24} \cdot x_2 - \lambda_{34} \cdot x_3.$$

These equations have 24 parameters: μ_1, μ_2, μ_3, μ_4, p_1^2, p_2^2, p_3^2, p_4^2, g_1^2, g_2^2, g_3^2, g_4^2, q_1, q_2, q_3, q_4, λ_{13}, λ_{31}, λ_{12}, λ_{21}, λ_{24}, λ_{42}, λ_{43}, λ_{34}. The influence of these parameters on the properties of the oscillators is very complex, because of numerous couplings between them.

The exact analytical solution describing the behaviour of the above systems of coupled oscillators is not known. Approximate solutions can be found (e.g. for coupled chemical oscillators). Approximate methods first assume the type of solution and later, by substitution, the parameter values of the assumed solution are found. Approximate solutions are useful in the analysis of the general features of equations, such as the limit cycle or the oscillation generation condition. Unfortunately those solutions do not picture the generated oscillations fully. Due to that, in the generation of gait rhythm, numerical methods of solving equations were used.

We solved the equations numerically applying Runge-Kutta method of second order. Besides the problems arising from using numerical methods (integration scheme, numerical stability), the fundamental condition for properly picturing oscillations is to determine equation parameters in such a way that stable oscillations (limit cycle) result [41]. Coupled oscillators are characterized by many locally stable cycles, the aim of our tuning is to obtain such cycle which resembles human gait.

The scaled values of x_i as a function of time, resolving the coupled equations of these oscillators describe the changes of angles in the hip and knee joints. The angles X_1, X_2, X_3, X_4 (see Fig. 13.7) correspond to the adequately scaled x_1, x_2, x_3, x_4. It is assumed that the angles X_i have positive values if the thigh or shank are in front of their respective axes and negative in the opposite direction. In the presented investigations of oscillators it has been assumed that the variables x_i and X_i represent the values expressed in degrees.

From the data obtained from over a dozen recordings of slow human walk (data from the literature, e.g. [12,38]) it was found that the angles in the knee joint change (on average) in the range $(-30°, 15°)$ and in the hip joint in the range $(-10°, 25°)$. To obtain the solutions of equations in the above ranges the following scaling has been done:

$$X_1 = x_1 * 10.0 - 22.0, \qquad X_2 = x_2 * 5.0 + 30.0, \qquad (13.10)$$
$$X_3 = x_3 * 10.0 - 22.0, \qquad X_4 = x_4 * 5.0 + 30.0.$$

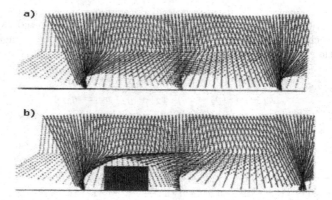

Fig. 13.9. Gait generated for a smooth surface – a), gait with avoiding obstacle – b)

The initial conditions have been determined by using the stick diagrams (side view of the image of leg motion recorded at discrete time instants) and the tables containing the joint angle values recorded during an experiment with a walking human [38]. The following have been assumed:

$$X_1 = 8.8°, \qquad\qquad X_2 = 22.9°, \qquad\qquad (13.11)$$
$$X_3 = -27.0°, \qquad\qquad X_4 = 17.0°.$$

For solution of (13.9) by numerical methods initial angular velocities were approximated by difference quotients of the first order, taking into account the angles at two consecutive instants. In the stick diagram (and table) from which the data was read the limb positions had been recorded every $\frac{1}{120}s$ ($0.008s$). Oscillator equation integration step was set at 0.01. When setting the initial angular velocities, it was assumed that this integration step was equivalent to $0.008s$ time interval. The initial angular velocities were equal to:

$$\dot{X}_1 = -1.0°/s, \qquad\qquad \dot{X}_2 = 1.0°/s, \qquad\qquad (13.12)$$
$$\dot{X}_3 = -31.0°/s, \qquad\qquad \dot{X}_4 = 43.0°/s.$$

The oscillators coupling parameters determine the type of gait – they are the motion coordinators between the joints for both limbs. They influence the phase shift between joint trajectories according to the coupling illustrated in Figure 13.7. After the initial selection of coupling parameters resulting in stable oscillations and in satisfactory coordination of joint motion, the remaining parameters were tuned in such a way that the required gait was obtained – the character of variability of x_i was similar to the changes of knee and hip joint angles [41]. A satisfactory accuracy of human gait estimation was obtained. The joint trajectories and phase diagrams for human and for values generated by the generators were compared to validate the parameters tuning. At that point is must be noted that the mathematical evaluation of estimation accuracy is here not a right approach. In the walk of a healthy person the joint trajectories from step to step can differ 5° or even more under constant conditions. The aim of designing the gait generator is not to get a perfect accuracy, which is not possible. Despite of using some reference data, the target is not to imitate ideally a walk of this person. In the research on gait generators it is expected that the synergy hidden in the changes of joint positions will be „learned". This, between others, is offered by coupled oscillators. Once this kind of synergy is detected, the gait changes similar to real can be obtained. The final confirmation of the correctness of oscillator parameters evaluation was, in our case, the application of generated joint trajectories to the motion animation of biped planar model. The observed gait included short double support phase, and each leg single support phases with equal or almost equal length. The walk was with constant speed adequate to human walking speed, the motion animation resembled well human gait. This all positively validated the parameters tuning. Figure 13.8a) presents stick diagram of human gait (according to the data from [38] and Figure 13.8b) displays gait generated by the discussed coupled oscillators [39].

The relationship between coupling parameters (λ_{ik}) and the gait pattern was analyzed in details. This relation is complex – it is not a linear dependence between the value and the phase shift, but in general the changes in coupling parameters cause changes in the phases of variables x_1, x_2, x_3, x_4 which result in changes of gait type (transfer from slow gait to run, jumps, etc.). The changes of phase plots and joint trajectories depending on other oscillator parameters were identified. The alteration of p_i^2, g_i^2, μ_i influences the shape of the trajectory X_i, and introduces a phase shift, so it is extremely useful in gait generation. The parameters p_i^2, g_i^2, q_i, μ_i, first of all, determine the shape of the trajectory of a single limb. To a lesser extent they influence the type of gait (jumping, pathological gait). During gait rhythm generation these parameters can be altered to correct the limb trajectory tuning it to obstacles or increasing the gait velocity.

By changing oscillator parameters the velocity and type of motion was changed in our tests – e.g. a transition from slow walking to running or jumping was caused [39]. Figure 13.9 contains stick diagrams for rhythmic gait and a gait when one leg avoids obstacles. The change of leg trajectory was obtained by temporary change of λ_{34}, and λ_{43} from 0.2, 0.2 do 0.6, 0.6, the duration of such change was 1/15 of gait period and this change was started just after the appropriate leg detached from the ground.

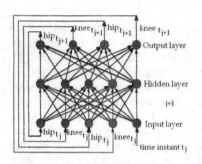

Fig. 13.10. Structure of recursive back-propagation neural network generating the gait

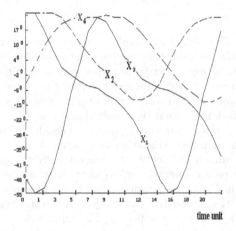

time unit

Fig. 13.11. Real gait: changes of joint angles

Another gait pattern generator was developed using the recursive back propagation artificial neural network with structure shown in Fig. 13.10. As the teaching examples were used the human gait data recorded in known time instants. After the learning process (it is a self-learning network) the network was stimulated to work by applying time to the input and outputs were closed to the inputs. The generated gait was close to human. By the change of selected parameter of threshold functions in hidden layer the human like gait changes were obtained. Figures 13.11 and 13.12 illustrate the typical time course of joint angles during human walk [38] and angles generated by a neural network.

13.3.2 Recursive Formula of Gait Generation Considering Joint Feedback

Gait Rhythm Generator produces a pattern of motion (reference trajectory). The van der Pol equations in differential form can be used when implementing the control law. This formula can be used to produce additional feedback on gait generation level. In humanoidal bipeds the control signals adjusting the robot posture are generated using the sensory feedbacks from leg force sensors. The pure model based

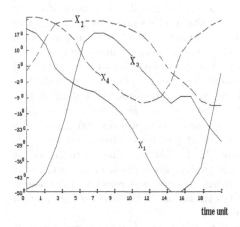

Fig. 13.12. Joint angles generated by artificial neural network

Fig. 13.13. Prototype of NUSBIP-II

control is not possible due to the existence of closed kinematic chain closed by ground in double support phase. The so-called „zero moment point" method [35] utilizing the measurement of reaction forces is widespread in real applications. The advantage of utilization of oscillator based control loop is the recursion of calculations and possible reference to the real (already realized) positions and velocities, this will bring the possibility of disturbed trajectories correction in a similar way as a human does. The oscillator based gait generation recursive formula is:

$$^{i+1}x_j^g = {}^i x_j + {}^i v_j[\Delta t + \Delta t^2 \, {}^i\mu_j({}^i p_j - {}^i x_{con}^2)] + \Delta t^2({}^i q_j - {}^i g_j x_{con}), \qquad (13.13)$$

$^{i+1}x_j^g$ is the angular position of j-th joint, generated by CPG in instant $i+1$, $^i x_j$ is the real (measured) position of j-th joint in instant i, $^i v_j = ({}^i x_j - {}^{i-1} x_j)/\Delta t$ is the angular velocity of j-th joint, Δt is the control step ($t_{i+1} = t_i + \Delta t$), and $^i x_{con} = {}^i x_j - {}^i \lambda_{kj}^i x_k - {}^i \lambda_{lj}^i x_l$, where $con = a, b, c, d$ (see rel. (13.9)).

The formula similarly to the original (13.9) keeps the synergy of joint angle changes. We tested it; the trajectories obtained were as those generated using Runge-Kutta integrating scheme. Next, we simulated the disturbances in joints by adding a white noise to $^{i+1}x_j^g$, and using it as $^i x_j$ (measured value). The generated trajectories were returning properly (human like walking posture was kept) into a regular rhythmic shape. This was tested for disturbances causing the change up 20% of the expected exact value and acting only at one time instant (one generation step), or for a time range not longer than 10% of the gait period.

13.4 Summary

In legged locomotion it is very important to obtain the joint trajectories resulting in well-coordinated displacements of body parts when postural stability is preserved. In multi-legged free gait generation (in free gait the leg transfers are not periodic) diverse methods are used for searching the graphs of possible leg movements, or possible sequences of leg transfers (e.g. [5,6]) for the motions preserving stability and assuring obstacle avoidance. Researchers assume here that the terrain map is known a priori, and gait is pre-planned rather than planned on-line (during motion). The two-stage method of gait generation introduced in this publication can be used in real time considering the actual state of environment. The motion planning for all the legs is performed simultaneously. Therefore the method is convenient for implementation using real-time based parallel computing. Until now only one feature of this method – the variable leg-end support time – was used by the authors in real walking machines for coordination of legs motion in the presence of obstacles. They were hexapods GROVEN following firstly build quadrupeds [42, 44]. All machines were moved by statically stable gaits. Considering a more complete application of the method it must be noticed that the real actuators have acceleration and velocity limits, which complicates the reasoning for leg-end velocity evaluation. In the discussed simulation example only velocity limits were considered.

Concluding the considerations dedicated to the humanoidal gait generation it must be underscored that the gait similar to human gait will result in postural stability when being applied to the robot with kinematics and mass distribution similar to human. But it does not mean that perfect similarity to the human body, human motion actuation, and control is required. With proper limb coordination, a wide range of motions is possible when postural stability is kept – this is true referring not only to humans but to robots – many of them successfully walk with reference trajectories differently produced. The sense of using human reference was proved first by HONDA robots using the recorded human motion data. This also confirms that our approach has applicable potential. It is obvious that true validation of our method is not possible without real device. We performed rough quasi-dynamic simulations using generated trajectories applied to a simplified model of the human body – with two planar legs and a point mass as the remaining body; the masses and size of limb parts, as well as the mass of the body, were taken from the person whose recorded gait was used as a pattern in our research. Considering hard surface, the changes of calculated leg-end reaction forces were similar to those of a human, which confirms that with generated gait the postural stability will be preserved. The coupled oscillators can be used in the control systems as the generators of motion patterns similar to the real gait. The generated limb position prior to execution should be adequately corrected taking into account the actual stability requirements (e.g. measured leg-end reaction forces), and other information gathered by sensors. It is expected that the presented here methods of two-legged gait generation will be applied in humanoidal biped. The robot prototype NUSBIP-II modeled after a 10-years old child is currently under development at NUS–MPE (Fig. 13.13).

References

1. Allemand S, Blanc F, Burnod Y, Dufosse M, Lavayssiere L. *A Kinematic and Dynamic Robotic Control System Based On Cerebro-Cerebellar Interaction Modelling*. In Neural Networks in Robotics. Ed. G.A.Bekey, K.Y.Goldberg. Kluwer Academic Publishers, USA 1993.

2. Arena P. *The Central Pattern Generator: a Paradigm for Articficial Locomotion.* Soft Computing 4, pp.251-265, 2000.
3. Ayers J, Zavracky P, McGruer N, Massa D, Vorus V, Mukherjee R, Currie S. *A Modular Behavioral-Based Architecture for Biomimetic Autonomous Underwater Robots.* Proc. of the Autonomous Vehicles in Mine Countermeasures Symposium. USA, 1998.
4. *Two-legged Walking* (in Russian), Moscov, Science, 1884
5. Shaoping Bai, Low K.H., Zielinska T.:*Quadruped free gait generation based on the primary/secondary gait.* ROBOTICA Journal, vol.17, 1999, pp.405-412
6. Shaoping Bai, K.H.Low, T.Zielinska: *Quadruped free gait generation for straight-line and circular motion.* Advanced Robotics Journal. vol.13, no.5, pp.513-538, 1999
7. Bassler U. *The Walking- (and Searching) Pattern Generator of Stick Insects, a Modular System Composed of Reflex Chains and Endogenenous Oscillators.* Biological Cybernetics, vol.69, pp.305-317, 1993.
8. Bay J.S, Hemami H. *Modelling of a Neural Pattern Generator with Coupled Nonlinear Oscillators.* IEEE Trans. on Biomedical Engineering, Vol. BME-34, No.4, pp.297-306,1987.
9. Brooks R.A, Flynn A.M. *Fast, Cheap and Out of Control: A Robot Invasion of the Solar System.* J. of the British Interplanetary Society, vol.42, pp.478-485, 1989.
10. Brooks R.A. and Stein L.A. *Building Brains for Bodies, Autonomous Robots.* vol.1, pp.7–25, 1994.
11. Budanov V.M, Lavrowsky E.K. *Design of the Program Regime for Biped Walking Antropomorphic Apparatus.* Theory and Practice of Robots and Manipulators. *Proc. of Ro.Man.SY'10.* Ed. Morecki A., Bianchi G., Jaworek K., pp.379-386. Springer Verlag 1995.
12. Capozzo A., Marchetti M., Tosi V. eds. *Biolocomotion: A Century of Research Using Moving Pictures.* Int. Society of Biomechanics Series - vol.1, Promograph, Rome, Italy 1992.
13. Cohen A.H., Rossignol S., Griller S. *Neural Control of Rhythmic Movements in Vertebrates.* John Wiley and Sons, New York, Chichester, Brisbane, Toronto, Singapore 1988.
14. Collins J.J, Steward I. *Hexapodal Gaits and Coupled Nonlinear Oscillator Models.* Biological Cybernetics, vol.68, pp.287-298, 1993.
15. Collins J.J, Stewart I. *A Group-Theoretic Approach to Rings of Coupled Biological Oscillators.* Biological Cybernetics, vol.71, pp.95-103, 1994.
16. Collins J.J, Richmond S.A. *Hard-wired Central Pattern Generators for Quadrupedal Locomotion.* Biological Cybernetics, vol.71, 375-385, 1994.
17. Cruse H, Kindermann Th, Schumam M, Dean J, Schmitz J. *Walknet – a Biologically Inspired Network to Control Six-legged Walking.* Neural Network 11, pp.1435-1447, 1998.
18. Cruse H. *The Functional Sense of Central Oscillations in Walking.* Biological Cybernetics, 86, pp.271-280, 2002.
19. Dutra M.S, de Pina Filho A.C, Romano V.F. *Modelling of a Bipedal Locomotor Using Coupled Nonlinear Oscillators of Van der Pol.* Bilogical Cybernetics, 88, pp.286-292. 2003.
20. Fagg A.H. *Developmental Robotics: A New Approach to the Specification of Robot Programs.* In *Neural Networks in Robotics.* Ed. G.A.Bekey, K.Y.Goldberg. Kluwer Academic Publishers, USA 1993.

21. Formalsky A.M. *Impulsive Control for Antropomorphic Biped.* Theory and Practice of Robots and Manipulators. *Proc. of RoManSy'10.* Ed. Morecki A., Bianchi G., Jaworek K, pp.387-394. Springer Verlag 1995.

22. Gao X.C, Song S.M, Zheng C.Q. *A Generalized Stiffnes Matrix Method for Force Distribution of Robotics System with Indeterminancy. Journal of Mechanical Design,* vol.115, no.3, pp.585-591, 1993.

23. Grillner S. *Control of Locomotion in Bipeds, Tetrapods and Fish. Hanbook of Physiology.* Ed. by Brookhat J.M., Mountcastle V.B., 1179-1236, American Physiological Society, 1981.

24. Kandel E.R, Schwartz J.H, Jessel T.M. *Principles of Neural Science.* Elsevier, New York 1991.

25. Kaske A, Winberg G, Coster J. *Emergence of Coherent Traveling Waves Controllong Quadruped Gaits in a Two-dimensional Spinal Cord Model. Biological Cybernetics,* 88, pp.20-32, 2003.

26. Kimura S, Yano M, Shimizu H. *A Self Organizing Model of Walking Patterns of Insects. Biological Cybernetics,* vol.69, pp.183-193, 1993.

27. Kooij H, Jacobs R, Koopman B, Helm F. *An Alternative Approach to Synthesizing Bipedal Walking. Biological Cybernetics,* 88, pp.46-59, 2003.

28. Massone L.L.E. *A Biollogically-Inspired Architecture for Reactive Motor Control.* In *Neural Networks in Robotics.* Ed. G.A.Bekey, K.Y.Goldberg. Kluwer Academic Publishers, USA 1993.

29. Muller-Wilm U. *A Neuron-Like Network with the Ability to Learn Coordinated Movement Patterns. Biological Cybernetics,* vol.68, 519-526, 1993.

30. Nishii J., Uno Y, Suzuki R. *Mathematical Models for the Swimming Pattern of a Lamprey. Part I: Analysis of Collective Oscillators with Time Delayed Interaction and Multiple Coupling. Part II: Control of the Central Pattern Generator by the Brainstem. Biological Cybernetics,* vol.72, pp.1-9, pp.11-18, 1994.

31. Raibert M.H, Brown H.B, Murthy S.S. *3-D Balance Using 2-D Algorithms? The Int. Journal of Robotics Research,* no.1, MIT Press, pp.215-224, 1984.

32. Shuuji Kajita, Tomio Yamaura, Akira Kobayashi *Dynamic Walking Control of a Biped Robot Along a Potential Energy Conserving Orbit. IEEE Trans. on Robotics and Automation,* vol.8, no.4, pp.431-438, 1992.

33. Taga G, Yamaguchi Y, Shimizu H. *Self-Organized Control of Bipedal Locomotion by Neural Oscillators in Unpredictable Environment. Biological Cybernetics,* vol.65, pp.147-159, 1991.

34. Todd D.J. *Walking Machines: An Introduction to Legged Robots.* Kogan Page, 1985.

35. Vukobratovic J. *Legged Locomotion and Anthropomorphic Mechanism.* Michailo Pupin Institute. Beograd 1975.

36. Vaughan Ch.L., Davis B.L., O'Connor J. *Dynamics of Human Gait.* Human Kinetics Publishers, USA 1992

37. Willer B.E, Miranker W.L. *Neural Organization of the Locomotive Oscillator. Biological Cybernetics,* vol.68, pp.307-320, 1993.

38. Winter D.A. *Biomechanics of Human Movement.* John Wiley, New York 1979.

39. Zielińska T. *Utilization of Human and Animal Gait Properties in the Synthesis of Walking Machines Motion* (in Polish). Polish Academy of Sciences, no.40, 1995.

40. Zieliński C., Zielińska T. *Sensor-Based Reactive Robot Control*, Proc. 10-th CISM-IFToMM Symp. on Theory and Practice of Robots and Manipulators, *RoManSy'94*, Ed. A. Morecki and G. Bianchi and K. Jaworek, Springer Verlag, 1995.
41. Zielińska T. *Coupled Oscillators Utilised as Gait Rhythm Generators of a Two-Legged Walking Machine*. *Biological Cybernetics* 74, pp. 263-273, 1996.
42. Zielińska T., Heng J. *Development of walking machine: mechanical design and control problems*. *Mechatronics*, 12, pp. 37-754, 2002.
43. Zieliński C. *By How Much Should a General Purpose Programming Language be Extended to Become a Multi-Robot System Programming Language?* *Advanced Robotics*, Vol. 15 no. 1, pp. 71–95, 2001.
44. Zielińska T., Heng J.: *Mechanical design of multifunctional quadruped*. Mechanism and Machine Theory. Pergamon. No. 38, 2003, pp. 463-478
45. Zielińska T.: *Walking Machines. Fundamentals, design, control and biological patterns*. Polish Scientific Edition PWN 2003. ISBN 83-01-13925-0 (in Polish)
46. Zielińska T.: *Motion Synthesis*. In: *Walking: Biological and Technological Aspects, CISM Courses and Lectures n. 467*. Ed. by F. Pfeiffer, T.Zielinska. Springer Verlag 2004, pp. 151-187

14

Control of an Autonomous Climbing Robot

Carsten Hillenbrand, Jan Koch, Jens Wettach, and Karsten Berns

Robotics Research Lab, University of Kaiserslautern, Germany

14.1 Introduction

Climbing robots using suction cups as the adhesion mechanism are examined in several projects worldwide. Examples of their application are window cleaning, painting, inspection of large concrete walls or steel tanks (see for example [1,4,5,7,9]). The wheel-driven or legged machines that are developed in these projects are able to move on vertical walls or overarm in normal situations, but their reliability is far from reaching 100%. For, in most cases, the reasons why the robot can move over a specific surface and why it falls down in nearly similar situations are not obvious. Seeking for an optimal climbing strategy one has to answer the following questions:

- How can the closed-loop control generate enough forces to hold the machine on the wall without preventing the sliding movements of the suction cups over the surface?
- In case of complete loss of vacuum, how long will it take to generate the needed negative pressure again (based on the characteristic of the vacuum engine)?
- How will a surface crack of a special size change the negative pressure of a suction cup?

The specific subject here is a climbing robot for the autonomous inspection of concrete walls of bridges and dams. It is wheel-driven because of the requirements of the inspection sensor system (see [8]) and its vacuum system consists of seven chambers as a result of theoretical preliminary examinations. Besides the adhesion mechanism, also the driving and navigation system have to be developed with regard to autonomously inspecting large concrete surfaces.

In Section 14.2, the thermodynamical fundamentals are presented that describe the changes of the negative pressure under specific surface conditions. This model serves as basis for a simulation tool of the seven chamber adhesion system, which calculates the actual pressure situation depending on the leakages of the sealing and the opening of valves connecting the chambers to the suction engine (detailed description can be found in [2] and in [9]). Afterwards, the controller for the vacuum chambers is introduced. Section 14.3 starts with a description of the omnidirectional wheel, that is used for the three wheel triangular drive of the robot. Then the inertial sensor system for solving the navigation task is introduced. The paper ends with

K. Kozłowski (Ed.): Robot Motion and Control, LNCIS 335, pp. 221–232, 2006.
© Springer-Verlag London Limited 2006

some tests of a one-chamber prototype, used for pressure measurements while moving on vertical walls and over different surfaces.

14.2 Closed-loop Control of the Vacuum System

In the following the thermodynamical model of the adhesion system and the control strategy for dependably generating a suitable vacuum situation are developed.

14.2.1 Dynamic Model of the Adhesion System

For examining the pressurization processes in the vacuum chambers, the pressure change due to air flows from or to the outside (another chamber or the environment) and the opening areas (valves and leakages) have to be modelled. Then the resulting pressure force that keeps the robot on the wall can be derived.

State Change of Gases and Modelling of Pressure Changes in a Vacuum Chamber

The first fundamental theorem of thermodynamics is:

$$\underbrace{\frac{d}{dt}(U + E_{mech.})}_{\text{change in the system}} = \underbrace{\sum_i \dot{W}_i + \sum_j \dot{Q}_j + \sum_k \dot{m}_k \left(h_k + \frac{c_k^2}{2} + gz_k \right)}_{\text{energies over the system limits}},$$

where U is the internal energy of the gas, E_{mech} is the mechanical energy of the gas volume (i.e. position, velocity, ...), $\dot{W}_i$ is the power of volume work i, $\dot{Q}_j$ is the heat capacity j, $\dot{m}_k$ means mass flow k, h_k is the specific enthalpy of $\dot{m}_k$, $\frac{c_k^2}{2}$ is the specific kinetic energy of $\dot{m}_k$ and gz_k is the specific potential energy of $\dot{m}_k$. In this application, the system is a vacuum chamber filled with air which is in contact with other chambers or the environment by sealings and chamber walls (i.e. system limits). From here on, the following simplifications and approximations are assumed:

$\frac{d}{dt}E_{mech.} \approx 0$, $\dot{m}_k gz_k \approx 0$, $\frac{c_{k,max}^2}{2} \ll h_k \Rightarrow \frac{c_k^2}{2} \approx 0$, $V_{chamber} = \text{const} \Rightarrow \sum_i \dot{W}_i = 0$, $T_{chamber} = \text{const} \approx T_{environment} \Rightarrow \sum_j \dot{Q}_j = 0$. With $\frac{d}{dt}U = c_v(\dot{m}T + m\dot{T})$, $pV = mRT \Rightarrow p\dot{V} + V\dot{p} = R\left(\dot{m}T + m\dot{T}\right)$, $h = c_P T$, and $\kappa = \frac{c_P}{c_v}$ (adiabatic exponent) the following holds:

$$\dot{p} = \frac{\kappa RT}{V} \sum_k \dot{m}_k. \tag{14.1}$$

Equation (14.1) describes the change of pressure in one chamber of volume V, filled with air of temperature T, adiabatic exponent κ_{air} and gas constant R_{air} caused by k mass flows $\dot{m}_k$ over the chamber limits.

Airflow Through a Control Area

Leakages in the chamber sealing and opening of the evacuation valve cause mass flows from or to the chamber. These air flows can be modelled by an air tube connecting the control volumes (for example the chamber and the environment). Figure 14.1 (a) illustrates this model: here c is the velocity of the air, A is the control area, p is the air pressure in A and t is the time stamp at which one air particle is in A.

With BERNOULLI's equation and neglecting the gravity pressure the flow of a medium with density ρ between A_0 and A_1 is

$$\frac{1}{2}\left(c_1^2 - c_0^2\right) + \int_{p_0}^{p_1} \frac{1}{\rho} dp + \int_0^1 \frac{\partial c\left(t,s\right)}{\partial t} ds = 0, \tag{14.2}$$

where p_0, p_1 and c_0 are known and c_1 has to be found. Assuming a stationary flow the inertial term $\int_0^1 \frac{\partial c(t,s)}{\partial t} ds$ can be neglected. Besides, the error introduced by assuming $\rho = const$ is about 6% for the relevant pressure differences between A_0 and A_1. As in each control volume there obviously exists a place where the air flow is 0 $(c_{air} = 0)$, it can also be assumed that $c_0 = 0$, which leads to

$$c_1 = \text{sgn}(p_0 - p_1) \cdot \sqrt{2\frac{\Delta p}{\rho}}, \tag{14.3}$$

where the sign of the difference (sgn-function) gives the direction of the air flow and the argument of the square root only takes the absolute value $\Delta p = |p_0 - p_1|$.

With $\dot{m} = \rho \cdot c_\perp \cdot A$ for a mass flow of velocity c perpendicular to control area A of a medium with density ρ and Eq. (14.3) – assuming $c_1 = c_\perp$ – the following holds:

$$\dot{m}_{incompressible} = \text{sgn}(p_0 - p_1) \cdot A_{in} \sqrt{2\rho\Delta p}. \tag{14.4}$$

Equation (14.4) describes the mass flow through a control area A_{in} caused by a pressure difference $\Delta p = |p_0 - p_1|$ between the control volumes connected by A_{in}.

Combined with Eq. (14.1) this yields

$$\dot{p} = \frac{\kappa RT}{V} \sum_k \left(\text{sgn}(p_{k_0} - p_{k_1}) \cdot A_{k_{in}} \sqrt{2\rho\Delta p_k}\right), \tag{14.5}$$

which describes the change of pressure in one vacuum chamber due to all incoming and outgoing mass flows.

Modelling of Leakage Areas in the Sealing

Airflow between a vacuum chamber and the environment is caused by leakages in the chamber's sealing. Concerning the robot's application area (concrete walls of buildings, bridges, and dams) the causes of leakages are grooves, holes, pores, cracks, steps between two concrete slabs and protrusions caused by out-pouring of concrete between paling-boards.

All these phenomena are summarized in the model as a crack of a certain width and depth, which is of infinite length and bounded by two parallel lines.

As the roughness of concrete and sealing surface causes a latent leakage, these "basic leakages" are approximated by a crack along the chamber edge of depth 0.1

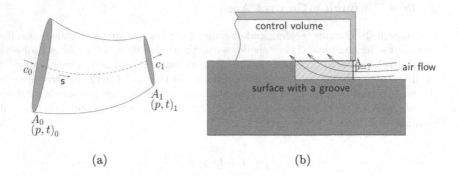

Fig. 14.1. Modelling of airflow by an *air tube* (a) and one possible leakage situation and corresponding flow paths (b). A is the effective leakage area, p is the pressure in A

mm (resulting in a leakage area of $l_{edge} \cdot 0.1$ mm). The leakage area caused by an arbitrary crack depends on the geometry of the crack and the base area of the chamber which has contact with the crack. As seen before, the effective leakage area (for calculating the pressure changes in the chamber according to Eq. (14.5)) must be perpendicular to the velocity of the air and therefore to the flow paths. One "possible" leakage area A, which is perpendicular to the flow path and parallel to the chamber wall is shown in Fig. 14.1 (b). For all following experiments this leakage area model is used.

Now all parameters in Eq. (14.5) are known since all leakage areas $A_{k_{in}}$ can be calculated, the valve areas are actively controlled and the pressures can be measured by sensors.

14.2.2 Realization of the Control System

The structure of the implemented control system, consisting of two control levels, is shown in Fig. 14.2. The top level control module is called *pressure force determination & driving strategy* and gets the overall forces and torque moments that affect the robot (i.e. gravity force, load of the manipulator) and its velocity as inputs. From these inputs the necessary value and working point of the total pressure force are calculated in order to compensate the forces and torque moments for keeping the robot dependably on the concrete surface. If the total pressure force is not as high as required the driving torques for the wheels must be reduced and eventually the moving path of the robot must be adapted. For the validation of the simulation system, a simple force model is used. This model will be refined with respect to different environmental conditions as the next step, but this task is decoupled from the assessment of the adhesion system. On the next level the *leakage estimation* module calculates the leakage area of each vacuum chamber based on the actual chamber pressures and valve areas. For this reason, Eq. (14.5) is transformed in order to calculate the opening area of a chamber to the environment based on the change of pressure $\dot{p}$ during the last control cycle and the pressure difference Δp between the inside and the outside of the chamber.

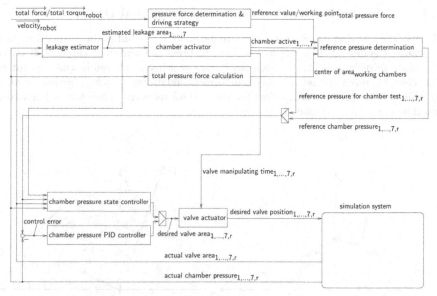

Fig. 14.2. Hierarchical structure of the control system

The *determination of active chambers* module computes which chambers cannot be evacuated any more due to a too big estimation value for their leakage area. Finally the *total pressure force calculation module* determines from the chamber pressures the actual value and working point of the total pressure force that affects the robot. From all these inputs the *reference pressure determination* module calculates the reference pressures for each working chamber, where non-active chambers get the outside pressure as the target. Naturally, the "lost" chambers prevent the control system from generating the desired pressure force. On the bottom level a PID based control loop calculates the reference valve areas for each vacuum chamber from the difference between the desired and the actual chamber pressures. The *valve actuator* module translates each valve area in a valve position as the input for the *plant simulation*, where the *valve simulation* module computes the new valve position and the corresponding valve area based on the reference input and the elapsed time (because a valve cannot be opened in "zero time"). The *vacuum system simulation* module calculates the new chamber pressures based on Eq. (14.5) from the actual valve areas. These values are the outputs of the plant simulation and come from a pressure sensor in each chamber in case of the real robot.

14.3 Drive and Navigation

In this section first the inverse kinematics of the omnidirectional driving wheel are developed. Afterwards the inertial navigation system is introduced: it is used for the correction of pose measurements derived from odometric data.

14.3.1 Kinematic Model of the Omnidirectional Drive

Beside the vacuum system there has to be a drive to fulfill the inspection task. Because it is difficult to generate the force to press the wheels to the wall, it makes no sense to develop a drive system with passive wheels. Therefore the concept for the climbing robot consists of three omnidirectional driven wheels. The disadvantage of this drive is the increased effort for the control software. Therefore a kinematics module must be composed.

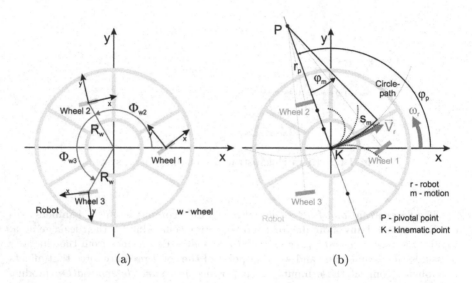

Fig. 14.3. Geometry description of the wheels (a) and the motion path of the robot as a circle (b)

The inputs are the velocity vector $\mathbf{v}_r$ and the rotation velocity ω_r of the robot (r). The given parameters are the radius R_w and the angle Φ_{wx} of the position of each ($x = 1,2,3$) omnidirectional wheel (w). The task of the inverse kinematics is the calculation of the velocity $v_{w1,2,3\_x}$ and the orientation $\varphi_{w1,2,3}$ of each wheel as a function of the inputs $f(\mathbf{v}_r, \omega_r)$. The path which the robot will drive can be described as a circle path with a pivotal point P. This point must be on the line that is perpendicular to the direction of the velocity vector $\mathbf{v}_r$ (see Fig. 14.3), because the given velocity represents the tangential direction of the circle at the kinematic point K.

The possible paths are dotted; the right one is defined by the rotation velocity ω_r. If the robot drives for a certain time Δt, the driven distance is $s_m = \Delta t \cdot |\mathbf{v}_r|$ and the angle of the segment of the circle is $\varphi_m = \Delta t \cdot \omega_r$, because the angular velocity of the robot itself is the same as the rotation velocity around the rotating point P. The radius for the path can be calculated by

$$r_p = \frac{s_m}{\varphi_m \cdot \left(\frac{\pi}{180}\right)} = \frac{\Delta t \cdot |\mathbf{v}_r|}{\Delta t \cdot \omega_r} \cdot \frac{180°}{\pi} = \frac{|\mathbf{v}_r|}{\omega_r} \cdot \frac{180°}{\pi} \qquad (14.6)$$

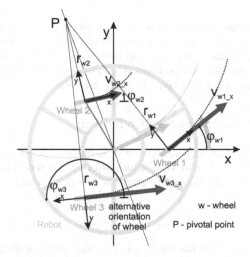

Fig. 14.4. Calculation of the wheel orientation

and the angle to the center point P by

$$\varphi_p = \arctan\left(\frac{v_{r\_y}}{v_{r\_x}}\right) + 90°.$$ (14.7)

With the calculated pivotal point the velocity and the orientation of each wheel can be calculated using the geometrical formulas (14.8), (14.9) and (14.10) (see Fig. 14.4):

$$v_{w1,2,3\_x} = |\mathbf{v}_r| \cdot \frac{r_{w1,2,3}}{r_p} = \frac{|\mathbf{v}_r|}{|\mathbf{v}_r|} \cdot r_{w1,2,3} \cdot \omega_r \cdot \frac{\pi}{180°} = r_{w1,2,3} \cdot \omega_r \cdot \frac{\pi}{180°},$$ (14.8)

$$r_{w1,2,3} = \sqrt{R_w^2 + r_p^2 - 2R_w r_p \cos\left(|\Phi_{w1,2,3} - \varphi_p|\right)},$$ (14.9)

$$\varphi_{w1,2,3} = \arcsin\left(\frac{r_p}{r_{w1,2,3}} \cdot \sin\left(|\Phi_{w1,2,3} - \varphi_p|\right)\right) + \Phi_{w1,2,3} + 90°.$$ (14.10)

Unfortunately there are two special cases for this calculation. In one case the robot is driving straight ahead without any rotation ($\omega_R = 0$). The wheels are pointing at the same direction and have the same velocity, but the result for r_p from Eq. (14.6) is infinite and therefore not reasonable. By examining the equations it can be seen that there exists the solution

$$\omega_r = 0 \;\Rightarrow\; \frac{r_{w1,2,3}}{r_p} = 1 \;\Rightarrow\; v_{w1,2,3\_x} = |\mathbf{v}_r| \;;\; \varphi_{w1,2,3} = \varphi_p + 90°.$$ (14.11)

In the other case the robot is only rotating around itself and there exists no velocity ($|\mathbf{v}_R| = 0$). This means that the wheels are tangential to a circle with its origin in the middle of the robot (P and K are at the same position). Also in this case a solution of the equations can be found:

$$|\mathbf{v}_r| = 0 \;\Rightarrow\; r_p = 0 \;;\; r_{w1,2,3} = R_w \;;\; \frac{r_p}{r_{w1,2,3}} = 0 \;;$$

$$\Rightarrow v_{w1,2,3\_x} = R_w \cdot \omega_r \cdot \frac{180°}{\pi} \; ; \; \varphi_{w1,2,3} = \Phi_{w1,2,3} + 90°. \qquad (14.12)$$

With the presented equations the inverse kinematics can be calculated. For odometric information the system needs additionally the direct kinematics to calculate the direction and velocity of the robot from the existing orientations and velocities of the wheel. As this problem is over-determined, it has to be solved by an optimization strategy. This is part of current research.

14.3.2 Pose Measurement for Navigation

Climbing robots can localize themselves on two sources. Global data can be obtained by landmarks, map matching or GPS information. Dead-reckoning data is acquired by measuring and calculating the movement of the robot itself.

Inertial Forces

To improve trajectory information from the odometry an inertial measurement system is used, see Fig. 14.5. Acceleration forces and angular velocity information can help to determine the trajectory of a system with a known initial state of motion, in theory. We chose the micro-mechanical sensors ADXRS150 and ADXL203 for angular rate and acceleration. After analog digital conversion of the sensor data the inertial calculations are done on a Motorola 56F803. Results are sent via a CAN bus.

Fig. 14.5. Inertial measurement unit

Algorithm

Our hardware provides discrete information on angular velocity and acceleration forces. The angular velocity information is mandatory to calculate the attitude of the robot. The attitude is required to determine the direction of the measured accelerations **a** as well as the direction of the speed from the odometry, see Fig. 14.6. The matrix B_w contains the attitude information of the system in the form of the unit vectors of the body frame from the view of the world frame. The commonly used strap-down approach, described in [6], is modified in this point. Using the unit matrix of the body frame shows the advantage that the calculation of the Euler angles has only to be done when needed [3]. The angle change σ in time step k is deduced by trapezoid integration of the current and last angular velocity measurement $\boldsymbol{\omega}$. Afterwards the angle $\boldsymbol{\sigma}$ is transformed to the world frame:

$$\boldsymbol{\sigma}_b^k = t^k \left(\boldsymbol{\omega}_b^k + \boldsymbol{\omega}^{k-1} \right) / 2, \tag{14.13}$$

$$\boldsymbol{\sigma}_w^k = B_w^k \boldsymbol{\sigma}_b^k. \tag{14.14}$$

The attitude information in B_w is updated by multiplication with a rotation tensor T. The rotation axis represented by this tensor is equal to $\boldsymbol{\omega}$. The rotation angle itself is derived as the length of $\boldsymbol{\omega}$:

$$B_w^{k+1} = T_{\boldsymbol{\sigma}^k, |\boldsymbol{\sigma}^k|} \cdot B_w^k. \tag{14.15}$$

B_w transforms from the body coordinate frame to the world frame.

Multiplying the measured acceleration vector with the matrix B_w translates it to the world frame. Continuous integration leads to velocity and again to position. But first the force of gravity has to be removed:

$$\mathbf{g}_w = \begin{pmatrix} 0 \\ 0 \\ 9,80665 \, \text{m/s}^2 \end{pmatrix}, \tag{14.16}$$

$$\mathbf{a}_w = B_w \mathbf{a}_b - \mathbf{g}_w, \tag{14.17}$$

$$\mathbf{v}_w^k = \mathbf{v}_w^{k-1} + t^k \left(\mathbf{a}_w^k + \mathbf{a}_w^{k-1} \right) / 2, \tag{14.18}$$

$$\mathbf{s}_w^k = \mathbf{s}_w^{k-1} + t^k \left(\mathbf{v}_w^k + \mathbf{v}_w^{k-1} \right) / 2. \tag{14.19}$$

Here we have a heavy source of errors as a wrong attitude causes an incorrect acceleration vector.

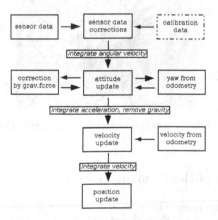

Fig. 14.6. Steps of the inertial calculations

Navigational applications typically need the orientation as three sequential rotations by the angles ϕ, θ and ψ around the world axes x, y, and z, or the body axes in opposite order (c: cos, s: sin):

$$D = \begin{pmatrix} c\psi c\theta & -s\psi c\phi + c\psi s\theta s\phi & s\psi s\phi + c\psi s\theta c\phi \\ s\psi c\theta & c\psi c\phi + s\psi s\theta s\phi & -c\psi s\phi + s\psi s\theta c\phi \\ -s\theta & c\theta s\phi & c\theta c\phi \end{pmatrix}. \tag{14.20}$$

By comparing coefficients in B_w these angles can be determined:

$$\phi = \arctan\left(\frac{b_{32}}{b_{33}}\right), \tag{14.21}$$

$$\theta = \arcsin(-b_{31}), \tag{14.22}$$

$$\psi = \arctan\left(\frac{b_{21}}{b_{11}}\right). \tag{14.23}$$

Using several heuristics like correcting the attitude by comparing the measured and expected gravitation leads to reduced positioning errors.

14.4 Experiments and Results

In this section the experiments performed with a one-chamber prototype robot are described and analyzed. This robot, shown in Fig. 14.7, has been built for the validation of the thermodynamical model and the corresponding simulation results and in order to test the control strategy while moving on a vertical wall. As the prototype and the validation tests are described in detail in [9], in the following only the wall driving tests are shown.

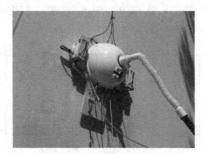

Fig. 14.7. Robot prototype driving on a vertical concrete wall

14.4.1 Validation of the Control System

First wall-driving experiments have been performed in the lab with the robot put on a vertically oriented smooth concrete plate. As could be anticipated from the results of pressure measurements on different surface conditions (see [9]) the sucking is no problem, but the sealing friction prevents the robot from moving reasonably. Wrapping with sellotape reduces the friction sufficiently and the robot moves quite well on the plate (sealing hose pressure: 17000 Pa, chamber pressure: 4000 Pa).

The control strategy has been tested outside on a common concrete wall of a building (see Fig. 14.7). Figure 14.8 shows the recorded pressure during this test.

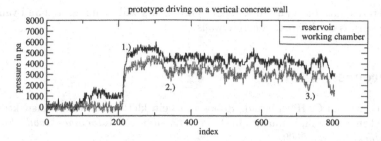

Fig. 14.8. Pressure situation in working and reservoir chamber of the prototype driving on a vertical concrete wall (index = time of measurement)

At point 1.) the suction engine is switched on and a suitable reference value for the total pressure force is given as input to the control system. According to the control strategy described in Section 14.2.2 the valves of the vacuum chambers are opened and the chamber pressures are adjusted. At point 2.) the pressure situation is as desired and the robot is able to start moving. During the run the chamber pressures change due to the dynamic leakage caused by the irregular concrete surface, and the valves are adjusted accordingly. At point 3.) the robot reaches a horizontal crack of about 1cm height and depth which causes a leakage that cannot be compensated with the single vacuum chamber: the control system is no longer capable to provide the needed pressure force, so the robot cannot clear the crack. At least the produced negative pressure is sufficient for keeping the robot on the wall, hence it does not fall down, but only slides downwards until the pressurization is suitable again.

The test run shows that the pressure situation can be controlled with the described strategy under normal surface conditions quite well. The pressure losses due to cracks and grooves are a challenge for the multi-chamber system, which is currently under construction.

14.5 Summary and Outlook

In this paper a climbing robot for autonomous inspections of large vertical concrete surfaces has been presented and its main components - the adhesion mechanism, the driving system and the inertial navigation unit - have been introduced. For keeping the robot dependably on the wall, a seven-chamber vacuum system has been developed and analyzed by a sound thermodynamical model. The main advantage of having multiple chambers are the capability to react to dynamically changing leakage situations by (de-)activating the affected chambers. The control strategy that exploits this fact has been introduced and validated during test runs with a prototype climber.

Besides, the kinematics for the triangular drive, consisting of three omnidirectional wheels, have been developed. The mechanical setup of this system as well as a suitable control algorithm are current research tasks.

Finally, the inertial system for solving the navigation and pose estimation problem has been presented. It will be integrated in the robot as soon as it is

232   C. Hillenbrand et al.

completely built. Then all the software components will be integrated and validated
with test runs on different concrete walls.

References

1. K. Berns and C. Hillenbrand. Robosense - ein kletterroboter zur inspektion von
   brücken und staudämmen. In *Informatik aktuell, Autonome Mobile Systeme,
   Springer Verlag*, 2000.
2. C. Hillenbrand, J. Wettach, and K. Berns. Simulation of climbing robots using
   underpressure for adhesion. In *7th International Conference on Climbing and
   Walking Robots (CLAWAR)*, Madrid, Spain, September 22-24 2004.
3. J. Koch, C. Hillenbrand, and K. Berns. Inertial navigation for wheeled robots in
   outdoor terrain. In *5th IEEE Workshop on Robot Motion and Control (RoMoCo)*,
   Dymaczewo, Poland, June 23-25 2005.
4. D. Longo and G. Muscato. Design of a single sliding suction cup robot
   for inspection of non porous vertical wall. In *ISR 2004: 35th International
   Symposium on Robotics. Proceedings.*, Paris, France, March 2004. International
   Federation of Robotics.
5. R. D. Schraft and F. Simons. Concept of a miniature window cleaning robot -
   development potentialities for a mass product. In *ISR 2004: 35th International
   Symposium on Robotics. Proceedings.*, Paris, France, March 2004. International
   Federation of Robotics.
6. D. H. Titterton and J. L. Weston. *Strapdown inertial navigation technology*. Peter
   Peregrinus Ltd. London, 1997.
7. S. K. Tso, J. Zhu, and B. L. Luk. Prototype design of a light-weight climbing
   robot capable of continuous motion. In *8th IEEE Conference on Mechatrinics
   and Machine Vision in Practice*, pages 235 – 238, Hong Kong, 2001.
8. F. Weise, J. Köhnen, H. Wiggenhauser, C. Hillenbrand, and K.Berns.
   Non-destructive sensors for inspection of concrete structures with a climbing
   robot. In *CLAWAR 2001, Karlsruhe, Germany, September 2001*, 2001.
9. J. Wettach, C. Hillenbrand, and K. Berns. Thermodynamical modelling and
   control of an adhesion system for a climbing robot. In *20th IEEE International
   Conference on Robotics and Automation (ICRA)*, Barcelona, Spain, April 18-22
   2005.

15

Computation of Optimum Consumption of Energy for Anthropomorphic Robot Driven by Electric Motor

Houtman P. Siregar[1] and Yuri G. Martynenko[2]

[1] Mechanical Engineering Department, Faculty of Industrial Technology
Indonesia Institute of Technology, Jl. Raya Puspiptek-Serpong-Tangerang 15320,
Indonesia, phone/fax (021)7561091 Aspirantsir@yahoo.com
[2] Department of Theoretical Mechanics, Moscow Power Engineering Institute
Ul. Krasnokazarmennaya, 11250 Moscow, Russia Yurim@termech.mpei.ac.ru

15.1 Introduction

The optimization problem for robot engineering systems according to the criterion of minimum consumed work is one of the important and complicated problems in robotics. Difficulties of this problem are caused by multi-parametrization, large number of degrees of freedom of the system, high order and bulkiness of differential equations systems, non-smoothness functional of energy consumption, etc.

In spite of a large number of works, which deal with this problem, there still arise many important questions in design of systems with practically important types of drives. These demands stimulate new approaches and solutions. From the point of view of consumed energy, walking robots need more energy than wheeled platforms. Consequently, walking robots require larger electric motors and larger sources of energy. Most of walking robots are autonomous devices, in which duration of functioning is determined both by the capacity of the onboard power supply and the speed of consumption of energy. Therefore, search of theoretically achievable estimations of energy consumption from the point of view of choice of optimum laws of control and development of appropriate mathematical models and software are very urgent.

In the paper, the method of numerical parametrical optimization problem for biped walking is proposed. As an object of the research, the model of the inverted mathematical pendulum, to which the moment is applied by the electric motor, and the model of a multi-link planar walking robot (biped walking robot) consisting of a heavy trunk and a pair of weightless identical legs are considered. Notice, that with the help of this model in monographs [1,4], estimation of a single support phase of a biped walking robot movement was received, when only one leg is supporting on the surface, while the other leg is in phase of moving. In comparison to [1,4], walking (gait) process is analyzed taking into account processes taking place in circuits of electric motors in the presented paper. The purpose of this work is to develop and improve methods of estimation of power consumption for walking robots and to

K. Kozłowski (Ed.): Robot Motion and Control, LNCIS 335, pp. 233–241, 2006.
© Springer-Verlag London Limited 2006

create the appropriate software with simulation of computer visualization of the studied process of walking (gait). The paper is a continuation of [5].

For research of power consumption in robot walking, methods of theoretical mechanics, the theory of optimization, the theory of ordinary differential equations, numerical methods and simulation of computer visualization are used.

15.2 Equation of Motion for Inverted Pendulum

The electromechanical system consisting of the planar inverted pendulum and the direct current electric motor, which applies a moment to the pendulum through reducer (Fig. 15.1), is considered in the paper. It is the model of the moment, which is produced in the foot of the walking robot.

Point O is the point of support of the walking robot. The position of the pendulum is determined by the angle of turn φ, which is counted counter-clockwise from the fixed horizontal X axis of the coordinate system Oxy. Let the weight of pendulum be equal to m, the distance between point O and the center of weight be h, the moment of inertia of the pendulum with respect to the suspension axis be J_0, and the moment of inertia of the motor anchor be J_m. The effective moment of inertia is J. The values of J_0 and J_m are included in J ($J = J_0 + J_m n^2$). Lagrange's function L and Rayleigh's function Φ for the considered electromechanical system are set in the form:

$$L = \frac{1}{2}J\dot{\varphi}^2 + \frac{1}{2}L_0 i^2 + c_1 n\varphi i - mgh\sin\varphi, \qquad \Phi = \frac{1}{2}\beta\dot{\varphi}^2 + \frac{1}{2}Ri^2. \qquad (15.1)$$

Here L_0, R – generalized inductance and resistance of the rotor windings, i – current in the external circuit of the rotor, c_1 – factor of the electromechanical interaction, n – gear ratio of the reducer, β – factor of the viscous friction.

The equation of the pendulum movement with the direct current electric motor may be written in the form of Lagrange-Maxwell equation:

$$\frac{d}{dt}\left(\frac{\partial L}{\partial \dot{\varphi}}\right) - \frac{\partial L}{\partial \varphi} + \frac{\partial \Phi}{\partial \dot{\varphi}} = 0,$$
$$\frac{d}{dt}\left(\frac{\partial L}{\partial i}\right) + \frac{\partial \Phi}{\partial i} = U. \qquad (15.2)$$

Here U – voltage, which is supplied to the motor. Substitution (15.1) into (15.2) leads to a system of two nonlinear differential equations as follows:

$$\begin{cases} J\ddot{\varphi} + \beta\dot{\varphi} + mgh\cos\varphi + c_1 ni = 0, \\ L_0\dfrac{di}{dt} + c_1 n\dot{\varphi} + Ri = U. \end{cases} \qquad (15.3)$$

After separating the current i from the first part of equation (15.3) and substituting into second part, the equation leads to one nonlinear differential equation of the third order as follows:

$$JL_0\frac{d^3\varphi}{dt^3} + (JR + L_0\beta)\frac{d^2\varphi}{dt^2} + (c_1^2 n^2 + R\beta - mghL_0\sin\varphi)\frac{d\varphi}{dt} +$$
$$+ mghR\cos\varphi = c_1 nU. \qquad (15.4)$$

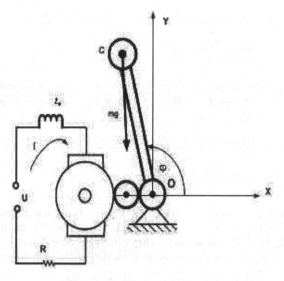

Fig. 15.1. Planar inverted pendulum

The length of the step is specified as L_s, the time of the step is t_k, and then we have two boundary-value conditions for the angle φ:

$$\left.\varphi\right|_{t=0} = \varphi_0, \qquad \left.\varphi\right|_{t=t_k} = \varphi_k. \tag{15.5}$$

For simplicity, the initial and final positions of the pendulum with respect to the vertical are supposed symmetric, so equations (15.5) become: $\varphi_k = \pi/2 - \arcsin \frac{L_s}{2h}$, $\varphi_0 = \pi/2 + \arcsin \frac{L_s}{2h}$.

For construction of periodic movement of the walking robot, we have to make equal the angular speed of the center of weight of the pendulum at the beginning and at the end of a single support phase:

$$\left.\dot{\varphi}\right|_{t=0} = \left.\dot{\varphi}\right|_{t=t_k}. \tag{15.6}$$

The programmed movement, which satisfies condition (15.5) is formulated in the form of the third-order Lagrange interpolation polynomial:

$$\varphi = \varphi_0 \frac{(t_1-t)(t_2-t)(t_k-t)}{(t_1-t_0)(t_2-t_0)(t_k-t_0)} + \alpha_1 \frac{(t-t_0)(t_2-t)(t_k-t)}{(t_1-t_0)(t_2-t_1)(t_k-t_1)} +$$
$$+ \alpha_2 \frac{(t-t_0)(t-t_1)(t_k-t)}{(t_2-t_0)(t_2-t_1)(t_k-t_2)} + \varphi_k \frac{(t-t_0)(t-t_1)(t-t_2)}{(t_k-t_0)(t_k-t_1)(t_k-t_2)}. \tag{15.7}$$

Here, α_1, α_2 are arbitrary constant values. They are functions of $\varphi(t)$ and lie between points t_1, t_2. Let us assume for simplicity $t_1 = t_k/3$, $t_2 = 2t_k/3$, $\varphi_0 = \pi/2 + \Delta$, $\varphi_k = \pi/2 - \Delta$, and $\Delta = \arcsin \frac{L_s}{2h}$. Taking into account boundary-value condition (15.5) we find that $\alpha_2 = \pi - \alpha_1$. So, for the programmed movement (15.7), we obtain:

$$\varphi(t, \alpha_1) = \frac{9(6\alpha_1 - 2\Delta - 3\pi)}{2t_k^3}t^3 - \frac{27(6\alpha_1 - 2\Delta - 3\pi)}{4t_k^2}t^2 +$$

$$+ \frac{(54\alpha_1 - 26\Delta - 27\pi)}{4t_k}t + \frac{\pi}{2} + \Delta. \tag{15.8}$$

Certainly, it is meaningful to consider only those programmed movements, at which angle φ monotonously decreases, i.e. at $0 \leqslant t \leqslant t_k$ the angular speed of the pendulum is negative:

$$\omega(t, \alpha_1) = \frac{27(6\alpha_1 - 2\Delta - 3\pi)}{2t_k^3}t^2 - \frac{27(6\alpha_1 - 2\Delta - 3\pi)}{2t_k^2}t +$$

$$+ \frac{(54\alpha_1 - 26\Delta - 27\pi)}{4t_k} < 0. \tag{15.9}$$

The inequality (15.9) is carried out for all $0 < t < t_k$, when parameter α_1 is inside the interval $\frac{\pi}{2} + \frac{1}{27}\Delta < \alpha_1 < \frac{\pi}{2} + \frac{13}{27}\Delta$. At $L_s = 0.8$ m, $h = 1.2$ m, $\Delta = \arcsin\frac{L_s}{2h} \approx 0.34 \approx 19.5°$ and the inequality (15.9) becomes $1.584 < \alpha_1 < 1.734$.

The energy consumption for the considered model is determined with the help of the functional

$$W = \int_{t_0}^{t_k} U(t)i(t)dt. \tag{15.10}$$

The subintegral expression in (15.10) is determined with use of equation (15.3) and (15.4) and has the following form:

$$U(t)i(t) = \frac{\left(\begin{array}{c} JL_0\dfrac{d^3\varphi}{dt^3} + (JR + L_0\beta)\dfrac{d^2\varphi}{dt^2} + \\ +(c_1^2n^2 + R\beta - mghL_0\sin\varphi)\dfrac{d\varphi}{dt} + \\ + mghR\cos\varphi)\,(J\ddot{\varphi} + \beta\dot{\varphi} + mgh\cos\varphi) \end{array}\right)}{(c_1^2n^2)}. \tag{15.11}$$

The energy consumption for the walking robot in one step, according to formulae (15.10) and (15.11) for the programmed movement (15.7) is a function of one parameter α_1:

$$W = W(\alpha_1). \tag{15.12}$$

Thus, the problem of energy optimization for movement of the inverted pendulum is reduced to the search of the minimum of the function (15.12). The result of calculation according to formula (15.12) is plotted in Fig. 15.2, and in this case the form of transients of optimum angle $\varphi = \varphi(t)$ and $\omega = \omega(t)$ is close to movement of free pendulum $J\ddot{\varphi} + mgh\cos\varphi = 0$.

In this case, the following numerical data [3,4] are used for calculating the energy consumption: $m = 75$kg, $J = 100.8$kg·m^2, $h = 1.2$m, $g = 9.8$m/sec^2, $L_s = 0.8$m, $L_0 = 0.03$H, $R = 0.8$ Ω, $c_1 = 0.05$H·m/volt, $n = 300$, $\beta = 150$kg·m^2/sec, time of step $t_k = 2/3$sec. The nodal points of the polynomial interpolation are $t_1 = 2/9$sec and $t_2 = 4/9$sec. In addition, the mean velocity of motion of the center of mass in one step is $V_s = 1.2$m/sec. With the given numerical data we have $\Delta = 0.34 \approx 19.5°$, and the mean angular velocity of the pendulum is $\omega(t, \alpha_1) \approx -1.02$/sec.

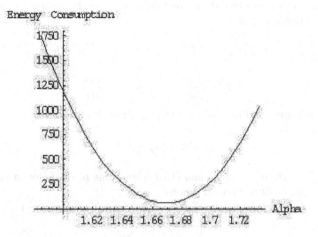

Fig. 15.2. Dependence of energy consumption on α

15.3 Exact Solution for Variational Problem

In this case, the transient in the circuit of the electric motor is neglected and it is assumed that in (15.4) $L_0 = 0$. Then for the case of a small deviation of the pendulum from the vertical, the problem minimization of the scalar energy criterion becomes

$$W = \int_0^1 F(u(\tau), x'(\tau))d\tau,$$

$$F = \frac{k}{2}u^2 - bu\frac{dx}{d\tau}.$$

(15.13)

The variables of the functional (15.13) satisfy restrictions in the form of a differential equation:

$$\frac{d^2x}{d\tau^2} + \beta_1\frac{dx}{d\tau} - \Omega^2 x = u,$$

(15.14)

where

$$x = \varphi - \pi/2, \qquad \beta_1 = \frac{\beta t_k}{J} + \frac{c_1^2 n^2 t_k}{JR}, \qquad \Omega^2 = \frac{mght_k^2}{J},$$

$$u = \frac{c_1 n t_k^2}{JR}U, \qquad k = \frac{2J^2 R}{c_1^2 n^2 t_k^4}, \qquad b = \frac{J}{t_k^3}.$$

For searching the optimal control $u(\tau)$, which minimizes functional (15.13), a method of classical calculus of variation of Euler-Lagrange [2] is used. As a preliminary, the equation (15.14) is written down in the normal form of Cauchy

$$x' = y, \qquad y' = \Omega^2 x - \beta_1 y + u,$$

(15.15)

where the accent sign denotes the derivative with respect to τ.

Let's write down the Hamilton function and then the system equations for the conjugated variable impulses

$$H = p_x y + p_y(\Omega^2 x - \beta_1 y + u) + \frac{k}{2}u^2 - buy, \qquad (15.16)$$

$$p_x' = \frac{\partial H}{\partial y}, \qquad p_y' = -\frac{\partial H}{\partial x}. \qquad (15.17)$$

From the minimum condition $\frac{\partial H}{\partial u} = 0$ of Hamilton function H

$$u = \frac{1}{k}(by - p_y). \qquad (15.18)$$

Substituting (15.18) into (15.15) and (15.17) we come to the following system of the fourth-order linear differential equations:

$$\mathbf{X}' = \mathbf{A}\mathbf{X}, \qquad \mathbf{X} = \begin{bmatrix} x & y & p_x & p_y \end{bmatrix}^T. \qquad (15.19)$$

Here T – transposition sign,

$$\mathbf{A} = \begin{bmatrix} 0 & 1 & 0 & 0 \\ \Omega^2 & \beta_{22} & 0 & -1/k \\ 0 & 0 & 0 & -\Omega^2 \\ 0 & b^2/k & -1 & -\beta_{22} \end{bmatrix}.$$

The eigenvalues of matrix $\mathbf{A}$ are determined after solving the following biquadratic equations:

$$\lambda^4 + \left(\frac{b^2}{k^2} - \beta_{22}^2 - 2\Omega^2\right)\lambda^2 + \Omega^4 = (\lambda^2 - \lambda_1^2)(\lambda^2 - \lambda_2^2) = 0. \qquad (15.20)$$

The solution of Eq. (15.19) is

$$\mathbf{F} = \mathbf{S}.\exp(\boldsymbol{\Lambda}\tau).\mathbf{S}^{-1} \qquad (15.21)$$

where

$$\exp(\boldsymbol{\Lambda}\tau) = \begin{bmatrix} e^{\lambda_1 \tau} & 0 & 0 & 0 \\ 0 & e^{-\lambda_1 \tau} & 0 & 0 \\ 0 & 0 & e^{\lambda_2 \tau} & 0 \\ 0 & 0 & 0 & e^{-\lambda_2 \tau} \end{bmatrix},$$

matrix $\mathbf{S}$ is formed from eigenvectors matrix $\mathbf{A}$, $\mathbf{S}^{-1}$ – inverse matrix of $\mathbf{S}$. Eigenvectors s_k are determined by formulas:

$$s_1 = \begin{bmatrix} \dfrac{-1}{k\delta_1} & \dfrac{-\lambda_1}{k\delta_1} & \dfrac{-\Omega^2}{\lambda_1} & 1 \end{bmatrix}^T, \qquad s_2 = \begin{bmatrix} \dfrac{-1}{k\delta_2} & \dfrac{\lambda_1}{k\delta_2} & \dfrac{\Omega^2}{\lambda_1} & 1 \end{bmatrix}^T,$$

$$s_3 = \begin{bmatrix} \dfrac{-1}{k\delta_3} & \dfrac{-\lambda_2}{k\delta_3} & \dfrac{-\Omega^2}{\lambda_2} & 1 \end{bmatrix}^T, \qquad s_4 = \begin{bmatrix} \dfrac{-1}{k\delta_4} & \dfrac{\lambda_2}{k\delta_4} & \dfrac{\Omega^2}{\lambda_2} & 1 \end{bmatrix}^T.$$

Here $\delta_1 = \lambda_1^2 - \lambda_1\beta_{22} - \Omega^2$, $\delta_2 = \lambda_1^2 + \lambda_1\beta_{22} - \Omega^2$, $\delta_3 = \lambda_2^2 - \lambda_2\beta_{22} - \Omega^2$, $\delta_4 = \lambda_2^2 + \lambda_2\beta_{22} - \Omega^2$.

The fundamental matrix (15.21) allows to write down the solution of Eq. (15.19) for the moment of time $\tau = 1$:

$$\mathbf{X}(1) = \mathbf{S} \cdot \exp(\boldsymbol{\Lambda}) \cdot \mathbf{S}^{-1} \mathbf{X}(0). \qquad (15.22)$$

Here $\mathbf{X}(0) = \begin{bmatrix} \Delta & \omega_0 & p_x^0 & p_y^0 \end{bmatrix}^T$, $\mathbf{X}(1) = \begin{bmatrix} -\Delta & \omega_0 & p_x(1) & p_y(1) \end{bmatrix}^T$ and, hence, the first term of Eq. (15.22) represents a system of two linear algebraic equations for initial values of impulses p_x^0, p_y^0. The solution of the indicated equations determines completely the behavior of the trajectories of system (15.19), and substituting formula (15.19) into (15.18) gives optimal control of energy for the electric motor.

15.4 Equation of Motion for Multi-link of the Walking Robot

The proposed method of parametric optimization of variational problem for searching for the minimum of a function of several variables in the first section is used in optimizing the problem of walking for an electromechanical biped walking robot. The model of the planar anthropomorphic walking robot consists of a heavy trunk and a pair of weightless identical legs is considered. The movement of the walking robot is provided with direct current electric motors, which create moments in the hinges of the robot.

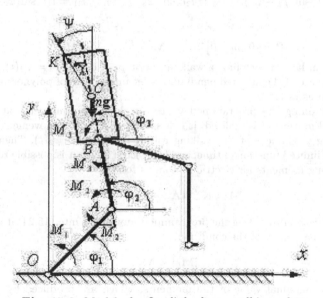

Fig. 15.3. Model of a five-link planar walking robot

In the given paper the walking process is analyzed taking into account the processes that run in the circuit of the electric motors. In a single support phase the considered electromechanical system of the walking robot has six degrees of freedom: three angles $\varphi = \begin{bmatrix} \varphi_1 & \varphi_2 & \varphi_3 \end{bmatrix}^T$, which specify the position of links OA, OB, BC and three currents $\mathbf{i} = \begin{bmatrix} i_1 & i_2 & i_3 \end{bmatrix}^T$, which flow in the circuits input of direct current electric motors, which create the moments in the hinges O, A, B (Fig. 15.3).

The equations of motion for the proposed electromechanical system in a single support phase represent a system of second order nonlinear differential equations for a vector of angular variables and a vector of currents

$$\mathbf{A}\ddot{\varphi} + \mathbf{B}\dot{\varphi}^2 - \mathbf{P} = \mathbf{K}.\mathbf{M},$$
$$\mathbf{L}\frac{di}{dt} + Ri + \mathbf{C}\dot{\varphi} = \mathbf{U}, \tag{15.23}$$

where

$$\mathbf{A} = m \begin{bmatrix} -l_1 \sin\varphi_1 & -l_2 \sin\varphi_2 & -l_3 \sin\varphi_3 \\ l_1 \cos\varphi_1 & l_2 \cos\varphi_2 & l_3 \cos\varphi_3 \\ 0 & 0 & \rho^2 \end{bmatrix},$$

$$\mathbf{B} = m \begin{bmatrix} -l_1 \cos\varphi_1 & -l_2 \cos\varphi_2 & -l_3 \cos\varphi_3 \\ l_1 \sin\varphi_1 & l_2 \sin\varphi_2 & l_3 \sin\varphi_3 \\ 0 & 0 & 0 \end{bmatrix},$$

$$\mathbf{K} = \frac{1}{s_{21}} \begin{bmatrix} l_2 \cos\varphi_2 & -l_1 \cos\varphi_1 - l_2 \cos\varphi_2 & l_1 \cos\varphi_1 \\ l_2 \sin\varphi_2 & -l_1 \sin\varphi_1 - l_2 \sin\varphi_2 & l_1 \sin\varphi_1 \\ s_{32} & -s_{32} - s_{31} & s_{31} + s_{21} \end{bmatrix},$$

$$s_{21} = l_1 l_2 \sin(\varphi_2 - \varphi_1), \quad s_{32} = l_2 l_3 \sin(\varphi_3 - \varphi_2), \quad s_{31} = l_1 l_3 \sin(\varphi_3 - \varphi_1),$$

$$\ddot{\varphi} = \begin{bmatrix} \ddot{\varphi}_1 & \ddot{\varphi}_2 & \ddot{\varphi}_3 \end{bmatrix}^T, \quad \dot{\varphi}^2 = \begin{bmatrix} \dot{\varphi}_1^2 & \dot{\varphi}_2^2 & \dot{\varphi}_3^2 \end{bmatrix}^T,$$

$$\mathbf{P} = \begin{bmatrix} 0 & mg & 0 \end{bmatrix}^T, \quad \mathbf{M} = \begin{bmatrix} M_1 & M_2 & M_3 \end{bmatrix}^T.$$

The movement law for each link of walking robot $\varphi_1 = \varphi_1(t)$, $\varphi_2 = \varphi_2(t)$, $\varphi_3 = \varphi_3(t)$ as well as in case of the inverted pendulum is set in the form of polynomial Lagrange interpolation as in (15.7).

In determining the programmed values of the motor moments at which the programmed movement is realized, it is assumed that in this movement there is no straightening of the legs of the walking robot, i.e. $\varphi_1(t) \neq \varphi_2(t)$. Then matrix $\mathbf{K}$ represents a limited function of time, and from Eq. (15.23) it is possible to determine the controlling moments for electric motors as follows

$$\mathbf{M} = \mathbf{K}^{-1} \left(\mathbf{A}\ddot{\varphi} + \mathbf{B}\dot{\varphi}^2 - \mathbf{P} \right). \tag{15.24}$$

In this case after substituting the programmed movement into (15.24) it is necessary to check up realization of the conditions

$$|\mathbf{M}_k| \leqslant \mathbf{M}_k^*, \tag{15.25}$$

where $\mathbf{M}_k^*$ – maximal value of the moments, which are produced by the electric drives of the walking robot.

The functional, which determines consumption of energy

$$W = \int_{t_0}^{t_k} \mathbf{U}^T i \, dt \tag{15.26}$$

is a function of a finite number of variables – angles of rotation of the links of the walking robot at the nodal points.

15.5 Conclusions

1. Computer simulation for movement of the considered walking robot shows that the offered technique allows to carry out a purposeful choice of design parameters of the walking robot, which use electric drives and programmed movement at walking.
2. Results of numerical experiments with respect to the proposed schema demonstrate the possibility of an essential reduction of energy consumption by the walking robot.

References

1. Beletsky V V (1984) Biped walking. Modeling problems of dynamics and control. Nauka, Moscow
2. Bryson A E, Ho Y C (1969) Applied optimal control: optimization, estimation, and control. Blaisdell, Waltham, MA
3. Chernousko F L, Bolotnik N N, Gradetsky V G (1989) Manipulation robots. Dynamics, control, and optimization. Nauka, Moscow
4. Formalsky A M (1982) Moving of the anthropomorphical mechanisms. Nauka, Moscow
5. Siregar H P, Martynenko Yu G (2002) Energy consumption for anthropomorphic robots driven by electromotor. In: Proc. 8th Int. Conf. for Engineering Sciences: Abstract of thesis, vol. 1, Moscow 370–371

Multi-agent Systems and Localization Methods

16

Real-Time Motion Planning and Control in the 2005 Cornell RoboCup System

Michael Sherback, Oliver Purwin, and Raffaello D'Andrea

Sibley School of Mechanical and Aerospace Engineering, Cornell University
Ithaca NY 14853 USA   {mas61|op24|rd28}@cornell.edu

16.1 Introduction

Since 1999 Cornell has competed in the RoboCup Small Size League. RoboCup is an annual international gathering featuring robot soccer competitions aimed at fostering research and education in robotics, autonomous systems, and artificial intelligence. In the small size league, autonomous teams of five cantaloupe sized robots play on a 4.9 × 3.4 m field with overhead global vision cameras. Teams are typically made up of a mixture of graduate and undergraduate students.

Virtually all teams have adopted a similar system architecture. The global vision camera captures an image and blob analysis is used to locate the robots and ball. This data is sent to a hub computer that makes decisions and determines trajectories. Information on desired velocities and ball control functions is communicated via wireless communication to the robots, who implement the motion and thus close the loop.

Cornell's architecture builds on this by having full duplex wireless to enable ball possession information to be shared, and by having x86 based PC104 computers on each robot to allow decision making functions to be distributed. Figure 16.1 should help readers understand our system and be a helpful reference for understanding each section of this paper.

This paper is organized in the following way: Section 16.2 explains how our motion-related methods and algorithms are used by the decision-making part of the system. A detailed description of the vehicles and their dynamics is in Section 16.3. Section 16.4 describes how the current state of the system is estimated in the presence of noise, latency, and vision problems. Section 16.5 is an overview of the trajectory generation method. Section 16.6 describes how trajectories are modified in the presence of obstacles. Section 16.7 presents our method of using an onboard gyro to improve trajectory execution. Section 16.8 describes the concepts used in our onboard motor control and their implementation. Section 16.9 briefly introduces several alternatives to our methods. Experimental results are presented in Section 16.10.

K. Kozłowski (Ed.): Robot Motion and Control, LNCIS 335, pp. 245–263, 2006.
© Springer-Verlag London Limited 2006

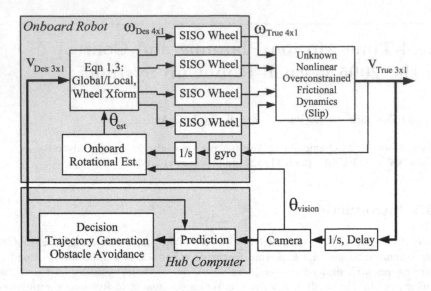

Fig. 16.1. System loop. Angle velocity control assumed. 1/s denotes integration

16.2 Decision

The decision algorithms (also known as artificial intelligence or AI) ultimately determine what sort of trajectories will be generated. At this level the following choices are made:

- *Trajectory Generation or Manual Velocities:* Either set a desired final state and employ our trajectory generation algorithms to find an optimal current velocity, or directly command the current velocity.
- *Obstacle Avoidance Override:* Under some conditions, such as playing goalie, we choose to tolerate collisions in trajectories.
- *Angle Control or Angular Rate Control:* Give the robot either a desired final angle, or a desired current angular velocity.

The decision algorithms are not the focus of this paper; see [7]. They can be briefly described as a hierarchical state machine, with both internal states and physical states (robot and ball positions and velocities).

16.3 Description of the Vehicles

Our system uses four-wheeled omnidirectional vehicles (Fig. 16.2). The wheels have rollers at their edges that allow unconstrained vehicle motion in the direction of the axle of the wheel. Each point of contact thus provides one translational constraint and negligible rotational constraint. Three such wheels, arranged to avoid singularities, are sufficient for omnidirectional holonomic control of the vehicle, as described in [5]. Since 2003 we have added a fourth wheel to aid control when accelerations cause weight shift that reduces friction at the wheels [6]. The equations describing this weight shift and its effects on the acceleration performance envelope are found in [8].

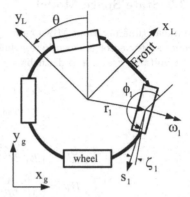

Fig. 16.2. 2005 Cornell RoboCup robot **Fig. 16.3.** Robot kinematics. Plan view.

16.3.1 Kinematics

The kinematics of our omnidirectional drive are straightforward, and their holonomic properties greatly simplify our motion panning. Variables used in this section are defined in Figure 16.3. We define the global coordinate system by x_g and y_g. The local frame of reference is defined by x_L and y_L where it is assumed that the vehicle remains in the plane and does not pitch or roll. The angle θ is the yaw rotation to x_L from x_g, defined as positive along $x_g \times y_g$, sometimes referred to as the "robot angle". The positions of the wheels with respect to the CM in the local frame of reference are defined by the vectors $\mathbf{r}_i$, and unit vectors in their driven directions by $\mathbf{s}_i$. Our convention is that the wheels' driven direction is negative in the direction that corresponds to positive theta motion.

Global velocity vectors are transformed to local coordinates using the standard rotation matrix:

$$\begin{bmatrix} \dot{x}_L \\ \dot{y}_L \\ \dot{\theta} \end{bmatrix} = \begin{bmatrix} \cos\theta & -\sin\theta & 0 \\ \sin\theta & \cos\theta & 0 \\ 0 & 0 & 1 \end{bmatrix} \begin{bmatrix} \dot{x}_g \\ \dot{y}_g \\ \dot{\theta} \end{bmatrix}. \tag{16.1}$$

Given a desired velocity vector in the local robot frame $(\dot{x}_L, \dot{y}_L, \dot{\theta}_L)$, the velocity u of any point i on the robot in the local frame can be calculated:

$$\begin{bmatrix} u_x \\ u_y \\ u_\theta \end{bmatrix}_i = \begin{bmatrix} 1 & 0 & |r_i \times x_L| \\ 0 & 1 & |r_i \times y_L| \\ 0 & 0 & 1 \end{bmatrix} \begin{bmatrix} \dot{x}_L \\ \dot{y}_L \\ \dot{\theta} \end{bmatrix}. \tag{16.2}$$

For control purposes we are interested in the scalar wheel velocities w_i obtained with the transformation T_ψ:

$$\begin{bmatrix} w_1 \\ \cdots \\ w_n \end{bmatrix} = \begin{bmatrix} \cos\phi_1 & \sin\phi_1 & |r_1|\cos\zeta_1 \\ \cdots & \cdots & \cdots \\ \cos\phi_n & \sin\phi_n & |r_n|\cos\zeta_n \end{bmatrix} \begin{bmatrix} \dot{x}_L \\ \dot{y}_L \\ \dot{\theta} \end{bmatrix}. \tag{16.3}$$

16.3.2 State Space Model

The linear time invariant MIMO state space system model for the robot with a six element state vector is given here, where $\dot{z} = Az + B_u u + B_w w, y = Cz + Du$, input motor voltages are u, and disturbance forces and torques are w:

$$z = (x_L, y_L, \theta, \dot{x}_L, \dot{y}_L, \dot{\theta}), \tag{16.4}$$

$$A = \begin{bmatrix} -\epsilon I_{3\times3} & I_{3\times3} \\ 0_{3x3} & M^{-1}(B_\mathrm{d} + M_\mathrm{emf}) \end{bmatrix}, \tag{16.5}$$

$$B_u = \begin{bmatrix} 0_{3\times3} \\ K_\mathrm{tv} M^{-1} T_\psi^T \end{bmatrix}, \tag{16.6}$$

$$B_w = \begin{bmatrix} 0_{3\times4} \\ M^{-1} T_\psi^T \end{bmatrix}, \tag{16.7}$$

$$C = \begin{bmatrix} 0_{4\times3} & \frac{N t_s CPR}{\pi d_w T_\psi} \\ 0_{5\times1} & K_\mathrm{gyro} \end{bmatrix}, \tag{16.8}$$

$$D = 0, \tag{16.9}$$

where ϵ can be used in A to avoid unobservable modes (due to the absence of an onboard absolute position reference) during estimator design, N is the gear ratio between the motors and wheels, d_w is the wheel diameter, CPR is the encoder counts per revolution, t_s is sample time, and

$$M^{-1} = \mathrm{diag}(1/M_{\mathrm{eff},x} \ \ 1/M_{\mathrm{eff},y} \ \ 1/M_{\mathrm{eff},\theta}), \tag{16.10}$$

$$C_\mathrm{emf} = \frac{4\beta\gamma N^2}{d_w \Omega}, \tag{16.11}$$

$$M_\mathrm{emf} = C_\mathrm{emf} T_\psi^T T_\psi, \tag{16.12}$$

$$K_\mathrm{tv} = \frac{2\beta N}{d_w \Omega}, \tag{16.13}$$

where γ is the back-EMF induced loss of torque per velocity, and the torque constant at stall is β. Effective masses $M_{\mathrm{eff},q}$ are described in Section 16.8. Process noise covariance must in this case account for disturbance torques and forces associated with drag and slipping wheels. This full vehicle model is included for completeness even though SISO loops are used in practice.

16.4 Translational State Estimation or Prediction

The principal state estimation problem in the 2005 system consists of taking camera measurements with a variable latency, at a variable rate, and constructing an estimate of the state of the game at the time when commands based on this data will reach the robot. It is customary at Cornell to think of oneself as doing calculations at the instant of frame grabbing, and so the problem is referred to as 'prediction.' These calculations are actually being done about halfway through the total latency, after the vision data is transferred and processed, and before the wireless packet is transmitted. This does not affect the calculations. We have no direct measure of latency during normal operation. Its value is determined off line using an LED visible to the camera so that there is no indeterminate return path latency to estimate.

There are three flavors of this problem: estimation of the ball state, opponent state, and teammate state. The rotational degree of freedom of the ball and opponents is neglected, and that of the teammates is handled by a separate method described in Section 16.7 due to the availability of reliable low-latency onboard gyro data.

16.4.1 Conceptual Basis

The ball state estimator consists of two Kalman filters, fast and slow. When their velocity estimates agree within a threshold, the slow filter is used. The slow filters' advantage is rejection of high frequency noise. The fast filter is used when the threshold is exceeded. This is most important when the ball is kicked, or a lost ball is found again.

The teammate state estimator consists of two sequential filters in order to make use of our knowledge of the commands that have been issued since the vision frame was captured. A Kalman filter is first applied to the vision data to generate an estimate of the robot state at the time of the most recent frame. This filter makes use of neither the dynamics of the robots nor their past commands but rather assumes that the robot is a mass driven by white noise force disturbances. Following this filter, the latency is compensated for by assuming that the velocity commands that had been issued were followed. This is in the spirit of a Smith Predictor [9]. There is a further correction where the phase lag in the reference tracking of the robot is represented as additional latency (see Section 16.8).

The opponent state estimator is the same as the teammate state estimator except that it is assumed that the velocity estimate from the first filter is the commanded velocity for all subsequent frames.

16.4.2 Complications

Non-ideal vision behavior is the source of most of the complexity of our approach. Variable vision frame rate is due to lack of sufficient computational power. Specifically, when a sought entity (typically the ball) is not being found by vision when it looks where the entity was last found, the algorithms require accessing the stored image of the entire field. On our system this brings the frame rate down from 60 Hz to as low as 30 Hz. We therefore re-discretize the filters for each new reading. This variable delay directly alters the total latency that must be compensated for. This is handled by putting time stamps on vision data. During implementation we found inaccuracies in our time stamp data that we were not able to eliminate.

There is also a delay associated with the execution time of the high level decision and control, especially obstacle avoidance. This is a 'chicken and egg' problem, where the state estimate is required to begin the decision and control, but where the decision and control execution time alters the latency and therefore the correct state estimate. We neglect this small variance.

If vision fails to find objects within a time limit, the object is marked 'lost' in the vision data. For a small number of frames this is dealt with as a simple case of increased latency. For a larger number of frames, different hand tuned logic is used depending on the lost object. Lost teammate robots are commanded to execute a slow arcing motion designed to exit any regions of poor vision performance.

16.4.3 Ball Occlusion Logic

Ball state estimation is complicated by the fact that the ball can be completely hidden or 'occluded' during normal game play for an arbitrary duration due to the geometry of the overhead cameras and tall robots. This is dealt with in two ways. If the IR sensor on the front of our robots is broken, possession is assigned, and the filters are overridden to locate the ball at that robot. Alternately, if the ball is not found and there are no broken IR sensors, we simultaneously assign two locations for the ball, attached to teammate and opponent robots near the last found ball, and functions use whichever is appropriate depending on whether they are focused on offensive or defensive behavior.

16.5 Trajectory Generation

Our system generates trajectories using a near optimal bang-bang method described in [8]. At each AI frame, trajectories are recomputed based on the current state estimate. Our application requires real-time calculation for a potentially large number of trajectories for use by obstacle avoidance. The method excels by having bounded and low execution times approximately four orders of magnitude shorter than the full nonlinear optimal method as implemented in RIOTS in MATLAB, while yielding trajectories with more than 85% of execution times being less than 18% suboptimal.

16.5.1 Simplification of Dynamics

Our vehicles have sufficient torque that it is a satisfactory assumption that their acceleration is limited by either friction or maximum motor speed for all trajectories that can be achieved within the space of our field. Any desired acceleration can be thought of as a vector in $\Re^3$, $(\ddot{x}_L, \ddot{y}_L, \ddot{\theta}_L)$. Following [8] we assume Coulomb friction and neglect the dynamic effects involving tipping of the vehicle on a compliant surface due to jerk, but retain quasi-static weight shift due to acceleration. More sophisticated friction models as described in [11] would require a more complicated algorithm. By assuming Coulomb friction the achievable acceleration in the three dimensional acceleration space is made invariant with respect to the state of the vehicle subject to the constraint

$$v(t) = \sqrt{\dot{x}^2(t) + \dot{y}^2(t)} \leqslant v_{\max}, \quad \forall t. \tag{16.14}$$

This constraint results from the loss of motor torque at high speed with fixed voltage. The selection of the constant $v_{\max}$ is left for the end of this subsection.

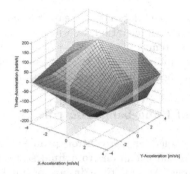

Fig. 16.4. Analytical performance envelope

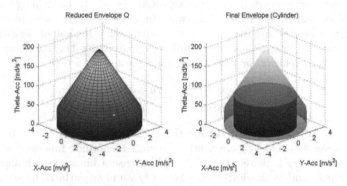

Fig. 16.5. Simplified performance envelope

Under these assumptions and the constraint of Eq. (16.14) the acceleration constraint can be expressed as a constant bounding surface about the origin in the three dimensional acceleration space as shown in Fig. 16.4 from [8]. Asymmetries result from weight shift interacting with wheel location and angle. Simplifications are required to yield a fast algorithm, as the boundary is nonlinear. Geometrically speaking, simpler surfaces within the original bounding surface are found. First it is convenient to make the bounding surface rotationally symmetric using the constraint

$$\sqrt{\ddot{x}^2(t) + \ddot{y}^2(t)} \leqslant a_{\max}, \quad \forall t. \tag{16.15}$$

Next it is convenient to make the bounding surface invariant with respect to $\ddot{\theta}_L$, which causes negligible loss of performance due to the small control efforts required for the $\ddot{\theta}_L$ desired in our application. The resulting surface is shown in Fig. 16.5. If bang-bang is used in θ as well, the operating space would be visualized as three thin stacked discs.

Similar simplifications are required for allowable velocities. The maximum velocity $v_{\max}$ in Eq. (16.14) for which the assumption of friction limited acceleration holds is a function of the direction of travel due to kinematics and motor properties, but it is expedient and reasonable to set the constraint as a constant rather than as a function of all velocities and accelerations. This simplification is analogous to the one that yielded Eq. (16.15). In this case the bound is the set of velocities for which the resulting motor speed does not decrease torque sufficiently to violate the assumption of friction limited acceleration. In practice v_{max} not derived analytically, but is tuned on the real system.

Ref. [10] describes a similar method for bang-bang control of a vehicle with constraints due to motor back-EMF rather than friction.

16.5.2 Application of Bang-bang Control

Having completely decoupled the θ control problem we seek minimum time trajectories from given initial conditions to a final state, typically with zero final velocity, in minimal time. At any time in an optimal single-DOF bang-bang trajectory the vehicle is either accelerating at its limit, or cruising at v_{max} with no acceleration, or stopped. In our case, we obtain two-DOF bang-bang trajectories with a binary search in the following way: first we generate decoupled x and y bang-bang trajectories with distinct $t_{f,x}, t_{f,y}$, with each using half of the total control effort. Second, we synchronize the solution by adjusting the maximum allowed control effort and velocity for both DOFs via a parameter $\alpha \in (0, \pi/2)$:

$$\ddot{x}_{max} = a_{max} \cos \alpha, \qquad \ddot{y}_{max} = a_{max} \sin \alpha, \qquad (16.16)$$

$$\dot{x}_{max} = v_{max} \cos \alpha, \qquad \dot{y}_{max} = v_{max} \sin \alpha. \qquad (16.17)$$

Equations (16.16) and (16.17) satisfy the constraints (16.14) and (16.15). The execution times $t_{f,x}$ and $t_{f,y}$ are continuous and monotonic functions of α (for a formal proof, see [19]), therefore it is possible to use a binary search algorithm to find an α which renders the difference between $t_{f,x}$ and $t_{f,y}$ arbitrarily small. In this manner the feasible near-optimal two-DOF trajectory is found.

During deceleration this algorithm is sensitive to any initial failure to follow the desired deceleration. This results in overshoot that will not be overcome unless deceleration is in excess of that used to calculate the distance at which deceleration was to begin. During deceleration we therefore calculate the required deceleration using known distance to target and current velocity, yielding $\ddot{q} = \text{sign}(D)v^2/2D$.

16.6 Obstacle Avoidance

It is essential for motion planning to include obstacle avoidance because of the need to infiltrate defenses and because the RoboCup rules penalize collisions between robots. There is an enormous and growing literature on obstacle avoidance. A 1991 book with information on the subject is [18]. The constraints of our application dictate that the method must work in a very dynamic environment, and must return a result in $\ll 1$ ms so that it can be executed for all robots by the central computer. We also require that it be compatible with our trajectory generation algorithm. These requirements lead us to a very simple method.

Any desired motion is classified by the decision algorithms as requiring obstacle avoidance or not. If obstacle avoidance is desired, the decision algorithm calls a function that seeks an obstacle free path to the current desired location at each frame. This function calculates a trajectory as described in Section 16.5 and steps through each point seeking collisions with the predicted locations of teammates and opponents. The future position of the objects is found by propagating current state estimates assuming no change in velocity. If a collision is detected, the desired final position is altered until either the function can return an obstacle free path, or it returns a failure value. The alternate final locations are determined by constructing a line perpendicular to that between the robot and its destination, and picking increasingly distant points from alternating sides. Fig. 16.6 depicts this process, but for the sake of clarity does not attempt to depict the fact that the obstacle may be in different locations depending on how far along the trajectory the collision occurs. The routine terminates in failure if this does not generate an obstacle free path in a certain number of iterations. The alteration of the desired final location is tolerable due to the fact that old trajectory solutions are not retained, and so the collision free path is recomputed at every frame until the 'corner is turned' and alterations are no longer required.

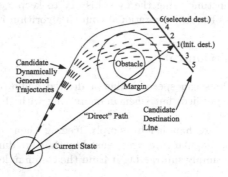

Fig. 16.6. Simplified obstacle avoidance diagram

16.7 Rotational Estimation and Control

The rotational degree of freedom of teammate robots is estimated and controlled in different way from the translational. Estimation is different because of the availability of low-latency gyro data on board the robots. Control is different because of greater control authority due to the relatively low apparent mass seen by the motors during rotation, as well as the availability of superior estimates. Similar methods could be applied to the translational state estimation described in Section 16.4 if one found analogous local translation sensors. We have not found them; in our experience, accelerometers cause problems due to unsensed tilt coupling in gravity, mouse-type sensors currently lack the required speed, and the motor rotation sensors are unreliable for estimates of the robot state due to slip during the aggressive maneuvers that motivate the problem.

16.7.1 Rotational Estimation

The accuracy of the rotational state estimate is crucial for robot control due to the appearance of the rotation angle θ in Eq. (1). By transmitting global velocities

$(\dot{x}_g, \dot{y}_g)$ to the robot and using in the global to local transformation an estimate $\hat{\theta}$ that makes use of onboard data, reference tracking is improved compared to past systems.

The $\hat{\theta}$ estimation is broken into three components: primary, catchup, and keepup. When each wireless packet is received (at 30 to 60 Hz if none are dropped), a primary estimator yields $\hat{\theta}_v$ at the time of vision frame capture, and a catchup estimator marches that forward in time to compensate for latency, yielding $\hat{\theta}_c$. The primary estimator is a simple feedthrough of raw vision data when it is available, because of the low noise in this measurement. The catchup estimator uses a circular gyro history buffer, integrating as many gyro readings as called for by a static latency estimate. At each local control frame (a consistent 300 Hz, see Section 16.8) the most recent gyro data is marched forward with a keepup function to yield the $\hat{\theta}$ used by the rotational control and by the global to local transformation.

If a packet is received in which the robot is lost, the last known vision data is marched forward in time using the gyro history to keep $\hat{\theta}_v$ current. For a large number of lost frames the entire rotational control algorithm is disabled.

16.7.2 Rotational Control

The decision algorithms can specify either a desired angle or velocity. Typically the desired angle is specified, but when maneuvering with the ball, it is useful to command velocities.

Desired velocities are handled by simply feeding them through to the local control algorithm. The angular rate error sensed by the gyro during any local control frame is corrected by simply subtracting it from the commanded angular rate of the next frame.

Desired angles are handled by a bang-dead-bang method similar to that used in trajectory generation. For a given angular error and angular velocity, it is determined whether the robot should be accelerating or decelerating. To avoid deviating into a state from which the bang-bang acceleration can not drive the angle monotonically to the desired, a safety factor is introduced into the decision that causes suboptimal premature deceleration. This is done by using a nominal bang-bang acceleration $\ddot{\theta}_{bb}$ less than what the robot is actually capable of. Notice that deceleration is at whatever rate is necessary to drive the angular rate to zero at zero angular error, as in translational trajectory generation. This means that we generally decelerate with a deceleration greater than $\ddot{\theta}_{bb}$. To avoid jitter due to sensor noise, a dead zone is introduced. A function 'NormalizeAngle' that normalizes angular errors into the range $(-\pi, \pi)$ is used. Pseudocode for the bang-dead-bang rotation control follows.

1. $\theta_{err} = \text{NormalizeAngle}(\hat{\theta} - \theta_{des})$

2. $\ddot{\theta}_{req} = \dfrac{(\dot{\hat{\theta}})^2}{2\theta_{err}}$

3. **if** $(|\ddot{\theta}_{req}| > \ddot{\theta}_{bb})$
   $\ddot{\theta}_{des} = \text{sign}(\theta_{err}) * \ddot{\theta}_{req}$ (case: decel.)
   else
   $\ddot{\theta}_{des} = -\text{sign}(\theta_{err}) * \ddot{\theta}_{bb}$ (case: acc.)

4. **if** $(|\theta_{\text{err}}| > \theta_{\text{deadZone}})$
$$\dot{\theta}_{\text{des}} = \dot{\theta}_{\text{des}} + t_s \ddot{\theta}_{\text{des}}$$
else
$$\dot{\theta}_{\text{des}} = 0$$

The acceleration is applied relative to the previous commanded velocity. When there is a transition from desired velocity control to desired angle control, the current gyro reading establishes the velocity from which the algorithm proceeds.

A limiting velocity is set to 8 rad/s to avoid the maximum gyro reading of 10 rad/s, and the limiting acceleration is set to 15 rad/s/s as a compromise between speed and jerk, which harms ball control.

16.8 Local Control

Local control is our name for the algorithms that attempt to execute the motions dictated by trajectory generation and rotational control. Its basic components are SISO PI loops on each wheel (Fig. 16.7) that track the references w_{des} generated by the kinematic calculations from Section 16.3. The loops execute at 300 Hz on the PC104. At each step k, we convert encoder signals to velocity and filter using a first order low pass filter parameter ψ, compute velocity error, and update integrated velocity error with a low frequency rolloff time constant Υ

$$w_{\text{raw},i,k} = \frac{\pi d_w}{t_s CPR} counts_k, \tag{16.18}$$

$$w_{\text{filt},i,k} = \psi w_{\text{filt},i,k-1} + (1 - \psi)w_{\text{raw},i,k}, \tag{16.19}$$

$$w_{\text{err},i,k} = w_{\text{filt},i,k} - w_{\text{des},i,k}, \tag{16.20}$$

$$w_{\text{errInt},i,k} = (1 - t_s\Upsilon)w_{\text{errInt},i,k} + w_{\text{err},i,k}t_s. \tag{16.21}$$

Integrated velocity error is capped at that corresponding to maximum available torque given the control gains. We calculate desired torques $\tau_{i,k}$:

$$\tau_{i,k} = K_p w_{\text{err},i,k} + K_i w_{\text{errInt},i,k}. \tag{16.22}$$

Torques are converted into motor voltages using the motor properties from the manufacturer, wheel velocity data, and continuously monitored battery voltage V_{cc}. The back-EMF induced loss of torque per velocity is γ, and the torque constant at stall is β. The desired voltage $V_{i,k}$ is then

$$V_{i,k} = \frac{\tau_i}{Vcc_k\beta - \gamma w_{\text{filt},i,k}}. \tag{16.23}$$

We pulse width modulate (PWM) by grounding one phase and alternately connecting the appropriate other phase to V_{cc} and ground. PWM is done at 20 kHz to allow RLC filtering of the pulses in the motors to result in an average voltage at motor i of $V_{i,k}$, and for inaudibility to humans.

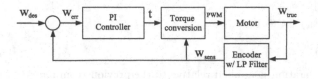

Fig. 16.7. SISO loop on each wheel

16.8.1 Reference Tracking Performance

Reference tracking performance is limited by process noise in the form of slip, and limited bandwidth. The robots have sufficiently powerful motors and high center of mass that weight transfer effects as described in [8] limit acceleration by causing some wheels to slip. Slip invalidates the kinematic transformation T_ψ in Eq. (16.3), and typically the robot both fails to follow the translational reference and rotates erratically if no correction is taken. Rotational control superposes corrective desired rotational velocities on the SISO loops to counteract this effect. The translational effects of skidding are not corrected locally, but only through trajectory generation.

16.8.2 Controller Bandwidth

The wheels' SISO bandwidth is limited by ringing of the wheels at high gain due to low encoder resolution. This is most apparent when the robot is on the carpeted playing surface, where small wheel motions are only weakly coupled to the robot through the compliance and backlash in the carpet, wheel mechanism, and gear mesh.

The effective MIMO bandwidth of the SISO PI controllers can be computed if we assume ideal kinematics in pure $x, y,$ or θ motion and neglect integrator leakage. First, the contribution of each of n loops to stiffness and damping in each direction is given by

$$K_x = \sum_{i=1}^{n} K_i |cos(\phi_i)|, \tag{16.24}$$

$$K_y = \sum_{i=1}^{n} K_i |sin(\phi_i)|, \tag{16.25}$$

$$K_\theta = \sum_{i=1}^{n} K_i |r_i| cos(\psi_i), \tag{16.26}$$

$$B_x = \sum_{i=1}^{n} K_p |cos(\phi_i)|, \tag{16.27}$$

$$B_y = \sum_{i=1}^{n} K_p |sin(\phi_i)|, \tag{16.28}$$

$$B_\theta = \sum_{i=1}^{n} K_p |r_i| cos(\psi_i). \tag{16.29}$$

The only external force on the robot is drag. Drag is actually strongly nonlinear with a static component, complicating tuning and necessitating feedforward control as described in the next subsection. For this linear analysis physical damping C is assumed to be viscous, and the equation of motion and transfer function in each DOF q are

$$M_{eq}\ddot{q} = F_q = -B_q(\dot{q} - \dot{q}_{des}) - C_q\dot{q} - K_q(q - q_{des}), \tag{16.30}$$

$$Q/Q_{des} = \frac{\frac{B_q}{M_{eq}s} + \frac{K_q}{M_{eq}}}{s^2 + \frac{B_q + C_q}{M_{eq}}s + \frac{K_q}{M_{eq}}}, \tag{16.31}$$

where M_{eq} is an effective mass resulting from both the robot body M and the moment of inertia of the motors J_m as amplified by the gear ratio N. The robot body contribution is simply mass for x, y and robot moment of inertia for θ:

$$M_{eq,x} = M + \sum_{i=1}^{n} \frac{2N^2 J_{motor}}{d_w} |\cos(\phi_i)|, \tag{16.32}$$

$$M_{eq,y} = M + \sum_{i=1}^{n} \frac{2N^2 J_{motor}}{d_w} |\sin(\phi_i)|, \tag{16.33}$$

$$J_{eq,\theta} = J_{robot} + \sum_{i=1}^{n} \frac{2N^2 J_{motor}}{d_w} |r_i|. \tag{16.34}$$

This analysis predicts a bandwidth of 47 rad/s. Experimental results for this are in Section 16.10.

The major impact of limited bandwidth on the system is apparent 'mechanical latency' that is actually the phase lag of the effective MIMO loop. Mechanical latency in rotation is negligible, but it is significant in translation. For convenience, the state estimators described in Section 16.4 treat these as pure delays added to the vision/hub/wireless delay.

16.8.3 Feedforward

Feedforward is used to improve reference tracking by compensating for unmodeled translational friction. The torque required to move the robot at a steady rate can be idealized as having a static friction component, a viscous friction component, and residual unmodeled components. The component analogous to viscous friction appears in the closed loop transfer function and provides damping, so we do not attempt to cancel it. The static component is not as benign, and leads to a 'lurching' behavior at low velocities as the integral error must 'wind up' to provide the required starting torque. To alleviate this we add a corrective torque at each wheel. The coefficient K_s is a hand tuned parameter.

$$\tau_{sFF,i} = \frac{K_s}{\dot{x}_L^2 + \dot{y}_L^2}(\dot{x}_L \cos(\phi_i) + \dot{y}_L \sin(\phi_i)) \tag{16.35}$$

Acceleration feedforward was tried to improve transient performance but resulted in high frequency robot motions that tended to cause the ball to lose contact with the robot.

16.9 Alternatives

Various alternative methods have been demonstrated in RoboCup or proposed in the literature.

16.9.1 Trajectory Generation

The most unique feature of our system is its dynamics based trajectory generation algorithm. Alternative path planning methods for omnidirectional vehicles described in [14] and [15] utilize combinations of simple shapes to produce dynamically feasible trajectories.

Rather than recomputing trajectories at every control frame, one could retain trajectories until either their destinations must change, or a collision is detected, as in [16]. The state-based method presented here is optimal for reasons found in [8]. The retained trajectories also represent an augmentation to the internal state of the system, complicating the architecture.

16.9.2 Neural networks applied to state estimation

The FU-Berlin team uses a neural network based nonlinear Smith Predictor for state estimation of its own robots as described in [13]. Its nonlinear elements improve system performance in the presence of slip. This is shown to result in lower typical errors than their linear predictor. No a priori knowledge of robot dynamics is used, and the system coefficients are learned by back-propagation.

16.9.3 MIMO approach to local control

Given that the robot as a whole is a MIMO system, and that optimal MIMO control is well understood, it is worth investigating. Linear quadratic gaussian (LQG) controllers were synthesized using process noise estimates and performance criteria. The glaring omission in this MIMO compensator is slip. When we implemented MIMO methods, we observed that the front wheels slipped at very low accelerations. The SISO approach implicitly does a form of traction control by allocating torque to wheels that are not slipping.

Adding four states to describe wheel velocities and coupling the torque input through the wheel states to the robot is attractive. Clearly the SISO approach is possible within this model if the wheel velocities are the sole performance measure. For improvements over SISO to be made, this coupling must be modelled well. Ultimately we decided to retain SISO loops for their combination of traction control, simplicity, and robustness.

16.9.4 Distributed behavior primitives

We observed at the competition that opponents had provided for sending instructions to their robots to carry out a behavior that required several decision frames to execute, essentially distributing decision making authority to the robot to circumvent game state knowledge latency. For example, the robot could be issued an instruction to turn to a certain angle and then shoot. In our system we must enable the kick manually, and so must either wait for the central hub computer to verify that the robot has turned to the proper angle, or must anticipate the behavior and enable the kicker without verification that the rotation is complete. Clearly in this case it is desirable to distribute behavior, and we already have an onboard rotational estimate appropriate for this. Whether any given decision or behavior should be distributed is an interesting question, and one that grows richer as the sensor environment and communication constraints become more complex.

16.10 Experiments

The hardware experiments presented here demonstrate the performance of the complete system and the performance of the local control loops.

16.10.1 Complete System Loop

Figures 16.8 and 16.9 show how the actual velocity history during a straight line motion with no rotation varies from that computed at the beginning of the motion. Recall that trajectories are recomputed in real time based on current state, so that there is no attempt to 'track' the nominal trajectory calculated at the beginning. Velocity data is simply differentiated position data with smoothing, which is noisier than the estimated velocity but excludes dynamic effects from the Kalman filter. There are two cases, one with low accelerations and velocities that corresponds to very low slip, and the other with the high accelerations and velocities that we use in competition.

The crucial non-ideal behaviors are visible in the more aggressive case of Fig. 16.9. We are willing to tolerate non-ideal behavior in exchange for speed in the competitive RoboCup environment. The robot is only intermittently able to sustain the full acceleration, though it is capable of deceleration at greater than the specified acceleration. This is because during the deceleration the robot actually tips until only two wheels and a nylon skid touch the field, and in that case the skid's friction is helpful rather than a hindrance. Acceleration also seems to be a function of velocity; the reduction in acceleration apparent at $v \geqslant 1.5$ m/s is not a result of controller dynamics. Also, the robot fails to sustain full speed. This is probably due to slip, given the presence of integral control. The integral leak time constant of 5 seconds is too slow to explain this effect. The overshoot can also be traced to slip; slip invalidates the assumption in translational state estimation that commands are executed, so that the calculated trajectories assume the robot is farther from the stopping point than it actually is.

Angular tracking during both trajectories is shown in Fig. 16.10. Angular tracking is very good during the slow trajectory, but becomes poor during skidding, especially when wheels lift off the ground during deceleration.

Figure 16.11 is a position history plot generated by commanding the robot to move in a square 1.5 meters on a side, using both slow and fast trajectory parameters as before. The reason for the erratic performance is that robot executes a half rotation on each leg. The robot always starts facing -x and finishes facing +x. This is challenging because the robot is rotating at high speed while translating at high

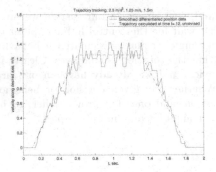

Fig. 16.8. Vel. Hist., Easy Traj.

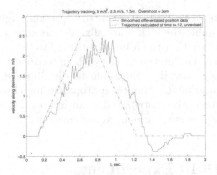

Fig. 16.9. Vel. Hist., Difficult Traj.

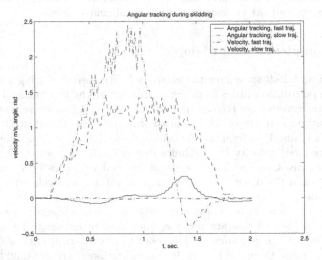

Fig. 16.10. Angular tracking, both trajectories, with vel. hist. for time ref.

speed. The global to local coordinate transformation is based on the rotational estimate, so that at high speed, small angular position estimate errors create large lateral deviations. Due to our state-based trajectory methods, the reader is again cautioned against interpreting the experiment as an attempt to track a box reference path. Severe wheel slip is observed during the fast trajectories but not reflected in the figure except in resulting position history deviations.

16.10.2 Local Control

Figure 16.12 shows the amplitude gain of the local control loop in reference tracking in translation. The reference signal generated consisted of sinusoidal motions with peak amplitudes of 5 m/s^2. The experimental translational bandwidth of 25 rad/s is significantly lower than the bandwidth of 47 rad/s implied by summing the effects of the SISO loops. Because this acceleration does cause slip, we could hypothesize that the two unweighted wheels are slipping and thus do not contribute to the transfer

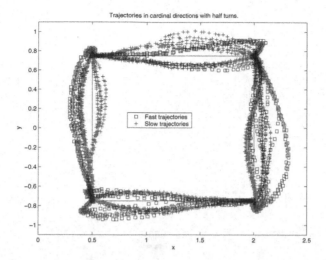

Fig. 16.11. Position histories for linear translations with half rotations

function. This analytical correction results in 27.5 rad/s. Another issue with the data is the absence of the resonant peak predicted by analysis, even with a static feedforward correction. Increasing the viscous friction C_q can correct for this, but the number used in the analysis can be verified in isolation by pulling the robot with a scale. This suggests the existence of an unmodeled dissipative process, possibly in the drive train.

The 25 rad/s experimental bandwidth agrees with the mechanical latency parameter that works well in our system, which corresponds to a 32ms delay. The small dead time approximation

$$e^{-Ts} = \frac{1}{Ts+1} \tag{16.36}$$

from [17] suggests that our compensation is appropriate to a system with a bandwidth of $1/T = 31$ rad/s. Another way to say this is that if we had set our mechanical latency parameter based on transfer function data rather than hand tuning, we would have obtained the same result. Better agreement is impossible given that mechanical latency is only adjusted in 16ms (1 frame) increments.

16.10.3 Obstacle Avoidance

The motivating case of a dynamic environment can not be conveyed here. We can only say that we were never penalized for collisions in our seven games at the competition.

16.10.4 Movies

Movie files of our system in competition can be found at
http://robocup.mae.cornell.edu.

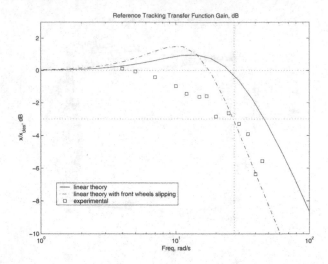

Fig. 16.12. Translational Reference Tracking Transfer Function Gain

16.11 Conclusion

The complete set of algorithms used by the 2005 Cornell RoboCup team is presented. The feasibility and quality of the algorithms is demonstrated by both competitive experience and experiment.

Future work includes automated parameter tuning in both the decision algorithms and the control algorithms. We are also interested in allowing human supervision of the decision algorithms. We do not intend to compete in RoboCup competitions in the near future in order to focus on these goals.

16.11.1 Acknowledgements

This work was supported by grants from NASA, sponsorship from Applied Materials, Microsoft, Lockheed, John Swanson, and UTC, and equipment donations and discounts from Keithley, Maxon, MicroMo, and others. Over the past seven years over one hundred and fifty undergraduates and M.Eng. students have built a new system every year as part of an interdisciplinary project team class, and without their work none of this would exist.

References

1. Kipphan H (2000) Handbook of printmedia. Springer, Berlin Heidelberg New York
2. Brandt J, Hein W (2001) Polymer materials in joint surgery. In: Grellmann W, Seidler S (eds) Deformation and fracture behavior of polymers. Engineering materials. Springer, Berlin Heidelberg New York
3. Che M, Grellmann W, Seidler S (1997) Appl Polym Sci 64:1079–1090

4. Ross DW (1977) Lysosomes and storage diseases. MA Thesis, Columbia University, New York
5. D'Andrea R., Kalmár-Nagy T., Ganguly P., Babish M.: "The Cornell RoboCup Team", in: Stone P., Balch T., Kraetzschmar (Eds), *Robocup 2000: Robot Soccer World Cup IV*, Springer, Berlin, 2001
6. Purwin O., D'Andrea R.: "Cornell Big Red 2003", in: Polani D., Bonarini A., Browning B., Yoshida K. (Eds), *Robocup 2003: Robot Soccer World Cup VII, Lecture Notes in Artificial Intelligence*, Springer, Berlin, 2003
7. D'Andrea et al. Cornell 2005 Team Description. To be published in *RoboCup 2005: Robot Soccer World Cup VIII, Lecture Notes in Artificial Intelligence*. Springer, Berlin, TBA.
8. Purwin O., D'Andrea R.: "Trajectory Generation for Four Wheeled Omnidirectional Vehicles", *Proceedings of the American Control Conference 2005*, 8-10 June 2005, pp. 4979-4984
9. O. J. M. Smith: "Closer control of loops with dead time", Chem. Eng. Progress, 53(5):217219, 1957.
10. Kalmár-Nagy T., D'Andrea R., Ganguly P.: "Near-Optimal Dynamic Trajectory Generation and Control of an Omnidirectional Vehicle", Journal of Robotics and Autonomous Systems, Vol. 46, 2004, pp. 47-64
11. Olsson H., Aström K.J., Canudas de Wit C., Gäfvert M., Lischinsky P.: "Friction Models and Friction Compensation"
12. Browning, B.; Bowling, M.; Veloso, M.M.: "Improbability Filtering for Rejecting False Positives". *Proceedings of ICRA-02, the 2002 IEEE International Conference on Robotics and Automation*, 2002.
13. Sven Behnke, Anna Egorova, Alexander Gloye, Raul Rojas, and Mark Simon: "Predicting away Robot Control Latency." *RoboCup-2003: Robot Soccer World Cup VII*, LNCS, Springer, 2004.
14. Faiz N., Agrawal S.K.: "Trajectory Planning of Robots with Dynamics and Inequalities", *Proceedings of the 2000 IEEE International Conference on Robotics and Automation*, Vol. 4, 2000, pp. 3976-3982
15. Moore K.L., Flann N.S.: "A Six-Wheeled Omnidirectional Autonomous Mobile Robot", Control Systems Magazine, IEEE, Vol. 20, Issue 6, Dec 2000, pp. 53-66
16. Watanabe K., Shiraishi Y., Tzafestas S.G., Tang J., Fukuda T.: "Feedback Control of an Omnidirectional Autonomous Platform for Mobile Service Robots", Journal of Intelligent and Robotic Systems, Vol. 22, Issue 3-4, 1998, pp. 315-330
17. Ogata, Katsuhiko. *Modern Control Engineering*. Prentice-Hall, Inc., Englewood Cliffs NJ, 1970.
18. J.C. Latombe. *Robot Motion Planning*. Kluwer Academic Publishers, 1991.
19. http://control.mae.cornell.edu/Purwin/PhDWork.htm

Transition-Function Based Approach to Structuring Robot Control Software*

Cezary Zieliński

Warsaw University of Technology, Faculty of Electronics and Information
Technology, Institute of Control and Computation Engineering
ul. Nowowiejska 15/19, 00-665 Warsaw, Poland C.Zielinski@ia.pw.edu.pl

17.1 Motivation

Whenever control software is to be coded for a new piece of equipment or for a new task that it has to execute, we notice that there is considerable similarity with the code that has been produced previously for alike devices and tasks. However, if the previously produced software does not have a structure facilitating its reuse it is usually very difficult to extract the useful portions. Moreover, any modification or extension of the old software might be hindered by its inadequate structure. This is especially true for robot control software. It is relatively easy to produce code for a specific device and a specific task, but when those change, it is sometimes easier to start coding from scratch than try to reuse the old pieces. The problem that we want to tackle in this paper is: whether we can give some insight into structuring the robot system control software without a priori knowledge of:

- the exact type of hardware (robot, sensors) that will be used,
- the number of controlled devices,
- the task that will be executed by the system,
- the control paradigm employed (e.g., behavioural, deliberative).

The above stated problem requires adequate tools for its solution. As we shall discuss general software structures, instead of specific solutions, a tool for describing the general system architecture is needed. Moreover, this tool must enable logical derivation of diverse system structures. Only formal methods have this property. Formalization imposes rigor and precision on the discussion. It provides a deeper understanding of the discussed topic and usually discloses otherwise hidden properties of the proposed solutions. Thus this paper will propose a formalism for describing diverse robot control systems.

Our aspiration is not only to provide a tool for the discussion of diverse software structures of the controllers, but also to provide a specification tool facilitating the implementation of those structures. Thus the formalism will use pseudocode that, on the one hand, can easily be translated to any imperative programming language, and on the other hand, will be implementation independent.

* This work was supported by Polish Ministry of Science and Information Technology grant: 4 T11A 003 25.

K. Kozłowski (Ed.): Robot Motion and Control, LNCIS 335, pp. 265–286, 2006.
© Springer-Verlag London Limited 2006

The practical result of the theoretical discussion will be the tool for specifying robot programming frameworks. **Programming frameworks** [15] are application generators with the following components: a library of software modules (building blocks out of which the system is constructed), a pattern according to which ready modules can be assembled into a complete system, and a method for designing new modules that can be included in the library. It should be underscored that, if the functionality of the programming framework is not adequate, it might not be possible to implement some control systems and to execute some tasks. The proposed formalism introduces a unified notation (symbolic denotations) and a transition-function-based approach to describing the state and control regardless of the control paradigm employed (e.g., deliberative [22], behavioral [2,5,6], fuzzy [8]).

This research was motivated by the necessity of quickly producing several controllers for diverse robots executing significantly differing tasks. Controllers for the following systems were designed using the methodology described in this paper:

- a serial-parallel robot exhibiting high stiffness and having a large work-space [18, 33], thus well suited to milling and polishing tasks [16],
- a direct-drive robot without joint limits [19], hence applicable to fast transfer of objects,
- a system of two IRp-6 robots acting as a two-handed manipulation system [25].

The speed with which we can produce a new controller strongly depends on the quantity of readily available software that can be reused from former projects and the extent to which this software can be modified. Programming frameworks are meant to facilitate quick design and reuse of software.

Many robot programming frameworks have been created: RCCL [11], KALI [10], PASRO [3], MRROC++ [28, 29], G^{en}oM [1], DCA [20], TCA [24], TDL [23], Generis [17], to mention only some of them. Depending on the implementation language (e.g., C or C++), procedural or object oriented approach to programming is fostered by those frameworks. Component based approach is also being considered (e.g., DCOM or CORBA [21]), but in this case the overhead of communication between distributed objects (localization and providing of services) usually is an obstacle to the implementation of the hard real-time portion of the software, thus those problems have to be solved within a component. However, component based software is a viable alternative for implementing systems composed of cooperating embodied agents needing coordination or for implementing soft real-time portion of the software within an agent. Currently efforts are being made to produce public domain generic robot control software (e.g., the OROCOS project [7] or the Player/Stage suite of software [9, 26]).

Although many robot programming frameworks have been implemented, no formal tool for their specification has been devised. Their quality, to a large extent, depends on the ingenuity and experience of their designers, rather than on systematic derivation and the discussion of many design possibilities. Because of the lack of a formal method of their specification, it is hard to compare the properties of those frameworks or to ascertain that implementation of a system executing a predefined task is possible. This paper proposes a formal tool that can solve the above problems.

17.2 Agents

The paper will treat robots as embodied agents, but also virtual agents will be discussed. A broad definition of an agent is assumed here. An **agent** is anything that can be viewed as perceiving its environment and acting upon that environment. However, we shall be considering the physical environment, and not virtual environments (e.g., the Internet). Thus, **embodied agents**, i.e. entities that can be viewed as perceiving the physical environment through sensors and acting upon that environment through effectors, will be at the focus of our attention. The distinction between the two types of agents is based on whether they posses an effector. Embodied agents have material bodies (i.e., effectors) and thus can influence the environment directly. **Virtual agents** do not have material bodies and thus cannot influence the environment directly – they have to do this indirectly through embodied agents. However, both kinds of agents have their own sensing capabilities.

In general a multi-agent system composed of agents a_j will be considered:

$$A = \{a_0, a_1, \ldots, a_{n_a}\}, \tag{17.1}$$

where $j = 0, \ldots, n_a$. The state of the agents, for a fixed structure system, will be represented by:

$$s = \langle s_0, s_1, \ldots, s_{n_a} \rangle, \tag{17.2}$$

where s_j is the state of agent a_j. However, if the structure of the system changes in time, due to communication links between agents appearing and disappearing, the state (17.2) does not represent the state of the multi-agent system as a whole. In that case the representation of the current structure of the system must be included in the discussion, thus a directed-graph representing the connections must be defined. Let the set of directed arcs of a graph be:

$$L = A \times A. \tag{17.3}$$

Hence the communication links between the agents, at a certain instant of time i (we assume discrete time), will be represented by:

$$L_a^i \subset L. \tag{17.4}$$

The structure of the communication network between the agents, at an instant i, is a directed graph:

$$G^i = \langle A, L_a^i \rangle. \tag{17.5}$$

Thus the state of a multi-agent system with a changing structure is:

$$s_s^i = \langle s^i, G^i \rangle. \tag{17.6}$$

This represents an *ad hoc* network of communicating agents (e.g., [12]). This is a standard representation used in telecommunications, and thus methods developed within that field of knowledge can be used for the analysis of such networks of agents. Further on we shall concentrate on the behaviour of a single agent.

17.3 State of a Single Agent

A system consisting of n_e embodied agents is considered. The state of an embodied agent a_j, $j = 1, \ldots, n_e$ is composed of:

$$s_j = \langle c_j, e_j, V_j, T_j \rangle \tag{17.7}$$

c_j – state of the **control subsystem** (e.g., internal data structures),
e_j – state of the **effector**,
V_j – bundle of **virtual sensor** readings,
T_j – information transmitted to/from the other agents.
To be brief, and because of contextual obviousness, the denotations assigned to the subcomponents of the considered system and their state are not distinguished.

Moreover, the system may contain zero or more virtual agents a_j, $j = 0$ or $j = n_e + 1, \ldots, n_a$. Their state is represented by:

$$s_j = \langle c_j, V_j, T_j \rangle. \tag{17.8}$$

Quite often the system contains only a single virtual agent, the coordinator. To underscore its role in the system it is appended to the list of agents at its beginning rather than at its end, and it is distinguished by the subscript $j = 0$. All the other virtual agents are appended to the end of the list of embodied agents ($j = n_e + 1, \ldots, n_a$). In this paper we shall consider only a single virtual agent a_0 (thus $n_e = n_a$), which will perform either coordinating activities for several embodied agents or deliberations for one or more of those agents.

As in living creatures, **receptors** in technical systems can be divided into three categories, depending on the source of stimuli that they respond to.

- **Exteroceptors** – receptors that detect stimulus external to the body (e.g., vision, smell, touch).
- **Proprioceptors** – receptors that detect stimulus from inside of the limbs. They enable perception of position of the limbs and body.
- **Interoceptors** – receptors that detect the stimulus from the internal organs of the body – they are instrumental in maintaining homeostasis. In technical systems interoceptors supply information about the internal condition of the control subsystem (e.g., error detection – errors as a source of pain).

Both in animals and technical systems receptors can exhibit duality. The same sensor can be sensitive to stimuli coming from different sources (e.g., a force sensor can detect collisions, and thus it is an exteroceptor, or internal forces acting in the wrist, hence it is a proprioceptor [35]).

A bundle of **virtual sensor** readings, associated with the agent a_j, contains n_{v_j} individual virtual sensor readings:

$$V_j = \langle v_{j_1}, \ldots, v_{j_{n_{v_j}}} \rangle. \tag{17.9}$$

Each virtual sensor v_{j_k}, $k = 1, \ldots, n_{v_j}$, produces an aggregate reading from one or more hardware sensors – exteroceptors. The data obtained from the exteroceptors usually cannot be used directly in motion control, e.g., control of arm motion requires the grasping location of the object and not the bit-map delivered by a camera. In other cases a simple sensor would not suffice to control the motion (e.g., a single touch sensor), but several such sensors deliver meaningful data. The process of extracting meaningful information for the purpose of motion control is named **data aggregation** and is performed by virtual sensors. Thus the kth virtual sensor reading obtained by the agent a_j is formed as:

$$v_{j_k} = f_{v_{j_k}}(c_j, R_{j_k}),\qquad(17.10)$$

where R_{j_k} is a bundle of exteroceptor readings used for the creation of the kth virtual sensor reading:

$$R_{j_k} = \langle r_{j_{k_1}},\ldots,r_{j_{k_{n_r}}}\rangle.\qquad(17.11)$$

where n_r is the number of exteroceptor readings $r_{j_{k_l}}$, $l = 1\ldots,n_r$, taken into account in the process of forming the reading of the kth virtual sensor of the agent a_j. It should be noted that (17.10) implies that the reading of the virtual sensor depends also on c_j. In this way the agent has the capability of configuring the sensor as well as delivering to the virtual sensor the relevant information about the current state of the agent (including its effector). This might be necessary in the case of computing the reading of a virtual sensor having its associated exteroceptors mounted on the effector (e.g. artificial skin).

There are diverse methods of expressing effector state e_j. How we express e_j depends both on the type of the embodied agent and the task that it has to execute. Manipulators, walking machines or underwater vehicles have different effectors, thus positions (and/or velocities) of different devices will have to be included in e_j. Moreover, diverse views of those devices might be used by the designers (e.g., joint versus end-effector space representation of manipulator position). If the task that is to be executed requires only the change of position and orientation of the device, then e_j will contain only pose information, but if the device has to exert forces on the environment, then this information will have to be included too.

The responsibility of the embodied agent's a_j control subsystem c_j is to:

- gather information about the environment through the associated virtual sensor bundle V_j,
- obtain the information from the other agents $a_{j'}$ ($j' \neq j$),
- monitor the state of its own effector e_j,

and to process all of this information to produce:

- a new state of the effector e_j,
- influence the future functioning of the virtual sensors V_j,
- communicate with the other agents $a_{j'}$.

As a side effect, the internal state of the control subsystem c_j changes. Thus three types of components of the control subsystem must be distinguished:

- input components providing the information about: the state of the effector, virtual sensor readings and the messages obtained from the other agents (they are distinguished by a leading subscript x),
- output components exerting influence over: the state of the effector, configuration of virtual sensors and the messages to be transmitted to the other agents (they are distinguished by a leading subscript y),
- other resources needed for data processing within the control subsystem (without a leading subscript).

Hence, the following subdivision results (Fig.17.1):

$_xc_{e_j}$ –   input image of the effector (a set of data conforming to the assumed input model of the effector in the control subsystem – it is produced by processing the input signals transmitted from the effector to the control subsystem, e.g., motor shaft positions, joint angles, end-effector location – they form diverse ontologies),

$_xc_{V_j}$ –   input images of the virtual sensors (current virtual sensor readings – control subsystem's perception of the sensors and through them of the environment),

$_xc_{T_j}$ –   input of the inter-agent transmission (information obtained from other agents),

$_yc_{e_j}$ –   output image of the effector (a set of data conforming to the assumed output model of the effector in the control subsystem – e.g., PWM ratios supplied to the motor drivers; thus the input and output models of the effector need not be the same),

$_yc_{V_j}$ –   output images of the virtual sensors (current configuration and commands controlling the virtual sensors),

$_yc_{T_j}$ –   output of the inter-agent transmission (information transmitted to other agents),

c_{c_j} –   all the other relevant variables taking part in data processing within the agent's control subsystem.

From the point of view of the system designer the state of the control subsystem changes at a servo sampling rate or a low multiple of that. If i denotes the current instant, the next considered instant is denoted by $i+1$. This will be called a **motion step**. The control subsystem uses:

$$_xc_j^i = \langle c_{c_j}^i, \, _xc_{e_j}^i, \, _xc_{V_j}^i, \, _xc_{T_j}^i \rangle \tag{17.12}$$

to produce:

$$_yc_j^{i+1} = \langle c_{c_j}^{i+1}, \, _yc_{e_j}^{i+1}, \, _yc_{V_j}^{i+1}, \, _yc_{T_j}^{i+1} \rangle. \tag{17.13}$$

For that purpose it uses **transition functions**:

$$\begin{cases} c_{c_j}^{i+1} = f_{c_{c_j}}(c_{c_j}^i, \, _xc_{e_j}^i, \, _xc_{V_j}^i, \, _xc_{T_j}^i), \\ _yc_{e_j}^{i+1} = f_{c_{e_j}}(c_{c_j}^i, \, _xc_{e_j}^i, \, _xc_{V_j}^i, \, _xc_{T_j}^i), \\ _yc_{V_j}^{i+1} = f_{c_{V_j}}(c_{c_j}^i, \, _xc_{e_j}^i, \, _xc_{V_j}^i, \, _xc_{T_j}^i), \\ _yc_{T_j}^{i+1} = f_{c_{T_j}}(c_{c_j}^i, \, _xc_{e_j}^i, \, _xc_{V_j}^i, \, _xc_{T_j}^i). \end{cases} \tag{17.14}$$

This can be written down more compactly:

$$_yc_j^{i+1} = f_{c_j}(_xc_j^i). \tag{17.15}$$

Formula (17.15) is a prescription for evolving the state of the system, thus it has to be treated as a program of the agent's behavior. For any agent exhibiting useful behaviors this function would be very complex, because it describes the actions of the system throughout its existence. The complexity of this function renders impractical the representation of the program of agent's actions as a single function. The function (17.15) has to be decomposed to make the specification of the agent's program of actions comprehensible and uncomplicated.

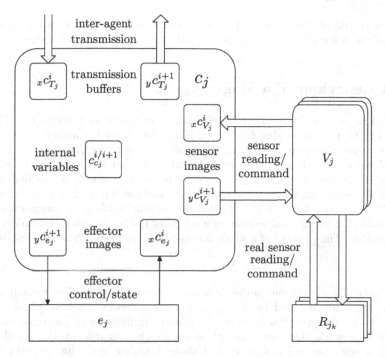

Fig. 17.1. A single embodied agent a_j, $j = 1, \ldots, n_e$ (thick arrows represent the connection to many entities, e.g., virtual sensors or other agents, while thin arrows to a single entity, e.g., the effector)

In the process of producing the output values $_y c_j^{i+1}$ the values of inputs $_x c_j^i$ and internal variables $c_{c_j}^i$ are used. The internal variables change their values, thus $c_{c_j}^{i+1}$ is created – those will be the values of internal variables at the onset of motion step $i + 1$. All of the mentioned quantities are stored in the agent's memory, thus their values form the total state of the agent's control subsystem. Obviously the process of computing the output values from the values of the input and internal agent's variables takes time, so the valid total state is obtained after the result of those computations is ready. This result is transferred to the other components of the system. Hence, the control subsystem total state holds the data obtained from the other components of the system and values of its own internal variables, both valid at the start of step i, and the output values computed during the initial part of the motion step i (those will be transmitted to the other components of the system at the end of motion step i, thus controlling the system in motion step $i + 1$). All of those values are memorized within the agent during the later part of motion step i, thus they compose the total internal state of the control subsystem:

$$c_j^i = \langle c_{c_j}^{i/i+1}, \, _x c_{e_j}^i, \, _y c_{e_j}^{i+1}, \, _x c_{V_j}^i, \, _y c_{V_j}^{i+1}, \, _x c_{T_j}^i, \, _y c_{T_j}^{i+1} \rangle. \tag{17.16}$$

The **total state** contains the input values, the internal variables, and the produced output (control). All of those are valid after the computations of (17.14) are completed. The focus of our attention is the creation of $_y c_j^{i+1}$ from $_x c_j^i$, i.e., the form of transition functions (17.14). The symbol $c_{c_j}^{i/i+1}$ underscores the change of state of the internal variables due to computations of (17.14) within step i.

17.4 Behaviour of a Single Agent

Internal functioning of an agent is defined by the transition function (17.15). The flexibility of a programming framework is attributed to the ability of expressing diverse approaches to programming the actions of each agent, and so the proposed formal description should enable easy formulation of diverse control strategies. In the case of a robot programming framework one should concentrate on the definition of motion commands. One feasible approach is: instead of providing a single function (17.15), describing the motion of an agent throughout its life, many simpler functions are specified, defining small motion segments, and the final result is obtained by their composition. Thus instead of a single function f_{c_j}, n_f partial functions are defined:

$$ _y c_j^{i+1} = {}^m f_{c_j}(_x c_j^i), \qquad m = 1, \ldots, n_f. \tag{17.17} $$

Variability of agents is due to the diversity of those functions. The more functions of this type are provided by a programming framework (its library), or can be expressed by the programmer using the general facilities (e.g., patterns) provided by the framework, the more types of agents can be constructed. Besides that the means of selection and composition of those functions must be provided by the framework. The actions brought about by the selected functions can be composed either sequentially (concatenated time slices) or superposed (composed within a single time slice). Obviously a mixture of the two is possible. The multiplicity of functions (17.17) requires some means of their selection. For that purpose predicates $^q p_{c_j}$, $q = 1, \ldots, n_p$, will be used. Predicates, as defined in first-order predicate calculus (e.g., [22]), are relations between objects of the domain of discourse, that can be evaluated to true, when they hold, or false, if not. In other words, this are expressions (conditions) which are either true or false depending on the current values of their arguments. A very general form of selection based on $_x c_j^i$ (i.e., on (17.12)) is assumed here. Predicate-based selection instructions of the following form will be used:

$$ \textbf{if } {}^q p_{c_j}(_x c_j^i) \textbf{ then } _y c_j^{i+1} := {}^m f_{c_j}(_x c_j^i) \textbf{ endif} \tag{17.18} $$

In actual systems an endless loop containing the conditional instruction (17.18) must be constructed. As many such instructions will appear within such a loop, two cases can be considered: sequential or parallel, i.e.:

- single predicate $^q p_{c_j}(_x c_j^i)$ can be true at a time – i.e., sequential composition,
- several predicates $^q p_{c_j}(_x c_j^i)$ can be true simultaneously – i.e., parallel composition.

If several predicates are true simultaneously, we can compose the outcome by:

- competitive methods (selection of the best), e.g.:

$$ _y c_j^{i+1} = \max_m \{ {}^m f_{c_j}(_x c_j^i) \}, \tag{17.19} $$

- cooperative methods (superposition of all), e.g.:

$$_yc_j^{i+1} = \sum_{m=1}^{n_f} \frac{w_m}{w} \, {}^mf_{c_j}(_xc_j^i), \quad w = \sum_{m=1}^{n_f} w_m, \tag{17.20}$$

where w_m are the weights.

In general, composition of results for several true predicates either pertains to one or several components of $_yc_j^{i+1}$, where

$$_yc_j^{i+1} = \langle\, {}^1_yc_j^{i+1}, {}^2_yc_j^{i+1}, \ldots, {}^{n_c}_yc_j^{i+1}\,\rangle \tag{17.21}$$

and n_c is the number of components. Here we assume that each of the four components of (17.13) can be subdivided into several subcomponents, thus we end up with a total of n_c subcomponents without any particular functional designation (i.e., not explicitly pertaining to effectors, virtual sensors or transmission buffers). Hence, the following situations can be distinguished:

- assembly of disjoint calculations of components of $_yc_j^{i+1}$, where each component is calculated separately:

$$_y^m c_j^{i+1} = {}^mf_{c_j}(_xc_j^i), \quad m = 1, \ldots, n_c, \tag{17.22}$$

- superposition of partial results for the vector $_yc_j^{i+1}$ (17.21) treated as a single entity:

$$_yc_j^{i+1} = f'_{c_j}(_xc_j^i) + f''_{c_j}(_xc_j^i) + \ldots + f'^{\ldots'}_{c_j}(_xc_j^i), \tag{17.23}$$

- superposition for each component and assembly of components (combination of the above two methods):

$$_y^m c_j^{i+1} = {}^mf'_{c_j}(_xc_j^i) + {}^mf''_{c_j}(_xc_j^i) + \ldots + {}^mf'^{\ldots'}_{c_j}(_xc_j^i), \quad m = 1, \ldots, n_c, \tag{17.24}$$

where the $+$ operator in (17.23) and (17.24) stands for any form of composition, e.g. addition or weighted superposition (17.20). In the formulas (17.19–17.24) it is assumed that all of the partial functions produce compatible values, so that the proposed mathematical operations can be performed.

For systems, where only one predicate can be true (i.e., in the case of sequential composition), the pseudocode will assume the following form:

```
loop
      // Determine the current state of the agent
      e_j ↦ xc^i_{e_j};  V_j ↦ xc^i_{V_j};  c_{T_{j'}} ↦ xc^i_{T_j};
      // Compute the next state of the agent
      if  ¹p_{c_j}(xc^i_j)  then  yc^{i+1}_j := ¹f_{c_j}(xc^i_j)  endif
      if  ²p_{c_j}(xc^i_j)  then  yc^{i+1}_j := ²f_{c_j}(xc^i_j)  endif
      ..............................................
      if  ^{n_p}p_{c_j}(xc^i_j)  then  yc^{i+1}_j := ^{n_p}f_{c_j}(xc^i_j)  endif
      // Transmit the results of computations
      yc^{i+1}_{e_j} ↦ e_j;  yc^{i+1}_{V_j} ↦ V_j;  yc^{i+1}_{T_j} ↦ c_{T_{j'}};
      i := i + 1;
endloop
```
$$\tag{17.25}$$

where n_p is the number of predicates used, and thus the number of the if instructions employed. The double slash precedes the comments and the symbol "$\rightarrowtail$" denotes transmission of data between components of the system. Those transmissions result in data input, execution of motion by the effectors, configuration of virtual sensors, and transmission of messages to the other agents. The instruction $i := i + 1$ synchronizes the loop iterations with the agent's motion steps (i.e., implements the agent's heartbeat).

Allowing several predicates to be true simultaneously, we need to modify (17.25) to enable composition of the obtained partial results:

loop
// Determine the current state of the agent
$$e_j \rightarrowtail {}_x c^i_{e_j}; \quad V_j \rightarrowtail {}_x c^i_{V_j}; \quad c_{T_{j'}} \rightarrowtail {}_x c^i_{T_j};$$
$$\text{for } q = 1, \ldots, n_p : \text{clear}({}^q_y c^{i+1}_j);$$

// Compute the next control subsystem state
$$\text{if } {}^1 p_{c_j}({}_x c^i_j) \quad \text{then } {}^1_y c^{i+1}_j := {}^1 f_{c_j}({}_x c^i_j) \quad \text{endif}$$
$$\text{if } {}^2 p_{c_j}({}_x c^i_j) \quad \text{then } {}^2_y c^{i+1}_j := {}^2 f_{c_j}({}_x c^i_j) \quad \text{endif}$$
$$\cdots\cdots\cdots\cdots\cdots\cdots\cdots\cdots\cdots\cdots\cdots\cdots\cdots$$
$$\text{if } {}^{n_p} p_{c_j}({}_x c^i_j) \quad \text{then } {}^{n_p}_y c^{i+1}_j := {}^{n_p} f_{c_j}({}_x c^i_j) \quad \text{endif}$$
// Compute the aggregate control   (17.26)
$$_y c^{i+1}_j := \text{composition}({}^q_y c^{i+1}_j); // \text{ for } q = 1, \ldots, n_p$$
// Transmit the results of computations
$$_y c^{i+1}_{e_j} \rightarrowtail e_j; \quad _y c^{i+1}_{V_j} \rightarrowtail V_j; \quad _y c^{i+1}_{T_j} \rightarrowtail c_{T_{j'}};$$
$$i := i + 1;$$
endloop

This pseudocode assumes sequential execution of its instructions. The involved computations should take significantly less time than one motion step. However, a concurrent version of this pseudocode is also possible (see further on: (17.33) and (17.34)).

Both the pseudocode (17.25) and (17.26) produce frequently switched systems, as the transition functions are chosen for just one motion step i. The selection of a function ${}^m f_{c_j}(\bullet)$ can be made once for a certain number of motion steps, all the more so as functions ${}^m f_{c_j}(\bullet)$ have the capability of testing all the components of ${}_x c^i_j$ in each motion step. Thus evaluation of the predicates ${}^m p_{c_j}$ in each motion step usually is too frequent. Having said that, we need to decide whether the number of motion steps, for which a transition function is chosen, should be fixed or variable, and what should be the criterion determining the number of steps for which this transition function is valid.

First let us consider behaviours with a fixed number of motion steps, i.e., fixed duration behaviours:

$$^{m'} b^i_j = \{c^{i+1}_j, c^{i+2}_j, \ldots, c^{i+n_s}_j\}, \quad m' = 1, \ldots, n_b, \quad (17.27)$$

where n_b is the number of behaviours, and n_s is the number of motion steps within a behaviour. Within $^{m'} b^i_j({}_x c^i_j)$ transitions $c^{i+\epsilon}_j \to c^{i+\epsilon+1}_j$, $\epsilon = 0, \ldots, n_s - 1$, are computed using one of the transition functions $^m f_{c_j}(\bullet)$, i.e., the behaviour defining function. Moreover, in each step of the behaviour ${}_x c^{i+\epsilon}_j$ must be determined and the result of computations ${}_y c^{i+\epsilon+1}_j$ must be transmitted for execution.

As $^{m'}b_j^i$ is multi-step, i in the definition of the behaviour looses significance, hence $^{m'}b_j^i = {}^{m'}b_j$. The modified form of the predicate based selection instruction assumes the form:

$$\text{if } {}^q p_{c_j}({}_x c_j^i) \text{ then } {}^{m'}b_j({}_x c_j^i) \text{ endif} \tag{17.28}$$

In the sequel, pseudocode for a simple behavioural system with **fixed duration behaviours** and a single predicate being true has the following form:

```
loop
    // Select and execute the next behavior
    if  ¹pcⱼ(ₓcⱼⁱ)    then  ¹bⱼ(ₓcⱼⁱ);  continue;  endif
    if  ²pcⱼ(ₓcⱼⁱ)    then  ²bⱼ(ₓcⱼⁱ);  continue;  endif
    ...................................
    if  ⁿᵇpcⱼ(ₓcⱼⁱ)   then  ⁿᵇbⱼ(ₓcⱼⁱ); continue;  endif
    //i := i + nₛ;
endloop
```

$$(17.29)$$

The comment containing the instruction $i := i + n_s$ signals that each iteration of the loop consists of the execution of n_s steps, but the actual incrementation of the counter i takes place within the code of behaviours $^{m'}b_j$. In the following pieces of pseudocode this remainding comment will be omitted. It is assumed, that here and in subsequent pseudocodes, prior to entering the loop $_x c_j^i$ has been determined. In the literature (e.g., [2]) the behaviours $^{m'}b_j(\bullet)$ use specific names, such as: search, avoid, roam $etc.$

The pseudocode (17.29) can represent a single **complex behaviour** $^{m''}b_j'$. Thus, behaviours can be composed recursively, producing a complex system:

```
loop
    // Select and execute the next behavior
    if  ¹pcⱼ(ₓcⱼⁱ)    then  ¹bⱼ'(ₓcⱼⁱ);  continue;  endif
    if  ²pcⱼ(ₓcⱼⁱ)    then  ²bⱼ'(ₓcⱼⁱ);  continue;  endif
    ...................................
    if  ⁿᵇpcⱼ(ₓcⱼⁱ)   then  ⁿᵇbⱼ'(ₓcⱼⁱ); continue;  endif
endloop
```

$$(17.30)$$

where $^{m''}b_j'({}_x c_j^i)$, $m'' = 1, \ldots, n_b'$, are either simple or complex behaviours. A hierarchy of reactions with a variable granularity results. The structures of pseudocodes (17.29) and (17.30) are the same, proving that recursive composition of complex behaviours from simpler ones does not require additional coding.

Usually fixed duration behaviours are over-constraining. Variable duration behaviours rely on the fact that the transition functions (17.17) have access to all of the components of $_x c_j^i$, because of (17.12). Thus, within those functions a terminal condition can be tested. The **terminal conditions** will be represented by Boolean functions $^m f_{\tau_j}$, $m = 1, \ldots, n_\tau$, where n_τ is the number of those functions. The pseudocode for a simple **variable duration behaviour** is:

$$
\begin{array}{l}
\textbf{loop}\\
\quad \text{// Check the terminal condition}\\
\quad \textbf{if } {}^{m}f_{\tau_j}(\,{}_{x}c_j^i\,) = \textbf{true}\\
\qquad \textbf{then break // Quit the loop}\\
\quad \textbf{endif}\\
\quad \text{// Compute the next control subsystem state}\\
\quad {}_{y}c_j^{i+1} := {}^{m}f_{c_j}(\,{}_{x}c_j^i\,);\\
\quad \text{// Transmit the results}\\
\quad {}_{y}c_{e_j}^{i+1} \rightarrowtail e_j; \quad {}_{y}c_{V_j}^{i+1} \rightarrowtail V_j; \quad {}_{y}c_{T_j}^{i+1} \rightarrowtail c_{T_{j'}};\\
\quad i := i+1;\\
\quad \text{// Determine the current state of the agent}\\
\quad e_j \rightarrowtail {}_{x}c_{e_j}^i; \quad V_j \rightarrowtail {}_{x}c_{V_j}^i; \quad c_{T_{j'}} \rightarrowtail {}_{x}c_{T_j}^i;\\
\textbf{endloop}
\end{array}
\qquad (17.31)
$$

This can be treated as a single thread variable duration behaviour. It is a simplified version of the `Move` instruction that `MRROC++` uses [28–30]. Treating (17.31) as the pseudocode of the `Move` instruction, complex variable duration behaviour assumes the following form:

$$
\begin{array}{l}
\text{//Determine the initial state of the agent}\\
e_j \rightarrowtail {}_{x}c_{e_j}^i; \quad V_j \rightarrowtail {}_{x}c_{V_j}^i; \quad c_{T_{j'}} \rightarrowtail {}_{x}c_{T_j}^i;\\
\textbf{loop}\\
\quad \text{// Select and execute the next behavior}\\
\quad \textbf{if } {}^{1}p_{c_j}({}_{x}c_j^i) \textbf{ then Move(} {}^{1}f_{c_j},\ {}^{1}f_{\tau_j}\textbf{)}; \ldots \textbf{endif}\\
\quad \textbf{if } {}^{2}p_{c_j}({}_{x}c_j^i) \textbf{ then Move(} {}^{2}f_{c_j},\ {}^{2}f_{\tau_j}\textbf{)}; \ldots \textbf{endif}\\
\quad \cdots\cdots\cdots\cdots\cdots\cdots\cdots\cdots\cdots\cdots\cdots\cdots\\
\quad \textbf{if } {}^{n_p}p_{c_j}({}_{x}c_j^i) \textbf{ then Move(} {}^{m}f_{c_j},\ {}^{m}f_{\tau_j}\textbf{)}; \ldots \textbf{endif}\\
\textbf{endloop}
\end{array}
\qquad (17.32)
$$

Both the fixed and variable duration behaviour pseudocodes, (17.29) and (17.32), assume that only a single predicate within the loop is true. However, we can also assume that several predicates are true, but that requires multi-thread implementation. The pseudocode for a multi-thread variable duration behaviour requires the following form of the `Move` instruction components:

$$
\begin{array}{l}
\textbf{loop}\\
\quad \text{// Check the terminal condition}\\
\quad \textbf{if } {}^{q}f_{\tau_j}(\,{}_{x}c_j^i\,) = \textbf{true}\\
\qquad \textbf{then } {}_{y}^{q}c_j^{i+1} := \text{null; break // Quit the loop}\\
\quad \textbf{endif}\\
\quad \text{// Compute the partial control subsystem state}\\
\quad {}_{y}^{q}c_j^{i+1} := {}^{q}f_{c_j}(\,{}_{x}c_j^i\,);\ \text{// This partial state is used}\\
\qquad\qquad\qquad\qquad\qquad\qquad \text{// by the composition thread}\\
\quad \texttt{Signal(CT)};\ \text{// that the partial result is ready}\\
\quad \texttt{Wait(CT)};\ \text{// for the composition thread (CT)}\\
\qquad\qquad\qquad\qquad \text{// to execute its iteration}\\
\textbf{endloop}
\end{array}
\qquad (17.33)
$$

Each of those components produces a partial value, that has to be dispatched to a **composition thread** (CT) to be assembled into a complete control value, that will be transferred to the other subsystems of the agent.

```
loop
    Wait (all Move); // for all partial results to be ready
    // Compose the next control subsystem state
```
$$_yc_j^{i+1} := \text{composition}(_y^qc_j^{i+1}); \quad // \text{ for } q = 1, \dots, n_p$$
```
    // Transmit the results
```
$$_yc_{e_j}^{i+1} \rightarrowtail e_j; \quad _yc_{V_j}^{i+1} \rightarrowtail V_j; \quad _yc_{T_j}^{i+1} \rightarrowtail c_{T_{j'}};$$
$$i := i + 1;$$
```
    // Determine the current state of the agent
```
$$e_j \rightarrowtail {}_xc_{e_j}^i; \quad V_j \rightarrowtail {}_xc_{V_j}^i; \quad c_{T_{j'}} \rightarrowtail {}_xc_{T_j}^i;$$
```
    Signal (all Move); // that the composition thread
                       // iteration is complete
endloop
```
$$(17.34)$$

The composition thread and the Move instruction components need to be synchronized with each other. The Wait and Signal instructions are responsible for that. The $i := i + 1$ instruction is responsible for synchronization with the agent's heartbeat. A complex behaviour can be formed by (17.32), if the composition of (17.33) and (17.34) is treated jointly as the code of the Move instruction.

At the level of behaviors the system gradually changes its image from synchronous (time-driven) into asynchronous (event-driven), thus the index i has been discarded in (17.28), (17.29) and (17.30). This is due to the fact that only at runtime the value of predicates is known (an event is detected by the fact that the predicate is true) and so only then we know which of the fixed duration behaviours will be chosen for execution, and thus only then the number of steps needed to complete it becomes known (each fixed duration behaviour might need a different number of motion steps to complete). In the case of variable duration behaviours (17.31) or (17.32) it is even harder to predict the number of steps that are necessary for their completion. The function ${}^mf_{\tau_j}$ is treated as an event detector, so a fully event-driven system results. This shows how a fully synchronous view of the system (17.25) gradually changes to a completely asynchronous, event-driven, one (17.32).

17.5 Types of Systems

17.5.1 Deterministic versus Indeterministic Systems

A **deterministic system**, being in a certain state, under a specific external influence, always produces the same behaviour. On the contrary, an **indeterministic system**, being in a certain state, under the same external influence, may produce different behaviours at different instances. For deterministic systems neither the predicates nor the transition functions contain random values. For indeterministic system either the predicates or transition functions contain random values. Here we shall consider only predicates with random values. Introduction of indeterminism, supplemented by addition of a critique, is interesting to us, because it enables the creation of learning systems. However, this problem will not be discussed in this paper.

Let us distinguish a certain random component of c_{c_j} by the superscript φ. Comparison of a random number $^\varphi c_{c_j} \in [0, 1]$ to the threshold probability $^\theta P$ forms a predicate:

$$^\varphi c_{c_j} \leqslant {}^\theta P. \tag{17.35}$$

In the case of deterministic agents each behavior $^{m'} b_j(_x c_j^i)$, $m' = 1, \ldots, n_b$, is executed only when an associated predicate is *true*. Likewise, with indeterministic systems, if a randomly chosen number $^\varphi c_{c_j}$ from the range $[0, 1]$ is below a threshold value $^\theta P$, the associated behaviour is executed. The higher the threshold value $^\theta P$, the higher the probability of executing the associated behaviour. With such predicates any of the algorithms (17.25), (17.26), (17.29), (17.30) or (17.32) can be used to construct an indeterministic system.

A system using (17.35) has a fixed threshold. A more realistic system can be produced if the probability of performing a certain action is associated with the level of stimulus. Swarm intelligence systems [4] mimicking the behavior of ants or bees frequently rely on such an approach. The stimulus can come from the environment (in this case $_x c_{V_j}^i$ is used) or from the other agents (then $_x c_{T_j}^i$ is utilized). Let us assume that one of the the the components of $_x c_{V_j}^i$ or $_x c_{T_j}^i$ is used, so let σ denote V or T, and let that component be singled out by the symbol φ, as in (17.35), so $^\varphi_x c_{\sigma_j}^i$ is being considered. The probability in (17.35) of executing a certain behavior can be expressed as [4]:

$$^\theta P = \frac{(^\varphi_x c_{\sigma_j}^i)^{n_\theta}}{(^\varphi_x c_{\sigma_j}^i)^{n_\theta} + (\theta_p)^{n_\theta}}, \quad n_\theta \in \mathcal{N}, \tag{17.36}$$

where θ_p is the stimulus threshold and n_θ is an appropriately chosen positive integer – the simulation experiments in [4] were carried out with $n_\theta = 2$. In such a system, when the stimulus is low (i.e., $^\varphi_x c_{\sigma_j}^i \ll \theta_p$) the probability of executing an associated action is next to nil, but if the stimulus is high (i.e., $^\varphi_x c_{\sigma_j}^i \gg \theta_p$) the action is executed almost certainly. Sometimes actions are performed only when stimulus is within a certain range. To express this fact the Gauss function can be used:

$$^\theta P = e^{-[\psi(^\varphi_x c_{\sigma_j}^i - \theta_p)]^2}, \tag{17.37}$$

where θ_p is the mid value in the range and ψ governs the steepness of the rise and fall of probability P_θ and the size of the range.

17.5.2 Crisp versus Fuzzy Systems

Besides the possibility of producing deterministic or indeterministic systems, one can design either crisp or fuzzy systems. The systems discussed so far are **crisp systems** – all the considered values are crisp. The structure of the **fuzzy systems** differs in that the decision process is not binary, i.e., it does not rely on predicates. The decision process in the fuzzy case takes into account the fuzzy rule base, and transforms the crisp input values into fuzzy ones and through aggregation, accumulation and defuzzyfication forms the crisp output values. Thus the decision process (i.e., the choice of appropriate transition functions) and the evaluation of those transition functions, which in the case of crisp systems were disjoint, are combined in the case of fuzzy systems. This influences the structure of the software.

Fuzzy computations are based on the following procedure. First, the measurements are fuzzified, i.e., a crisp value is converted into a degree of membership in every fuzzy set within the domain of that value. Next, aggregation is performed, where the firing strength of each condition in the rule base is computed. The respective firing strengths are then used to scale or trim the conclusions of each rule. This is called the activation of the rules, i.e., deduction of the conclusion reduced by the firing strength of the associated condition. The scaled output fuzzy sets are then accumulated to produce the resulting output fuzzy set. Finally this set is subject to the defuzification process in which a crisp output value is obtained. For that purpose, e.g., the abscissa of the centre of gravity of the composition of output fuzzy sets can be used. For each of the enumerated steps diverse computational methods have been elaborated [8,13]. The above procedure is presented more formally below.

Let the current sensor readings $_x c^i_{V_j}$ contain $_x \hat{c}^i_{V_j}$ – some of the readings that are of interest to us. Moreover, let $_x \hat{c}^i_{V_j}$ belong to a sensor reading subspace $_x \hat{C}_{V_j}$ – a part of the sensor reading space $_x C_{V_j}$. Each dimension of the space $_x \hat{C}_{V_j}$ forms a universe $_x^u \hat{C}_{V_j}$, $u = 1, \ldots, v$, where v is the number of those dimensions (universes). Each universe $_x^u \hat{C}_{V_j}$ can be further subdivided into ρ_u regions $_x^{uh} \hat{C}_{V_j}$, $h = 1, \ldots, \rho_u$. Each of those regions forms a support for a fuzzy set $_x^{uh} F$. To create a fuzzy set $_x^{uh} F$ an adequate membership function $_x^{uh} \mu$ must be provided. The functions $_x^{uh} \mu$ map $_x^u \hat{C}_{V_j}$ onto $[0, 1]$. Figure 17.2 presents an example showing the relationships between the above defined entities.

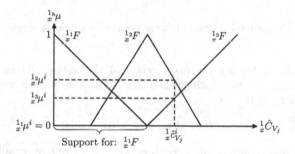

Fig. 17.2. An example of fuzzy sets, where: $_x^1 \hat{C}_{V_j}$ – universe 1; $_x^1 \hat{c}^i_{V_j}$ – measurement at instant i (crisp value); ρ_1 – number of fuzzy sets $_x^{1h} F$ in universe 1 (here $\rho_1 = 3$); $_x^{1h} F$ – fuzzy set h, $h = 1, \ldots, \rho_1$; $_x^{1h} \mu$ – degree of membership; $_x^{1h} \mu^i$ – degree of membership of a crisp value (measurement $_x^1 \hat{c}^i_{V_j}$) in the fuzzy set $_x^{1h} F$

Moreover, let either $c^{i+1}_{c_j}$ or $_y c^{i+1}_{e_j}$ contain the so-called output value $_y \hat{c}^{i+1}_j$. Beside that let that value belong to a space $_y \hat{C}_j$. Here a one dimensional space is assumed. The space $_y \hat{C}_j$ will form an output universe. It must be divided into regions $_y^g \hat{C}_j$, $g = 1, \ldots, \rho_y$, which will form the supports for the output fuzzy sets $_y^g F$, where ρ_y is the number of those regions and thus the number of the output fuzzy sets. Again, to create the fuzzy sets $_y^g F$ adequate membership functions $_y^g \mu$ must be provided. They map $_y \hat{C}_j$ onto $[0, 1]$.

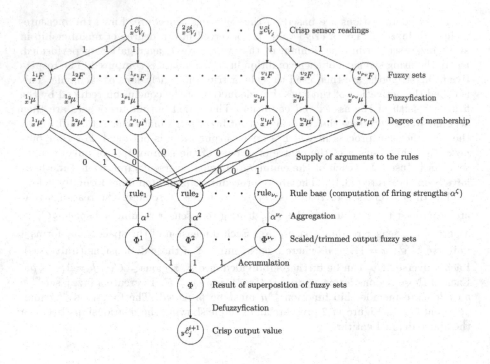

Fig. 17.3. The general scheme of fuzzy computations of a single output value

For the sake of brevity the fuzzy sets and the values of linguistic variables representing them are not distinguished here. The rule base is created by using many instructions of the type:

$$\text{if } \textit{proposition } \textbf{then } \textit{fuzzy conclusion } \textbf{endif} \tag{17.38}$$

where *proposition* consists of a conjunction of primitive conditions (atomic expressions; fuzzy predicates) in the form: *linguistic variable* **is** *term*, i.e., with the above notational simplification: *value* **is** *fuzzy set*.

Now the rule base can be constructed out of conditions formed by using the *AND* connective between the v fuzzy predicates (again linguistic variables are not used):

$$\text{if } (^1_x \hat{c}^i_{V_j} \text{ is } ^{1\kappa_1}_x F) \ AND \ (^2_x \hat{c}^i_{V_j} \text{ is } ^{2\kappa_2}_x F) \ AND \dots AND \ (^v_x \hat{c}^i_{V_j} \text{ is } ^{n_{u\kappa_v}}_x F)$$
$$\text{then } (_y \hat{c}^{i+1}_j \text{ is } ^\zeta_y F) \tag{17.39}$$

where:

$$\kappa_1 = 1, \dots, \rho_1,$$
$$\kappa_2 = 1, \dots, \rho_2,$$
$$\dots$$
$$\kappa_v = 1, \dots, \rho_v,$$
$$\zeta = 1, \dots, \nu_r.$$

The rule base consists of ν_r rules of the form (17.39). The fuzzification process converts each crisp value $_x^u \hat{c}_{V_j}^i$ into a degree of membership in $_x^{u\kappa_\xi} F$. As a result we obtain:

$$_x^{u\kappa_\xi} \mu^i = {}_x^{u\kappa_\xi} \mu({}_x^u \hat{c}_{V_j}^i), \tag{17.40}$$

where $\xi = 1, \ldots, \upsilon$. Aggregation consists in computation of the firing strength of the condition of each rule. In the case of the AND connective used in (17.39) the firing strength of the rule ζ will be, e.g.:

$$\alpha^\zeta = \min_{u=1}^{\upsilon}\{{}^{u\kappa_\xi} \mu^i\}, \quad \zeta = 1, \ldots, \nu_r, \tag{17.41}$$

where in each case the adequate κ_ξ for the rule ζ are used. The firing strengths α^ζ are used to scale or trim the output membership functions $_y^\zeta \mu$ associated with the corresponding rules. Those scaled membership functions form the conclusions (fuzzy sets) Φ^ζ. The resulting fuzzy set is obtained through accumulation, e.g., performing a max operation on a member by member basis on the fuzzy sets Φ^ζ:

$$\Phi = \{\max_{\zeta=1}^{\nu_r} \Phi^\zeta\}. \tag{17.42}$$

The defuzzyfied crisp value of $_y\hat{c}_j^{i+1}$ is finally obtained by, e.g., finding the abscissa of the centre of gravity of the set Φ. The thus computed value $_y\hat{c}_j^{i+1}$ is an element of either $_y c_{c_j}^{i+1}$ or $_y c_{e_j}^{i+1}$. The procedure described above can be presented graphically (Fig.17.3). The resulting pseudocode is as follows:

```
loop
      // Determine the current state of the agent
      ej ⟼ ₓcⁱₑⱼ;   Vj ⟼ ₓcⁱᵥⱼ;   cTⱼ' ⟼ ₓcⁱTⱼ;
      // Compute the next state of the agent
      ĉ'cⱼ := fuzzyfy(ₓĉⁱᵥⱼ);
      ĉ''cⱼ := accumulate(ĉ'cⱼ);
      ĉ'''cⱼ := aggregate(ĉ''cⱼ);
      yĉⱼⁱ⁺¹ := defuzzyfy(ĉ'''cⱼ);
      // Transmit the results of computations
      ycⁱ⁺¹ₑⱼ ⟼ ej;   ycⁱ⁺¹ᵥⱼ ⟼ Vj;   ycⁱ⁺¹Tⱼ ⟼ cTⱼ';
      i := i + 1;
endloop
```
$$\tag{17.43}$$

17.5.3 Behavioural versus Deliberative Systems

Historically two different approaches to intelligent behaviour of robots have been followed. The **deliberative** approach, stemming from traditional artificial intelligence [22], assumed that three components are necessary: sensing the environment, planning the actions taking into account the information obtained from the sensors, and acting, i.e., executing the plan. As planning is a computationally intensive activity, by the time the right plan is deduced, the state of the environment might change to such an extent that the plan is invalid. Therefore, an alternative approach has been proposed, the **behavioural** approach linking directly sensing and acting [2, 5, 6]. By including a multitude of reactions to diverse stimuli, intelligent behaviour should emerge. However, although emergent behaviour has been presented, it rarely attained a significant level of intelligence. Thus, currently there are many advocates for a hybrid approach [2] integrating planning and reactivity. Following the biological inspiration provided by the evolution of animal brain it is reasonable to distribute the reactive and deliberative components between two system parts. In mammals the cortex is distinct from the evolutionarily older parts of the brain. Thus, the behavioural component can be located in the embodied agent (e.g., a_j) and the deliberative one in a virtual agent (e.g., a_0). The structure of the system as it is implemented in `MRROC++` robot programming framework is presented in Fig. 17.4. The internal structure of each of the agents can be defined by any of the already described means. If predicates ${}^q p_{c_j}(\bullet)$ and transition functions ${}^m f_{c_j}(\bullet)$ use only ${}_x c^i_{V_j}$, a purely reactive system results, and if only $c^i_{c_j}$ is used, a purely deliberative system results. If both are used a hybrid system results. In the proposed solution exchange of information through transmission channels is employed, thus the virtual agent a_0 uses $c^i_{c_0}$ and $c^i_{T_0}$, and the virtual agent mainly relies on $c^i_{V_j}$ and $c^i_{T_j}$. The virtual agent a_0 within $c^i_{c_0}$ contains the model (i.e., the representation of the problem solution state) and the operators (i.e., the transition functions transforming one problem solution state into another). The generated plan

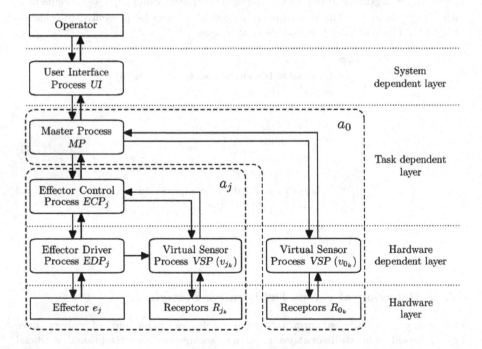

Fig. 17.4. Hybrid deliberative-behavioural system composed of a virtual agent a_0 and an embodied agent a_j implemented in `MRROC++` robot programming framework

is transmitted by the virtual agent a_0 to the embodied agent a_j through ${}_x c^i_{T_j}$ and stored in $c^{i/i+1}_{c_j}$, hence c_{c_j} acquires the effect of deliberation of a virtual agent a_0. The plan stored in the embodied agent forms a goal pursuing behaviour.

Deliberation assumes the use of artificial intelligence techniques [14] to find a plan (i.e., a sequence) of actions leading to the execution of a task (goal) set forth before the system. This is implemented within the virtual agent a_0 by search techniques. A search requires the following entities:

- a search space (i.e., a problem domain) composed of search space states – not to be mistaken with the state of the environment or the agent itself,
- an initial state, belonging to the search space – from this state the search commences,
- operators, which transform the current search space state into the next states (those operators may result either from production rules or be the side-effect of application of predicate logic [14]),
- a data structure accumulating the generated states (i.e., the search tree or graph),
- a goal test deciding whether the generated state is the goal state
- a path cost function, which evaluates the quality of the obtained search space state – usually it takes into consideration the cost of both the path traversed so far and the remaining path to the goal state (e.g., A^* algorithm or its derivatives).

In the case of deliberative systems c_{c_0} must contain the data structures accumulating the generated states (i.e., the problem solution). Deliberation is a search process starting in the initial state of the problem solution. This state includes a partial description of the current state of the agent, but also other search related information. As the operators are applied, new problem solution states are generated. The operators are equivalent to transition functions transforming one problem solution state into another. Heuristics are included in the path cost function and help in discarding the produced states that either do not lead to a solution or are along a far from optimal search path, thus avoiding a combinatorial explosion in the search process. The path leading from the initial state and ending in a goal state describes a plan of actions that the virtual agent a_0 sends to the embodied agent a_j for execution.

17.6 Conclusions

A transition function based formalism used to describe the functioning of agents captures all of the usually considered system structures: deterministic and indeterministic, crisp and fuzzy, behavioural and deliberative. Moreover, it can be used for both single or multi-agent systems composed of embodied and virtual agents. It is especially well suited to the specification of behavioural systems, where fixed and variable duration action sequences can be formulated, sequential or parallel forms of behaviour composition can be defined, and systems with diverse frequency of behaviour selection based on predicates can be implemented. Recursive composition of simple behaviours into complex ones is also possible.

The proposed internal structure of the agent is such that the transition functions (17.14) have access to all of the input and internal data structures. Moreover, there are as many transition functions within (17.14) as there are output and internal data structures. Hence, there is no structural reason, from the point of view of the data flow, for stopping the implementation of any control structure. The functions (17.14) have access to all the available information and can influence every component of the system.

Although the formalism is implementation independent, its transformation into code is straightforward, because it is expressed in pseudocode. Last, but not least, it enables logical reasoning about structures of agents' software. MRROC++ robot programming framework is an example of its utilization.

Usually MRROC++ based controllers are composed of at least a single embodied agent a_j and one virtual agent a_0 (Fig. 17.4). Each embodied agent controls a single effector. In the case of multi-effector systems the virtual agent is responsible for the coordination of several embodied agents. The coordination may be continuous (e.g., simultaneous transfer of a rigid object by two robots), sporadic (e.g., handing of an object by a robot to another one) or absent (i.e., robots act completely independently of each other). In the case of single effector systems the virtual agent is either dormant or is responsible for performing deliberation (e.g., planning future actions, which usually takes considerable time with respect to servo sampling rate). Each embodied agent is composed of two processes controlling its effector: Effector Control Process ECP and Effector Driver Process EDP. The former is agent task dependent and the latter is effector hardware dependent. Moreover, zero or more Virtual Sensor Processes VSPs may be used by an embodied agent. The virtual agent is composed of the Master Process MP and zero or more VSPs.

Both ECPs and MP contain program objects derived from an abstract class called the motion generator [30]. They are used by the Move instructions which encode, within the bodies of MP and ECPs, the behavior of the effectors. In MRROC++ each function $^m f_{c_j}$ is decomposed into two functions [28–30, 32]: $^m f'_{c_j}$ and $^m f''_{c_j}$, the former determining the behavior of the agent in the first step of the execution of the Move instruction, and the latter in all the following steps. The two function pairs $(^m f'_{c_j}, {}^m f_{\tau_j})$ and $(^m f''_{c_j}, {}^m f_{\tau_j})$ are embedded within the two methods of this class. The programmer derives descendant classes from this class and simultaneously defines the code of those function pairs. The code of the Move procedure invokes those methods causing the agent to execute a part of its task. Sequences of Move instructions in conjunction with the standard C++ loops and conditional statements enable the definition of the whole task. For this purpose the program templates described in this paper are used (e.g., (17.32)). Thus the programmer concentrates only on two aspects of control: the motion generator of each Move instruction, and the insertion of the Move instructions into a program template executing the demanded task. Other aspects of program implementation are provided by the framework. Details of different aspects of MRROC++ can be found in [27, 28, 30, 31, 34]. MRROC++ has been used in diverse applications [35].

References

1. R. Alami, R. Chatila, S. Fleury, M. Ghallab M., and Ingrand F. An architecture for autonomy. *Int. J. of Robotics Research*, 17(4):315–337, 1998.
2. R. C. Arkin. *Behavior-Based Robotics*. MIT Press, Cambridge, Mass., 1998.

3. C. Blume and W. Jakob. *Programming Languages for Industrial Robots.* Springer-Verlag, Berlin, 1986.
4. E. Bonabeau, M. Dorigo, and G. Theraulaz. *Swarm Intelligence: From Natural to Artificial Systems.* Oxford University Press, New York, Oxford, 1999.
5. R. A. Brooks. A robust layered control system for a mobile robot. *IEEE Journal of Robotics and Automation,* RA-2(1):14–23, March 1986.
6. R. A. Brooks. Intelligence without representation. *Artificial Intelligence,* (47):139–159, 1991.
7. H. Bruyninckx, P. Soetens, and B. Koninckx. The real-time control core of the OROCOS project. In *Proceedings of the IEEE International Conference on Robotics and Automation, Taipei, Taiwan,* pages 2766–2771. September, 14–19 2003.
8. D. Driankov, H. Hellendoorn, and M. Reinfrank. *An Introduction to Fuzzy Control.* Springer-Verlag, Berlin, Heidelberg, 1993.
9. B.P. Gerkey, R.T. Vaughan, and A. Howard. The Player/Stage project: Tools for multi-robot and distributed sensor systems. In *Proceedings of the International Conference on Advanced Robotics, ICAR'03, Coimbra, Portugal,* pages 317–323. June 30 – July 3 2003.
10. V. Hayward, L. Daneshmend, and S. Hayati. An overview of KALI: A system to program and control cooperative manipulators. In K. Waldron, editor, *Advanced Robotics,* pages 547–558. Springer-Verlag, Berlin, 1989.
11. V. Hayward and R. P. Paul. Robot manipulator control under unix RCCL: A robot control C library. *Int. J. Robotics Research,* 5(4):94–111, Winter 1986.
12. X. Hong, K. Xu, and M. Gerla. Scalable routing protocols for mobile ad hoc networks. *IEEE Network,* pages 11–21, July/August 2002.
13. J. Jantzen. Design of fuzzy controlers. Technical University of Denmark, Department of Automation, September 30 1999.
14. G. F. Luger and W. A. Stubblefield. *Artificial Intelligence and the Design of Expert Systems.* Benjamin-Cummings, Redwood, 1989.
15. M. E. Markiewicz and C. J. P. Lucena. Object oriented framework development. *ACM Crossroads,* 7(4), 2001. Also: http://www.acm.org/crossroads/xrds7-4/frameworks.html.
16. K. Mianowski, K. Nazarczuk, M. Wojtyra, and S. Zitarski. Application of the UNIGRAPHICS system for milling and polishing with the use of rnt robot. In *Proceedings of the Workshop for the users of UNIGRAPHICS system, Frankfurt, November,* pages 98–104. 1999.
17. E. R. Morales. Generis: The EC-JRC generalised software control system for industrial robots. *Industrial Robot,* 26(1):26–32, 1999. Also: http://www.erxa.it/Eng/GENERIS/description.html.
18. K. Nazarczuk, K. Mianowski, A. Olędzki A., and C. Rzymkowski. Experimental investigation of the robot arm with serial-parallel structure. In *Proceedings of the 9-th World Congress on the Theory of Machines and Mechanisms, Milan, Italy,* pages 2112–2116. 1995.
19. K. Nazarczuk, K. Mianowski, and S. Łuszczak. Development of the design of POLYCRANK manipulator without joint limits. In *Proceedings of the 13-th CISM-IFToMM Symposium on Theory and Practice of Robots and Manipulators Ro.Man.Sy 13, Zakopane, Poland, 3–6 July,* pages 285–292. 2000.
20. L. Petersson, D. Austin, and H. Christensen. DCA: a distributed control architecture for robotics. In *Proc. Int. Conference on Intelligent Robots and Systems IROS'01.* 2001.

21. J. Pritchard. *COM and COBRA Side by Side: Architectures, Strategies, and Implementations.* Addison-Wesley, Reading, 1999.
22. S. Russell and P. Norvig. *Artificial Intelligence: A Modern Approach.* Prentice Hall, Upper Saddle River, N.J., 1995.
23. R. Simmons and D. Apfelbaum. A task description languagefor robot control. In *International Conference on Itelligent Robots and Systems IROS'98. Victoria, Canada.* October 1998. Also: http://www-2.cs.cmu.edu/ tdl/.
24. R. Simmons, R. Goodwin, C. Fedor J., and Basista. Task control architecture: Programmer's guide to version 8.0. Carnegie Mellon University, School of Computer Science, Robotics Institute, May 1997. Also: http://www-2.cs.cmu.edu/afs/cs.cmu.edu/project/TCA/www/tca.orig.html.
25. W. Szynkiewicz. Motion planning for multi-robot systems with closed kinematic chains. In *Proceedings of the 9th IEEE International Conference on Methods and Models in Automation and Robotics MMAR'2003, Midzyzdroje,* pages 779–786. August 25-28 2003.
26. R.T. Vaughan, B.P. Gerkey, and A. Howard. On device abstractions for portable, reusable robot code. In *Proceedings of the IEEE/RSJ International Conference on Intelligent Robots and Systems, IROS'03, Las Vegas, Nevada,* pages 2121–2427. October 2003.
27. T. Winiarski and C. Zieliński. Implementation of position–force control in MRROC++. In *Proceedings of the 5th International Workshop on Robot Motion and Control, RoMoCo'05, Dymaczewo, Poland,* pages 259–264. June, 23–25 2005.
28. C. Zieliński. The MRROC++ system. In *1st Workshop on Robot Motion and Control, RoMoCo'99, Kiekrz, Poland,* pages 147–152. June 28–29 1999.
29. C. Zieliński. By how much should a general purpose programming language be extended to become a multi-robot system programming language? *Advanced Robotics,* 15(1):71–95, 2001.
30. C. Zieliński. Motion generators in MRROC++ based robot controller. In G. Bianchi, J-P. Guinot, and C. Rzymkowski, editors, *14th CISM–IFToMM Symposium on Robotics, Ro.Man.Sy'02, Udine, Italy, CISM Courses and Lectures no.438,* pages 299–306. Springer, Wien, New York, July 1–4 2002.
31. C. Zieliński. Reaction to errors in robot systems. In *Third International Workshop on Robot Motion and Control, RoMoCo'02, Bukowy Dworek, Poland,* pages 201–208. November 9–11 2002.
32. C. Zieliński. Formal approach to the design of robot programming frameworks: the behavioural control case. *Bulletin of the Polish Academy of Sciences – Technical Sciences,* 53(1):57–67, March 2005.
33. C. Zieliński, K. Mianowski, K. Nazarczuk, and W. Szynkiewicz. A prototype robot for polishing and milling large objects. *Industrial Robot,* 30(1):67–76, January 2003.
34. C. Zieliński, A. Rydzewski, and W. Szynkiewicz. Multi-robot system controllers. In *Proc. of the 5th International Symposium on Methods and Models in Automation and Robotics MMAR'98, Midzyzdroje, Poland,* volume 3, pages 795–800. August 25–29 1998.
35. C. Zieliński, W. Szynkiewicz, and T. Winiarski. Applications of MRROC++ robot programming framework. In *Proceedings of the 5th International Workshop on Robot Motion and Control, RoMoCo'05, Dymaczewo, Poland,* pages 251–257. June, 23–25 2005.

18

Steps Toward Derandomizing RRTs

Stephen R. Lindemann and Steven M. LaValle

Dept. of Computer Science, University of Illinois, Urbana, IL 61801 USA
{slindema|lavalle}@uiuc.edu

18.1 Introduction

For over a decade, randomized algorithms such as the Randomized Potential Field Planner (RPP) [3], the Probabilistic Roadmap (PRM) family [1, 9, 21, 24, 25], Rapidly-Exploring Random Trees (RRTs) [10, 12, 13], and others [4, 8, 17], have dominated the field of motion planning. Recently, a great deal of attention has been given to comparing random versus deterministic sampling in the context of PRMs [6, 11]. A recent survey of this field, termed *sampling-based motion planning*, is given in [15]. In this paper, we discuss the role of randomization in RRTs, and introduce two new planners which move toward their derandomization.

Randomization is a common algorithmic technique, and it is of great value in many contexts. Sometimes, it is used to defeat an adversary who might gain an advantage from learning one's deterministic strategy (e.g., cryptographic or sorting algorithms). Randomization is also useful for approximation or in conjunction with amplification techniques (e.g., the randomized min cut algorithm). It also allows for probabilistic performance analysis, which can be very useful. For problems of numerical integration, randomization can sometimes defeat the "curse of dimensionality" [22].

In the context of the original PRM, the primary use of randomization is to uniformly sample the configuration space. Recently, the usefulness of randomization for this purpose has been challenged; proponents of deterministic sampling argue that there are deterministic sequences satisfying other uniformity measures (e.g., discrepancy and dispersion) which perform at least as well as random sampling. Furthermore, these methods give deterministic guarantees of convergence (such as resolution completeness). Currently, work is being done both to study the performance of various random and deterministic sampling sequences in the PRM [11], and to construct new deterministic uniform sequences with other properties that are useful for motion planning [14].

It cannot be denied that contemporary motion planning algorithms, many of which use randomization, are very efficient and able to solve many challenging problems. This might lead one to conclude that randomization is the key to their effectiveness; however, this is not necessarily the case. On the contrary, randomization can easily become a "black box" which obscures the reasons for an algorithm's success. Hence, attempts to derandomize popular motion planning algorithms do not reflect antipathy toward randomization, but rather the desire

K. Kozłowski (Ed.): Robot Motion and Control, LNCIS 335, pp. 287–300, 2006.
© Springer-Verlag London Limited 2006

BUILD_RRT(x_{init})
1 G_{sub}.init(x_{init});
2 **for** $k = 1$ **to** *maxIterations* **do**
3      $x_{rand} \leftarrow$ RANDOM_STATE();
4      $x_{near} \leftarrow$ NEAREST_NEIGHBOR(x_{rand}, G_{sub});
5      $u_{best}, x_{new}, success \leftarrow$ CONTROL($x_{near}, x_{rand}, G_{sub}$);
6   **if** success
7   G_{sub}.add_vertex(x_{new});
8   G_{sub}.add_edge($x_{near}, x_{new}, u_{best}$);
9 Return G_{sub}

Fig. 18.1. The basic RRT construction algorithm

to understand the fundamental insights of these algorithms. After studying an algorithm in both its randomized and derandomized forms, it will be possible to intelligently decide which to use, or whether some mixture of the two is appropriate. This approach might even result in algorithms that combine deterministic and randomized strategies in a way that achieves the benefits of both. For example, [20] uses a combined sampling strategy for constructing a Visibility PRM; they recursively divide the space into quadrants (a deterministic strategy) and choose a random sample within each quadrant. In the area of low-discrepancy sampling, Wang and Hickernell have constructed and analyzed randomized Halton sequences [23]. Geraerts and Overmars have also investigated randomizing Halton points [6].

We believe that a great deal of work remains to investigate deterministic variants of contemporary motion planning algorithms. We have already mentioned efforts to derandomize PRMs; namely, those which attempt to use deterministic uniform sampling methods. It is also interesting to consider derandomizing RRTs; this is a very challenging task, due to the way that randomization is used in RRTs.

18.2 Randomization in RRTs

In the case of RRTs, derandomization is more difficult than with PRMs. In the original PRM, the primary use of randomization is to produce a uniformly distributed sample sequence in order to cover the space; for RRTs, the use of randomization is more subtle. As opposed to the former case, simply replacing random samples with deterministic ones will not capture the essence of the exploration strategy of RRTs. In order to understand the best way to derandomize RRTs, we will outline the role of random sampling in the basic RRT algorithm.

The basic RRT algorithm operates very simply; the overall strategy is to incrementally grow a tree from the initial state to the goal state. The root of the tree is the initial state; at each iteration, a random sample is taken and its nearest neighbor in the tree computed. A new node is then created by growing the nearest neighbor toward the random sample. Pseudocode for basic RRT construction can be found in Figure 18.1; for in-depth description and analysis of RRTs, see [13].

In [13], it is argued that RRTs explore rapidly because samples "pull" the search tree toward unexplored areas of the state space. This occurs because the probability that a vertex is selected for expansion is proportional to the area of its Voronoi region. Hence, a node at the frontier of the tree is likely to be chosen to grow into previously unexplored territory. This argument can be made because the samples are independent and taken from a uniform random distribution. RRTs are consequently able to explore in a Voronoi-biased manner, with very low cost (generating random samples is inexpensive). This appears to be the primary use of randomization in RRTs. In this light, it may not be beneficial to simply replace random samples with deterministic ones in RRT planners. In a uniform deterministic sequence, samples are not independent from each other; consequently, one cannot argue about the probability of a vertex being selected in the same way as one can when using random samples. It is also clear that in the context of RRTs, the concepts of *Voronoi bias* and *deterministic sampling* are not necessarily related. In the case above, one could use deterministic sampling and have little or no Voronoi bias; as we will see later, one may freely vary the amount of Voronoi bias using random sampling.

It should be noted that for RRTs, Voronoi bias does not play the same role as in some other recent motion planning algorithms [5, 19]. These algorithms compute the discretized GVD (generalized Voronoi diagram) of the environment and use this information to sample near the medial axis of free space (for another medial axis-based approach, see [24]). In this case, the Voronoi diagram is based on the environment. In RRTs, however, the Voronoi bias occurs with respect to the Voronoi diagram of the nodes in the search tree; that is, the nodes in the search tree with the largest Voronoi regions tend to be selected for exploration.

One may use the concept of Voronoi bias to construct a deterministic RRT. Simply construct the d-dimensional Voronoi diagram of the nodes in the tree and use this information (along with a decision rule) to incrementally grow the search tree. Two possible decision rules are: grow toward the centroid of the largest Voronoi region (this requires calculation of the volumes of the Voronoi regions); or, attempt to reduce the size of the largest empty ball (the center of the largest empty ball is a Voronoi vertex). Either of these approaches is theoretically feasible, since it is well-known how to construct Voronoi diagrams in arbitrary dimensions. Practically, however, it is no simple task to robustly compute Voronoi diagrams in d dimensions, and implementing algorithms that do so is quite difficult. In addition to this, most construction methods use an L_2 (Euclidean) metric in Euclidean space; this is more restrictive than is appropriate for general motion planning problems, which may have different metrics and whose topologies are are often more complicated. Finally, the cost of explicitly computing this information may be prohibitive from a practical point of view.

While these Voronoi bias-maximizing approaches are worth exploring, we do not seek to do so in this paper. Instead, we propose two algorithms which lie between the original RRT and the fully Voronoi-biased approaches. These algorithms are more Voronoi-biased than the original RRT, but not as much so as those based on explicitly-constructed Voronoi diagrams. In related work, we introduce another derandomized RRT variant, which is based on incremental dispersion reduction [16]; however, that approach will not be discussed in this paper.

18.3 A Spectrum of RRT-like Planners

We wish to construct planners that take the idea of Voronoi bias and emphasize it more strongly than in the original RRT algorithm. However, to avoid the difficulties with explicitly constructing Voronoi diagrams, we desire an approximate, sampling-based approach. Hence, our new algorithms will lie in the middle of a spectrum of RRT-like planners, with the original RRT on one side and a deterministic Voronoi diagram-based method on the other.

We have seen that the RRTs grow in a Voronoi-biased manner due to the way they process the random samples drawn from the configuration space. What would happen, then, if instead of taking a single sample, one took k samples? One can sort the nodes in the tree according to how many samples they were the nearest neighbor for, and grow from the nodes which collected the most samples. We call this algorithm the Multi-Sample RRT (MS-RRTa), and pseudo-code for the basic algorithm is given in Figure 18.2. Note that during a particular iteration, a node grows toward the average of the samples it collected; this may be viewed as an estimate of the centroid of that node's Voronoi region (clearly, it is guaranteed to be within the node's Voronoi region). Also, as k approaches infinity, we probabilistically obtain the exact Voronoi volumes and Voronoi region centroids. This means that one may view k as a knob which changes the behavior of the algorithm from randomized to deterministic (hence we refer to it as partially randomized). We also see that the cost of running this algorithm grows linearly with k, since at each iteration k nearest-neighbor queries must be performed. Presumably, having more Voronoi bias will result in better exploration and consequently fewer nodes in the search tree; however, the cost of each node grows as the Voronoi bias is increased. Hence, the best performance is achieved by finding the best value of the parameter k, which is not necessarily a simple task.

BUILD_MS-RRTA(x_{init})
1 G_{sub}.init(x_{init});
2 **for** $x = 1$ **to** $maxIterations$ **do**
3   **for** $i = 1$ **to** k **do**
4          $x_{rand} \leftarrow$ RANDOM_STATE();
5          $x_{near} \leftarrow$ NEAREST_NEIGHBOR(x_{rand}, G_{sub});
6          $x_{near}.sampleCount$ += 1;
7          $x_{near}.sampleAverage$ += x_{rand};
8      $x_{best} \leftarrow$ max($x.sampleCount, x \in G_{sub}$);
9      $x_{next} \leftarrow x_{best}.sampleAverage / x_{best}.sampleCount$;
10     $u_{best}, x_{new}, success \leftarrow$ CONTROL($x_{best}, x_{next}, G_{sub}$);
11   **if** $success$
12   G_{sub}.add_vertex(x_{new});
13   G_{sub}.add_edge($x_{near}, x_{new}, u_{best}$);
14   CLEAR_SAMPLE_INFO(G_{sub});
15 Return G_{sub};

Fig. 18.2. The basic MS-RRTa construction algorithm

Our second algorithm is similar to the first Multi-Sample RRT. Its differences are based on two key observations. First, if one can obtain approximate Voronoi information from a set of k uniformly distributed random samples, one may also obtain it from k uniformly distributed deterministic samples (the Voronoi information depends only on uniformity, not randomness). Hence, as long as one has a sequence with good incremental quality (i.e., the set of samples $\{s_i, \ldots, s_{i+k}\}$ from the sequence is uniformly distributed for all i, k), one may take k samples from that deterministic sequence and expect the algorithm to work as well as with random samples (in our experiments so far, we have used Halton points [7]). Second, if some set of k samples gives a good approximation, then there is no need to pick k new samples during the next iteration. Instead, the old samples may be used again, saving the cost of doing k nearest-neighbor queries at each iteration. At most, one must do k metric evaluations per new node, which can yield significant savings. To distinguish the algorithm resulting from these observations from the previous one, we denote this one as MS-RRTb; pseudo-code is given in Figure 18.3. In both of these algorithms, one may reach a point where the k samples no longer provide enough resolution to make a good choice about where to extend the tree; in the case of the MS-RRTa, one may decide to take $k + k_2$ samples each iteration instead of k. In the case of MS-RRTb, one can add k_2 new samples to the previous set and continue as before.

BUILD_MS-RRTB(x_{init})
1 G_{sub}.init(x_{init});
2 **for** $k = 1$ **to** K **do**
3   ADD_NEW_SAMPLE($sample, sampleList$);
4 **for** $x = 1$ **to** $maxIterations$ **do**
5      $x_{best} = \max(x.sampleCount, x \in G_{sub})$;
6      $x_{next} \leftarrow x_{best}.sampleAverage/x_{best}.sampleCount$;
7      $u_{best}, x_{new}, success \leftarrow$ CONTROL($x_{best}, x_{next}, G_{sub}$);
8   **if** success
9   G_{sub}.add_vertex(x_{new});
10   G_{sub}.add_edge($x_{near}, x_{new}, u_{best}$);
11   REDISTRIBUTE_SAMPLES($sampleList, G_{sub}$);
12 Return G_{sub}

Fig. 18.3. The basic MS-RRTb construction algorithm

Each of these algorithms takes the key feature of RRTs, Voronoi bias, and uses sampling techniques to increase that bias. Another approach, based on dispersion reduction, is given in [16]. In the next section, we discuss a few implementation details of our algorithms, and some experimental results.

18.4 Implementation Details and Experimental Results

Under certain situations, the performance of RRT algorithms can be degraded by local minima. For example, imagine the scenario where a node has a large Voronoi region but is prevented from growing due to proximity to an obstacle. In many cases, this causes little trouble because eventually another node will be chosen to expand. However, both of the MS-RRT algorithms exhibit greedy, Voronoi-biased behavior. This causes problems with local minima to be magnified. To address this problem, both of our planners include *obstacle nodes*, which are nodes in the search tree representing configurations in the obstacle region of C-space. They are generated when an obstacle is encountered during a connection attempt. These nodes are not candidates for expansion (i.e., they are all leaves in the tree), but they are allowed to "own" samples. As a result, nodes which are selected for expansion do not repeatedly grow toward a local minimum. Unfortunately, obstacle nodes can have problems of their own. For certain difficult problems (e.g., those with narrow corridors), many obstacle nodes can be created, which can significantly degrade performance because more samples are required to decide how to grow the tree. Consequently, we are currently investigating alternative approaches which combine the local minimum-avoiding effects of obstacle nodes without their disadvantages.

Each of our new algorithms has bottlenecks that do not appear in standard RRT planners. As seen from Section 18.3, MS-RRTa does a large number of nearest-neighbor queries; consequently, the performance of this algorithm depends both on the number of samples taken per iteration and on the efficiency of the nearest-neighbor calculation. Doing many nearest-neighbor queries is unavoidable for this method, but it is possible to reduce the cost of these queries to a manageable level. We do this by using a nearest neighbor package based on Kd-trees [2, 18]. Likewise, MS-RRTb has a bottleneck as well: updating the nearest neighbor for each sample after adding new nodes to the tree. Currently, we use the naive approach which calls the metric function for each new node and each sample, updating the nearest neighbor where appropriate. It should be possible to accelerate this by using an appropriate data structure (most likely, some form of a Kd-tree); developing a way to do this is not trivial, however, and we have not yet done so. Once this is accomplished, however, MS-RRTb should speed up significantly, particularly for difficult problems which require large numbers of samples.

A few other implementation details will suffice to introduce some experiments. First, one of the best RRT planners is RRTConCon, a variant which uses two trees (one starting at the initial state, the other at the goal state) and is more greedy than the basic RRT (see [13] for details). Hence, our experiments use corresponding versions of the MS-RRT planners (except where otherwise noted). Also, MS-RRTa defaults to random sampling for a particular iteration if it is unable to grow the tree in the attempted Voronoi-biased manner. In experiments using the basic algorithms, we use the basic RRT which has been modified to attempt to connect to the goal after each iteration (we denote this variant as ModRRT). Finally, our experiments below are all holonomic. This is a departure from typical RRT applications, since RRTs can easily be applied to systems with dynamics. However, our new algorithms are somewhat metric-sensitive (which is to be expected, since they are strongly Voronoi-biased, and the Voronoi diagram depends on the metric), and for systems with dynamics finding an appropriate metric is difficult. We are currently considering ways to resolve this difficulty.

First, we present two two-dimensional examples to illustrate how emphasizing Voronoi bias affects the growth of the search tree. The first example consists of a simple local minimum separating the initial and goal states (see Figure 18.4). Both MS-RRT planners initially grow into the local minimum and create an obstacle node upon encountering the obstacle; they then grow the other direction and around the local minimum. Observe the effects of randomization in MS-RRTa: while the deterministic MS-RRTb always chooses a single direction to grow, randomization in the MS-RRTa sometimes causes it to grow in both directions (sometimes, it grows in a single direction like MS-RRTb). Also, the modified RRT planner (ModRRT) requires significantly more planning iterations to solve the problem than either MS-RRT algorithm. This is primarily due to the effect of obstacle nodes, which cause MS-RRTs to avoid obstacles more than an ordinary RRT. Our second example is shown in Figure 18.5. The algorithms behave in a manner similar to the previous example. Third, in our description of the MS-RRTa algorithm, we mentioned that one could view the number of samples taken per iteration as a knob which changes behavior from a small degree of Voronoi bias (as in the basic RRT) to a large degree of Voronoi bias. In Figures 18.6 and 18.7, we show how varying the value of k affects the growth patterns.

Finally, we give two six-dimensional problems to illustrate our algorithms' performance for these problems. The algorithms presented are not capable of outperforming RRTConCon with respect to solution time; however, they do represent reasonable approaches to planning. An alternative RRT-based approach, which is based on incremental dispersion reduction, and uses many of the ideas from this paper, has led to improved performance over RRTConCon, based on our recent experiments [16].

18.5 Conclusions and Future Work

In conclusion, we have discussed the role of randomization in RRTs and introduced two new algorithms which increase Voronoi bias. By studying these algorithms, insight may be gained into the reasons for RRT algorithms' effectiveness at solving motion planning problems. Decreasing the effect of randomization allows us to isolate certain aspects of the algorithms' behavior, without the inherent "sloppiness" that results from randomization. Understanding the key reasons for RRTs' effectiveness is the first step toward making more efficient planners, which may or may not utilize randomization.

Our long-term goals include developing efficient planners based on insights gained from studying the algorithms presented in this paper, as well as other algorithms from the spectrum of RRT-like planners. In the near future, we plan to implement and study other RRT-like planners similar to those presented here; we hope to examine both sampling-based approaches and those based on explicit Voronoi computations. We also would like to study the performance of different sampling techniques (randomized and deterministic) in these different planners; this will show whether or not randomization is of value in RRT algorithms. We believe that studies of this type will greatly increase our understanding of efficient motion planning and configuration space exploration, and will enable us to develop better-performing algorithms.

Acknowledgments

This work was funded in part by NSF Awards 9875304, 0118146, and 0208891.

References

1. N. M. Amato and Y. Wu. A randomized roadmap method for path and manipulation planning. In *IEEE Int. Conf. Robot. & Autom.*, pages 113–120, 1996.

2. A. Atramentov and S. M. LaValle. Efficient nearest neighbor searching for motion planning. In *Proc. IEEE Int'l Conf. on Robotics and Automation*, pages 632–637, 2002.

3. J. Barraquand and J.-C. Latombe. Robot motion planning: A distributed representation approach. *Int. J. Robot. Res.*, 10(6):628–649, December 1991.

4. D. Challou, D. Boley, M. Gini, and V. Kumar. A parallel formulation of informed randomized search for robot motion planning problems. In *IEEE Int. Conf. Robot. & Autom.*, pages 709–714, 1995.

5. M. Garber and M. C. Lin. Constraint-based motion planning using voronoi diagrams. In *Proc. Workshop on Algorithmic Foundation of Robotics*, 2002.

6. R. Geraerts and M. H. Overmars. A comparative study of probabilistic roadmap planners. In *Proc. Workshop on the Algorithmic Foundations of Robotics*, December 2002.

7. J. H. Halton. On the efficiency of certain quasi-random sequences of points in evaluating multi-dimensional integrals. *Numer. Math.*, 2:84–90, 1960.

8. D. Hsu, J.-C. Latombe, and R. Motwani. Path planning in expansive configuration spaces. *Int. J. Comput. Geom. & Appl.*, 4:495–512, 1999.

9. L. E. Kavraki, P. Svestka, J.-C. Latombe, and M. H. Overmars. Probabilistic roadmaps for path planning in high-dimensional configuration spaces. *IEEE Trans. Robot. & Autom.*, 12(4):566–580, June 1996.

10. J. J. Kuffner and S. M. LaValle. RRT-connect: An efficient approach to single-query path planning. In *Proc. IEEE Int'l Conf. on Robotics and Automation*, pages 995–1001, 2000.

11. S. M. LaValle and M. S. Branicky. On the relationship between classical grid search and probabilistic roadmaps. In *Proc. Workshop on the Algorithmic Foundations of Robotics*, December 2002.

12. S. M. LaValle and J. J. Kuffner. Randomized kinodynamic planning. *International Journal of Robotics Research*, 20(5):378–400, May 2001.

13. S. M. LaValle and J. J. Kuffner. Rapidly-exploring random trees: Progress and prospects. In B. R. Donald, K. M. Lynch, and D. Rus, editors, *Algorithmic and Computational Robotics: New Directions*, pages 293–308. A K Peters, Wellesley, MA, 2001.

14. S. R. Lindemann and S. M. LaValle. Incremental low-discrepancy lattice methods for motion planning. In *Proc. IEEE International Conference on Robotics and Automation*, 2003.

15. S. R. Lindemann and S. M. LaValle. Current issues in sampling-based motion planning. In P. Dario and R. Chatila, editors, *Proc. Eighth Int'l Symp. on Robotics Research*. Springer-Verlag, Berlin, 2004. To appear.

16. S. R. Lindemann and S. M. LaValle. Incrementally reducing dispersion by increasing voronoi biasin RRTs. In *IEEE IEEE International Conference on Robotics and Automation*, 2004. Under review.

17. E. Mazer, J. M. Ahuactzin, and P. Bessière. The Ariadne's clew algorithm. *J. Artificial Intell. Res.*, 9:295–316, November 1998.

18. D. M. Mount. ANN programming manual. Technical report, Dept. of Computer Science, U. of Maryland, 1998.

19. C. Pisula, K. Hoff, M. Lin, and D. Manocha. Randomized path planning for a rigid body based on hardware accelerated voronoi sampling. In *Proc. Workshop on Algorithmic Foundation of Robotics*, 2000.

20. B. Salomon, Maxim Garber, Ming. C. Lin, and Dinesh Manocha. Interactive navigation in complex environments using path planning. In *Proceedings of the ACM SIGGRAPH 2003 Symposium on Interactive 3D Graphics*, 2003.
21. T. Simeon, J.-P. Laumond., and C. Nissoux. Visibility based probabilistic roadmaps for motion planning. *Advanced Robotics Journal*, 14(6), 2000.
22. J. F. Traub, G. W. Wasilkowski, and H. Wozniakowski. *Information-Based Complexity*. Academic Press Professional, Inc., San Diego, 1988.
23. X. Wang and F. J. Hickernell. Randomized halton sequences. *Mathematical and Computer Modelling*, 32:887–899, 2000.
24. S. A. Wilmarth, N. M. Amato, and P. F. Stiller. Maprm: A probabilistic roadmap planner with sampling on the medial axis of the free space. In *IEEE Int. Conf. Robot. & Autom.*, pages 1024–1031, 1999.
25. Y. Yu and K. Gupta. On sensor-based roadmap: A framework for motion planning for a manipulator arm in unknown environments. In *IEEE/RSJ Int. Conf. on Intelligent Robots & Systems*, pages 1919–1924, 1998.

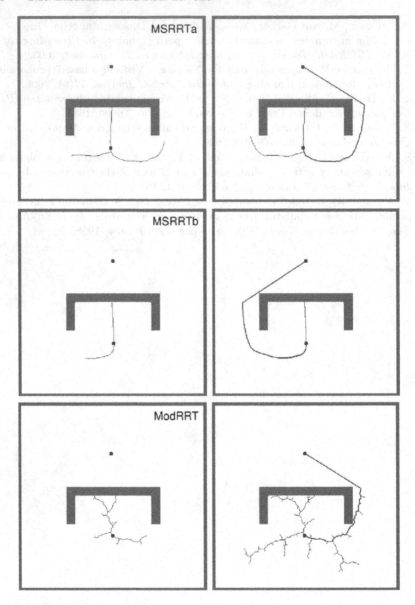

Fig. 18.4. A simple 2-d example. Top row, from left to right: MS-RRTa after 81 iterations (81 nodes, 1 obstacle node), MS-RRTa at completion (114 iterations, 114 nodes, 1 obstacle node). Second row: MS-RRTb after 41 iterations (41 nodes, 1 obstacle node), MS-RRTb at completion (81 iterations, 81 nodes, 1 obstacle node). Third row: ModRRT after 92 iterations (81 nodes), ModRRT at completion (249 nodes, 384 iterations).

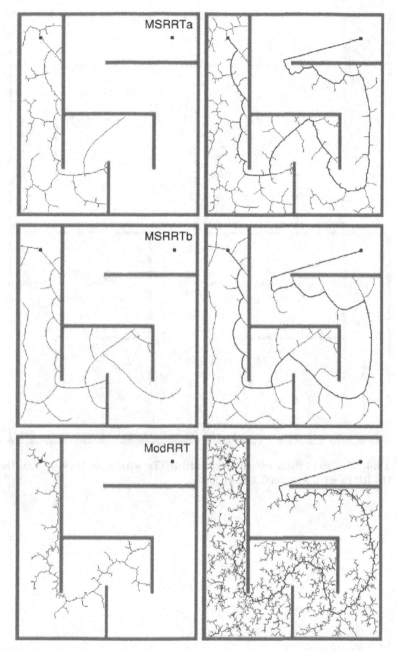

Fig. 18.5. A 2-d maze. Top row, from left to right: MS-RRTa after 400 iterations (400 nodes, 14 obstacle nodes), MS-RRTa at completion (989 iterations, 989 nodes, 39 obstacle nodes). Second row: MS-RRTb after 400 iterations (400 nodes, 14 obstacle nodes), MS-RRTb at completion (762 iterations, 762 nodes, 25 obstacle nodes). Third row: ModRRT after 858 iterations (400 nodes), ModRRT at completion (4869 iterations, 2896 nodes).

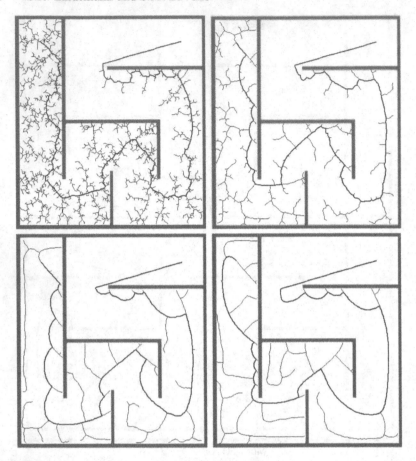

Fig. 18.6. Top row, from left to right: MS-RRTa with $k = 10, k = 100$. Bottom row, MS-RRTa with $k = 1000, k = 10000$.

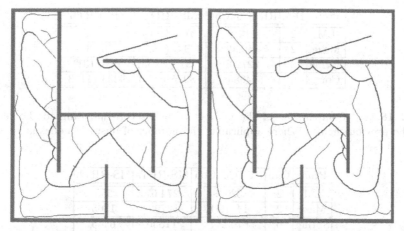

Fig. 18.7. From left to right: MS-RRTb with $k = 1000, k = 10000$

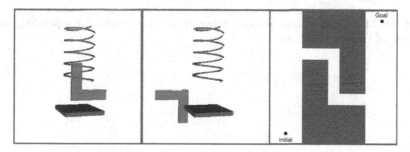

Fig. 18.8. The first two frames are the initial and goal configurations of a 3d rigid body example. The third is a two-dimensional representation of a 6-d bent corridor problem.

Prob.	Dim	RRTConCon	MS-RRTa	MS-RRTb
LM	2	0.01	0.01	0.01
Maze	2	0.19	0.21	0.17
Spring	6	0.7652	8.37	6.01
Corr.	6	101.18	1170	444.6

Fig. 18.9. Comparisons of the planning times required for our experiments. RRTConCon and MS-RRTa results are averaged over 100 trials. Implementations were done in Gnu C++ on a 2.0 GHz PC running Linux.

Prob.	Dim	RRTConCon	MS-RRTa	MS-RRTb
LM	2	113.4	91.1 (2.1)	88 (2)
Maze	2	456.49	389.3 (28.7)	450 (31)
Spring	6	2272.5	3392 (76.33)	1875 (31)
Corr.	6	14085	18975 (773)	29223 (1173)

Fig. 18.10. Comparisons of the number of nodes corresponding to the results of the previous figure. Where applicable, the number of obstacle nodes is given in parentheses.

Prob.	Dim	RRTConCon	MS-RRTa	MS-RRTb
LM	2	320.2	274.67	255
Maze	2	1411.41	1279.6	1365
Spring	6	6793.17	11850	6405
Corr.	6	42494	55686	80750

Fig. 18.11. Comparisons of the number of collision checks corresponding to the results of the previous figure

19

Tracking Methods for Relative Localisation

Frank E. Schneider, Andreas Kräußling, and Dennis Wildermuth

Research Institute for Applied Sciences (FGAN)
Neuenahrerstrasse 20, 53343 Wachtberg, Germany
{frank.schneider|a.kraeussling|dennis}@fgan.de

19.1 Introduction

The multi-robot relative localisation problem asks whether it is possible for an autonomous vehicle to start at an unknown location and incrementally estimate its own position with reference to the relative locations of the other robots based only on local sensor information. As shown in our previous papers [13] and [14] the topic of relative localisation (RL) for heterogenous multi-robot systems in unknown environments is a challenging task. We used the information of SICK laser scanner systems to estimate the relative positions between the robots. An Extended Kalman Filter was used to combine this information into continuously updated position estimations. All robots of a group utilized this data to generate one common co-ordinate system. In this paper we would like to address the sub-problem of tracking which is inherent to RL.

Tracking denotes the estimation of the position of a moving object based on consecutive sensor measurements. In the area of robotics tracking is a well established research topic [4, 15]. Usually laser range scanners are the preferred sensor devices. A SICK laser range scanner for example can measure the distance to the next reflecting object with high resolution.

The problem of tracking moving robots in densely populated environments with a robot mounted laser scanner shows certain peculiarities. Most of the readings are from obstacles like walls or other static objects and only a few measurements originate from the tracked robots themselves. This fact is illustrated in Figure 19.1. It shows the measurements of a 360-degree laser scan in our testing environment. The observing robot is located in the centre and has two SICK lasers, each with a planar 180-degree field of view. Since the laser scanners are mounted back to back and have a 1-degree resolution, the scan results in 360 readings. There are two other robots and also two wall-like obstacles in the field of view of the observer. Most of the measurements originate from the walls of the laboratory. The problem of allocation of data obtained from the presently accounted target is called the data association problem [1]. As a solution to this problem, a tracking algorithm might use a validation gate which separates the signals belonging to the current target from other signals. A second characteristic of tracking robots with laser range scanners is the occurrence of several measurements from the same object. In contrast to

K. Kozłowski (Ed.): Robot Motion and Control, LNCIS 335, pp. 301–313, 2006.
© Springer-Verlag London Limited 2006

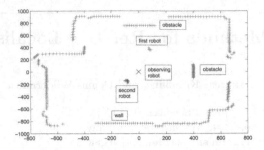

Fig. 19.1. Measurements of a 360-degree scan

common radar based tracking, robotic sensors like the SICK laser scanner have a much higher resolution and refresh rate. This leads to the fact that the tracked object generates several measurements. Therefore, we have to deal with what we call extended objects instead of punctiform objects like in the common radar tracking literature. Thereby, punctiform targets are the ones, which are just the origin of one measurement. A third characteristic of tracking in the field of multi-robot systems and especially in the case of relative localisation is the occurrence of crossing targets. This situation occurs when two targets get very close to each other, so that they cannot be separated by common tracking algorithms [3], [8]. This situation is very common in moving multi-robot systems and is one of the interesting problems in relative localisation.

In section two the underlying models for the dynamics and the observation process of the robot are proposed and the details of the validation gate are given. We then introduce two algorithms which can deal with tracking extended objects as long as they are not crossing. The first algorithm is based on the Viterbi algorithm [18], whereas the second makes use of the well known Kalman filter [6]. A brief comparison of the accuracy and the computation complexity of the algorithms is given.

In section three the performance of the algorithms under the condition of crossing targets is studied and it is shown, that none of the algorithms can handle this situation sufficiently. A new hybrid switching algorithm, which is the main contribution of this work, is proposed. It uses an improved Viterbi algorithm only when a crossing occurs and otherwise it just uses the Kalman filter. The performance of the switching algorithm is tested on real data.

Finally, in section 4 the summary and an outlook on future work are given.

19.2 Mathematical Background

19.2.1 Model

The dynamics of the object to be observed and the observation process itself are modeled by a hidden Gauß–Markov chain with the equations

$$x_k = Ax_{k-1} + w_{k-1} \qquad \text{and} \qquad z_k = Bx_k + v_k, \qquad (19.1)$$

where x_k is the object state vector at time k, A is the state transition matrix, z_k is the observation vector at time k and B is the observation matrix. Furthermore, w_k and v_k are supposed to be uncorrelated zero mean white Gaussian noises with covariances Q and R, i.e. it is $E(w_i w_j^\top) = Q\delta_{ij}$, $E(v_i v_j^\top) = R\delta_{ij}$ and $E(w_i v_j^\top) = 0$.

Since the motion of a target in the plane has to be described, a two dimensional kinematic model is used. Therefore, it is

$$x_k = \begin{pmatrix} x_{k1} & x_{k2} & \dot{x}_{k1} & \dot{x}_{k2} \end{pmatrix}^\top \tag{19.2}$$

with x_{k1} and x_{k2} the Cartesian coordinates of the target and $\dot{x}_{k1}$ and $\dot{x}_{k2}$ the corresponding velocities. This approach has the advantage that the orientation Θ of the observed robot can be calculated using the formula

$$\tan(\Theta) = \frac{\dot{x}_{k2}}{\dot{x}_{k1}}. \tag{19.3}$$

Vector z_k gives just the Cartesian coordinates of the target. For the coordinates the equations of a movement with constant velocity hold, i.e.

$$x_{k+1,j} = x_{kj} + \Delta T \dot{x}_{kj}. \tag{19.4}$$

For the progression of the velocities we use the equation

$$\dot{x}_{k+1,j} = e^{-\Delta T/\Theta} \dot{x}_{kj} + \Sigma \sqrt{1 - e^{-2\Delta T/\Theta}} u(k) \tag{19.5}$$

from [17] with the zero mean white Gaussian noise $u(k)$ with $E[u(m)u(n)^\top] = \delta_{mn}$. Thus the velocity is supposed to decline exponentially. The term

$$\Sigma \sqrt{1 - e^{-2\Delta T/\Theta}} u(k) \tag{19.6}$$

models the process noise and the accelerations. Thereby the parameters Σ and Θ model the dynamics of the observed robots. For these parameters we use the values $\Theta = 20$ and $\Sigma = 60$. For the details of the resulting matrices we refer to [17].

In the hereby proposed model the object is supposed to be punctiform and to produce only one measurement at time k. Nevertheless this model, at least with the extension given in the next section, will also be helpful for the description of an extended target. In this case x_k will get the state vector of a selected point of the target and z_k will get a prestigious measurement generated from all the measurements from the target. Details will be given below.

19.2.2 The Validation Gate

The validation gate is realised using the Kalman filter [6]. The Kalman filter calculates a prediction $y(k + 1|k)$ for the measurements $z_{k+1,l}$ from the actually handled target at time step $k + 1$ via the formulae

$$x(k + 1|k) = A \cdot x(k|k) \qquad \text{and} \qquad y(k + 1|k) = B \cdot x(k + 1|k). \tag{19.7}$$

Here $x(k|k)$ is the estimate for the position of the target at time step k. For every sensor reading $z_{k+1,l}$ of the time step $k+1$ ($l = 1, \ldots 360$), the Mahalanobis distance λ [10]

$$\lambda = (z_{k+1,l} - y(k+1|k))^\top \cdot [S(k+1)]^{-1} \cdot (z_{k+1,l} - y(k+1|k)) \qquad (19.8)$$

is computed. Matrix $S(k+1)$ is the innovations covariance from the Kalman filter [1]. In common filter applications this matrix is calculated from the predictions covariance $P(k+1|k)$ with the equation

$$S(k+1) = BP(k+1|k)B^\top + R \qquad (19.9)$$

with the given covariance matrix R of the measurement noise. The predictions covariance is derived from the equation

$$P(k+1|k) = AP(k|k)A^\top + Q. \qquad (19.10)$$

In tracking extended objects this approach is not sufficient, since there is an additional influence of the extendedness of the object on the deviation of the measurements from the prediction $y(k+1|k)$. To take care of this feature an accessory positive definite matrix E should be added in Eq. (19.9). Because the lateral dimension of present-day service robots usually shows a radius of an approximate range of $30\,cm$, the entries of E should be in the range of 900. Thus, after some optimization process (for details see [9]) we use

$$E = \begin{pmatrix} 780 & 0 \\ 0 & 780 \end{pmatrix} \qquad (19.11)$$

and

$$S(k+1) = BP(k+1|k)B^\top + R + E. \qquad (19.12)$$

Thereby the values of the entries of matrix E vastly exceed the values of the entries of matrix R, so that the main contribution in Eq. (19.12) comes from matrix E.

The points with the same Mahalanobis distance lie on the surface of an ellipse with the semi–major axis in the direction of the greater uncertainty. The Mahalanobis distance λ is χ^2–distributed with two degrees of freedom [1]. Thus a χ^2–test is performed to select the measurements from the target: all measurements with $\lambda > \lambda_{\max}$ with a given bound $\lambda_{\max}$ are excluded.

One characteristic of the model proposed in this paper is the fact that the sequence $\{K_k\}_{k=1}^\infty$ of the Kalman gains (see Eq. (19.22) for a definition) converges very rapidly to a limit. Thus the calculations of the matrices K_k can be omitted and instead it is sufficient to calculate and use the limit $K = \lim_{k \to \infty} K_k$. This limit can be calculated very easily, similar to the case of the α–β–filter described in [1].

19.2.3 The Viterbi-based Algorithm

The Viterbi algorithm has been introduced in [18]. A good description is given in [2]. It has been recommended for tracking punctiform targets in clutter in [11] and for tracking extended targets in [7].

Whereas the afterward introduced Kalman filter algorithm (KFA) uses all measurements in the validation gate as an unweighted mean, the Viterbi based algorithm (VBA) calculates a separate estimate $x(k+1|k+1)_i$ for every selected measurement $z_{k+1,i}$ with $i = 1, \ldots, m_{k+1}$, where m_{k+1} is the number of selected measurements at time step $k+1$. The selected measurements will be defined later. For the calculation of the estimates $x(k+1|k+1)_i$ the VBA uses a directed graph.

The nodes of this graph are the measurements in the validation gates or the selected measurements. Given the selected measurements $\tilde{Z}_k = \{z_{k,j}\}_{j=1}^{m_k}$ at time step k the selected measurements for the time step $k + 1$ are determined as follows: for every selected measurement $z_{k,j}$ the prediction $y(k + 1|k)_j$ is calculated based on the estimate $x(k|k)_j$. Then the corresponding validation gate is applied to the measurements of time $k + 1$. This results in a set $\tilde{Z}_{k+1,j}$ of measurements which have passed the particular validation gate for the measurement $z_{k,j}$ successfully. The set $\tilde{Z}_{k+1}$ of selected measurements at time $k + 1$ is then just the union of these sets, i.e.

$$\tilde{Z}_{k+1} = \cup_j \tilde{Z}_{k+1,j}. \tag{19.13}$$

The distance $a_{k+1,j,i}$ between nodes $z_{k,j}$ and $z_{k+1,i}$ is calculated using the formula

$$a_{k+1,j,i} = \frac{1}{2}\nu_{k+1,j,i}^\top [S_{k+1}]^{-1} \nu_{k+1,j,i} + \ln\left(\sqrt{|2\pi S_{k+1}|}\right), \tag{19.14}$$

where $\nu_{k+1,j,i}$ is the innovation defined as

$$\nu_{k+1,j,i} = z_{k+1,i} - y(k + 1|k)_j. \tag{19.15}$$

S_{k+1} and $y(k + 1|k)_j$ are, respectively, the innovation covariance and the previously defined prediction evaluated by the Kalman filter, based on the nodes $Z_{k,j} = \{z_{l,i(l,j)}\}_{l=1}^k$ belonging to the path ending in $z_{k,j}$. Set $Z_{k,j}$ is called the tracking history belonging to node $z_{k,j}$. The indices $i(l, j)$ are defined later on. This procedure gives

$$p(z_{k+1,i}|Z_{k,j}) = \exp(-a_{k+1,j,i}) \tag{19.16}$$

and therefore with Bayes' rule and

$$d_{k,j} = \sum_{l=2}^k a_{l,i(l-1,j),i(l,j)} \tag{19.17}$$

as well

$$p(Z_{k,j}) = \exp(-d_{k,j}). \tag{19.18}$$

Finding the tracking histories with the minimal distance corresponds therefrom strictly to determining the tracking history with the maximal likelihood.

Now the predecessor of the node $z_{k+1,i}$ in the graph is the node $z_{k,j}$ which minimises the length $d_{k,j} + a_{k+1,j,i}$ of the corresponding path in the graph. The corresponding index is referred to as $j(k, i)$. Thus by a recursive algorithm for every selected measurement at time $k+1$ the tracking history $Z_{k+1,i}$ is determined. Next, a Kalman filter is applied to calculate an estimate $x(k + 1|k + 1)_i$ for the measurement $z_{k+1,i}$ using the prediction $y(k + 1|k)_{j(k,i)}$ as the input in the following equation

$$x(k + 1|k + 1)_i = x(k + 1|k)_{j(k,i)} + K_{k+1}(z_{k+1,i} - y(k + 1|k)_{j(k,i)}) \tag{19.19}$$

with the predictions

$$x(k + 1|k)_{j(k,i)} = Ax(k|k)_{j(k,i)} \tag{19.20}$$

and

$$y(k + 1|k)_{j(k,i)} = Bx(k + 1|k)_{j(k,i)} \tag{19.21}$$

and the Kalman gain K_{k+1} derived from the equation

$$K_{k+1} = P(k + 1|k)B^\top[S(k + 1)]^{-1}. \tag{19.22}$$

Furthermore the state covariance is updated via the formula

$$P(k+1|k+1) = P(k+1|k) - K_{k+1}S(k+1)[K_{k+1}]^{\top}. \qquad (19.23)$$

It should be mentioned that the covariances and the Kalman gain do not depend on the measurements and thus are independent of index i.

We use the results of the Viterbi algorithm for the tracking process as follows. The tracking history with the shortest length of the path is selected, when the last scan is reached. The corresponding estimates are used as the estimates for the state and the position of the object. These estimates are further improved by use of the Kalman smoother [16].

19.2.4 The Kalman Filter Algorithm

The tracking algorithm applies well known Kalman filter [6]. At first it calculates an unweighted mean z_{k+1} of the m_{k+1} measurements $\{z_{k+1,l}\}_{l=1}^{m_{k+1}}$ that have been selected by the gate, i.e.

$$z_{k+1} = \frac{1}{m_{k+1}} \sum_{l=1}^{m_{k+1}} z_{k+1,l}. \qquad (19.24)$$

This mean is then used as the input in updating the Equation (19.19) of the Kalman filter. For further details of this rather simple algorithm we refer to [9].

19.2.5 Comparison of KFA and VBA

In this section we refer to two criterions that should be considered for the evaluation of the proposed algorithms. One is accuracy, which measures the deviation of the estimates of the position of the tracked object from the true position. The other is computational complexity, which is measured by the time needed for the performance of the calculations for the recursion step from time k to time $k+1$. The following tables give the results for the comparison of the two algorithms with respect to accuracy and computational complexity. For the comparison of the algorithms we use simulated data. A circular object with radius $27\,cm$ moves on a circle with radius d around the observing robot. The radius d has been varied from $1\,m$ to $8\,m$ and the standard deviation of the laser measurements has been varied from $1\,cm$ to $10\,cm$. Simulated data have been used, because we needed to know the true position of the target very accurately, a goal which is hard to achieve using data from a real experiment. This case has already been mentioned by other authors [19]. The values used for the distance and the standard deviation of the measurement noise are typical for real objects observed with a SICK laser. In this and all further experiments the robots had velocities below $50cm/s$. Tables 19.1 and 19.2 give the distances of the by the algorithm estimated position from the true position in cm. Tables 19.3 and 19.4 give the computation time needed for one time step in seconds. It is obvious from the tables that the KFA outperforms the VBA with respect to both criterions.

19.3 Crossing Targets

In this section we study another performance feature that should be considered when evaluating algorithms for relative localization. This feature is qualitative rather

Table 19.1
Distance for the KFA

d/m	1	2	4	6	8
$\sigma = 0cm$	0.96	0.36	0.21	0.55	0.24
$\sigma = 1cm$	0.96	0.36	0.22	0.56	0.25
$\sigma = 3cm$	0.97	0.38	0.29	0.57	0.34
$\sigma = 5cm$	0.98	0.42	0.41	0.65	0.53
$\sigma = 7.5cm$	1.04	0.56	0.56	0.78	0.78
$\sigma = 10cm$	1.10	0.63	0.74	0.96	1.01

Table 19.2
Distance for the VBA

d/m	1	2	4	6	8
$\sigma = 0\,cm$	9.09	5.87	16.12	14.19	11.68
$\sigma = 1\,cm$	12.84	8.60	6.67	14.34	11.76
$\sigma = 3\,cm$	13.66	9.90	8.46	12.73	11.35
$\sigma = 5\,cm$	13.91	10.37	10.81	11.93	11.57
$\sigma = 7.5\,cm$	15.15	13.22	11.12	11.43	11.12
$\sigma = 10\,cm$	17.28	12.75	11.77	11.24	11.06

Table 19.3
Computing time for the KFA

d/m	1	2	4	6	8
$\sigma = 0cm$	0.02	0.02	0.02	0.02	0.02
$\sigma = 1cm$	0.02	0.02	0.02	0.02	0.02
$\sigma = 3cm$	0.02	0.02	0.02	0.02	0.02
$\sigma = 5cm$	0.02	0.02	0.02	0.02	0.02
$\sigma = 7.5cm$	0.02	0.02	0.02	0.02	0.02
$\sigma = 10cm$	0.02	0.02	0.02	0.02	0.02

Table 19.4
Computing time for the VBA

d/m	1	2	4	6	8
$\sigma = 0cm$	1.45	0.41	0.17	0.10	0.08
$\sigma = 1cm$	1.46	0.41	0.17	0.10	0.08
$\sigma = 3cm$	1.46	0.42	0.17	0.10	0.08
$\sigma = 5cm$	1.47	0.41	0.16	0.10	0.08
$\sigma = 7.5cm$	1.47	0.41	0.17	0.10	0.08
$\sigma = 10cm$	1.48	0.42	0.17	0.10	0.08

than quantitative. It investigates the question whether an algorithm can deal with the complex situation of crossing targets. Crossing of two targets means that the validation gates of the two targets intersect, i.e. some measurements are lying in the validation gates of both targets. Figure 19.2 shows a typical situation.

For the evaluation of the algorithms with respect to the problem of crossing targets real data have been applied. Figure 19.3 shows the behaviour of the VBA when applied to the problem of crossing targets using real data originating from an experiment with two robots in our laboratory. It shows the estimates for the position of the objects calculated by the algorithm by use of ellipses. The estimated position is the centre of the ellipse, whereas the shape of the ellipse represents the actual geometry of the tracked object. The objects start in the left and move to the right as indicated in Fig. 19.2. The behaviour of the KFA is very similar and thus an additional graphics is omitted.

Obviously none of the algorithms can deal with the problem of crossing targets. They all locate both objects at the same position after the crossing. This behaviour is common to other algorithms for tracking extended targets, for instance a nearest–neighbour–based algorithm.

19.3.1 The Cluster Sorting Algorithm

Since the native Viterbi approach and the KFA cannot deal with the problem of crossing targets, an improved algorithm has been developed, which can handle this situation. Our approach to deal with two crossing expanded targets relies on a specific feature of the Viterbi algorithm. The Viterbi algorithm is able to cope with bimodal or even multimodal probability densities to some degree. It shares this feature with e.g. Schulz' SJPDAF algorithm proposed in [15]. But while Schulz'

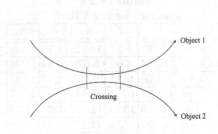

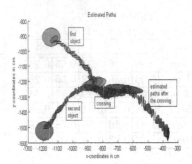

Fig. 19.2. Two crossing objects

Fig. 19.3. Application of the VBA to crossing targets, real data

algorithm uses particle filtering and thus has to deal with several hundreds of points, the Viterbi algorithm in our application only handles a few points. Additionally, these points contain some information about the surface of the tracked objects as proposed in [8]. If a crossing between the two targets occurs, the Viterbi algorithm shows the following behaviour. As soon as the crossing occurs, the algorithm tracks all points originated from both objects simultaneously. When the crossing is over, these points are again separated into two distinct clouds of measurements corresponding to the two targets. The two clouds or clusters of points are still tracked simultaneously. Our new approach is based on these observations from our experiments and consists of three different steps:

1. At first it is investigated whether a crossing between the two objects has occurred, i.e. at least one measurement lies in the gates of each object.
2. If a crossing between the two targets has been detected, it is examined whether the crossing has finished, i.e. the measurements have dispersed into two distinct clouds.
3. As soon as the end of the crossing has been observed, the two corresponding clouds or clusters are assigned to the two objects. By this procedure the two objects might be interchanged after the crossing.

The three steps are carried out based on geometrical considerations and can be viewed under the superordinate concept of data mining [5]. Finally, like for the VBA at the end of the tracking process for each object the path with the minimum length is determined and a Kalman smoother is applied. Since the introduced algorithm associates clusters of measurements with objects, the improved algorithm is called Cluster Sorting Algorithm (CSA).

It could be argued that there are already well established algorithms for tracking crossing targets like the JPDAF [3] or the Multiple Hypothesis Filter [12]. However, these algorithms have been developed for tracking punctiform targets in clutter and would probably fail when tracking extended targets. There might be several measurements from the same extended target. Thus two different measurements from the same target can be associated with the two targets. But the exclusion of the association of measurements from the same target to both objects is the essential core of these two algorithms. Thus they will fail to separate the two targets after the crossing in most of the cases, especially when an additional occlusion takes place.

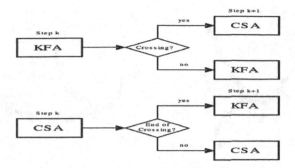

Fig. 19.4. Flowchart of the new switching algorithm

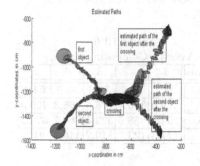

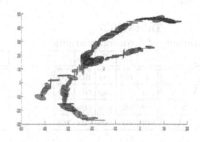

Fig. 19.5. Crossing targets, handled by the CSA, real data

Fig. 19.6. Real data, tracked with
the Cluster Sorting algorithm

In most of the cases the two tracks will coincide after the crossing like for the VBA
in figure 19.3. Moreover, the computational burden of applying these algorithms to
extended targets is very high.

19.3.2 A Switching Algorithm

Since the CSA is able to deal with crossing targets, it could be used for the whole
tracking process. However, since this algorithm is based on the VBA algorithm it is
not as accurate as the KFA as long as no crossing takes place and needs much more
computation time as will be shown in the next section. Therefore, we developed a
new switching or hybrid algorithm (SA), which uses the CSA only when a crossing
takes place. For the rest of the time it uses the very accurate and fast KFA. The
choice of the KFA is furthermore motivated by the fact, that it is more accurate and
faster than all other algorithms we have tested for tracking extended objects [9].
Crossings are detected as in the case of the CSA. Figure 19.4 shows the flowchart
of the SA.

19.3.3 Further Experiments

Figure 19.5 shows the application of the CSA to the previously used data. In the

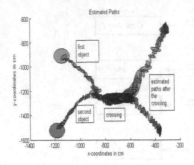

Fig. 19.7. Crossing targets, handled by the switching algorithm, real data

Fig. 19.8. Crossing targets, real data of scenario 1, handled by the new SA

Table 19.5
Handling of Crossing Targets

scenario	1	2	3	4	5
KFA	no	no	no	no	no
VBA	no	no	no	no	no
CSA	yes	yes	yes	yes	yes
SA	yes	yes	yes	yes	yes

Table 19.6
Average Computing Time

scenario	1	2	3	4	5
KFA	54.1	53.8	54.6	53.9	54.4
VBA	379.0	355.7	478.6	366.3	469.3
CSA	294.3	258.6	329.5	260.7	320.3
SA	171.3	123.7	119.2	110.8	91.5

second example given in Fig. 19.6, which uses real data of two moving robots, the assignment after the crossing is wrong. But again, the two objects are separated well after the crossing has finished.

Figure 19.7 shows that the SA can deal with crossing targets as well as the CSA using the same data from a real experiment as above.

To illustrate the power of the SA further experiments with real data were carried out, in which two robots moved around in our laboratory. The measurements were recorded with SICK lasers mounted on a mobile robot. The evaluation of the algorithms was performed by means of five similar scenarios. In each scenario the two robots moved separately for some time interval t_1 at the beginning of the experiment. Then the robots met and moved together for some time interval t_2, so that a crossing took place. Finally, the robots split again and moved alone for time interval t_3. Time interval t_2 was approximately 30 seconds for each scenario. Time intervals t_1 and t_3 were of the same length, varying from 30 seconds to 150 seconds. Figure 19.8 shows an example of the results for the estimated paths using the SA.

Like in the experiment examined above, KFA and VBA always failed, whereas CSA and SA worked well for all the five scenarios.

Table 19.6 shows the required computing time. It contains the average time t_a needed for the calculation of one time step in milliseconds. The table shows improvement that can be achieved using the SA in comparison to the CSA. Moreover, with growing intervals t_1 and t_3 the gain increases rapidly.

19.4 Conclusions

In this paper we have addressed the problem of tracking robots in typical RL situations. Two basic algorithms for the tracking process have been introduced: they are either using the Kalman filter (KFA) or the Viterbi algorithm (VBA). The comparison of the algorithms has shown that the KFA is faster and more accurate than the VBA. Thereafter, the problem of crossing targets has been introduced. It has been shown that both algorithms produce insufficient results under the constraints of crossing targets. Thus an enhancement of the VBA in the form of the CSA has been proposed, which can deal with crossing targets. Since the CSA is based on the VBA and thus is imprecise and slow, we finally developed the SA, which makes use of the CSA only when a crossing has been detected and otherwise uses the KFA. The power of the SA has been demonstrated on real data. It has been shown that the SA can handle crossing targets as well as the CSA but needs much less computing time. The proposed algorithms are capable of coping with occlusions to a certain limit.

In the future we will try to generalise our method for tracking crossing objects from the case of two objects to n objects (multi target tracking). Furthermore, the SA should be compared to algorithms like the SJPDAF [15].

Finally, further efforts have been made to improve the SA for real time applications. The effort goes mainly in two directions. First, the calculation of the Kalman gain in Eq. (19.22) and the inverse of the innovations covariance needed for the calculation for the Mahalanobis distance in Eq. (19.8) are substituted by the calculation of the corresponding limits as mentioned in Section 2.2. These limits can be calculated before the intrinsic tracking process is started. This reduces the computing effort, because matrix multiplication and inversion are omitted.

Another improvement is the introduction of filtering of the data before the application of the validation gate. This procedure is motivated by the fact that usually when applying the CSA the Mahalanobis distance has to be calculated for a few thousand measurements. This results from the fact that for every selected measurement of the last time step each of the 360 new measurements has to be tested in the validation gate. This requires the vast majority of the computation time when using this algorithm. In order to improve this, we have developed an extra filter step, which eliminates most of the new measurements before applying the validation gate. The new filter step uses the following sophisticated geometrical considerations.

One feature that the robots have in common is that they are located in some distance in front a wall. When investigating the measurements from the laser scan there should be two jumps in the scan line of the observing robot when there is an object in front of a wall. One jump results in an edge where the distance decreases, while the other jump leads to an edge where the distance increases. Figure 19.9 shows a typical situation.

Another feature comes from the fact that the number of measurements generated by the robot is very limited. The limit depends on the extension which is e.g. about 50cm for robots, and the distance to the robot. Using the combination of these two features in most cases nearly all measurements originating from the walls can be eliminated. These measurements are by far the major fraction of the 360 measurements from the scan.

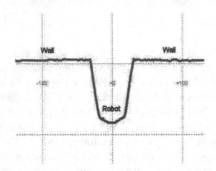

Fig. 19.9. A robot in front of a wall

First preliminary results show that by these two modifications the SA needs only about 30ms for the computation of one time step on a P4 with 3.5GHz. Since our SICK lasers have a frequency of 6Hz this means, that the SA is capable for real time applications after these modifications.

References

1. Bar-Shalom, Y. and Fortmann, T. (1988). *Tracking and Data Association.* Academic Press.
2. Forney Jr., G.-D. (1973). The Viterbi algorithm. *Proceedings of the IEEE,* 61(3):268–278.
3. Fortmann, T. E., Bar-Shalom, Y., and Scheffe, M. (1983). Sonar tracking of multiple targets using joint probabilistic data association. *IEEE Journal of Oceanic Engineering,* OE–8(3).
4. Fuerstenberg, K. C., Linzmeier, D. T., and Dietmayer, K. C. J. (2002). Pedestrian recognition and tracking of vehicles using a vehicle based multilayer laserscanner. In *Proceedings of IV 2002, Intelligent Vehicles Symposium,* volume 1, pages 31–35.
5. Han, J. and Kamber, M. (2001). *Data Mining.* Academic Press, London.
6. Kalman, R. E. (1960). A new approach to linear filtering and prediction problems. *Trans. ASME, J. Basic Engineering,* 82:34–45.
7. Kräußling, A., Schneider, F. E., and Wildermuth, D. (2004a). Tracking expanded objects using the Viterbi algorithm. In *Proceedings of the IEEE Conference on Intelligent systems, Varna, Bulgaria.*
8. Kräußling, A., Schneider, F. E., and Wildermuth, D. (2004b). Tracking of extended crossing objects using the viterbi algorithm. In *Proceedings of the 1st International Conference on Informatics in Control, Automation and Robotics (ICINCO).*
9. Kräußling, A., Schneider, F. E., and Wildermuth, D. (2005). Zur Verfolgung ausgedehnter Ziele — eine Übersicht über ausgewählte Algorithmen und ein Vergleich deren Güte. Technical report, FKIE/FGAN, Wachtberg, Germany (in German).
10. Mahalanobis, P. C. (1936). On the generalized distance in statistics. *Proceedings of the National Institute of Science,* 12:49–55.

11. Quach, T. and Farooq, M. (1994). Maximum likelihhod track formation with the viterbi algorithm. In *Proceedings of the 33rd Conference on Decision and Control, Lake Buena Vista, Florida*.
12. Reid, D. B. (1979). An algorithm for tracking multiple targets. *IEEE Transactions on Automatic Control*, AC–24:843–854.
13. Schneider, F. E., Moors, M., Wildermuth, D., and Kräußling, A. (2003). Relative position estimation in a group of robots. In *IEEE International Conference on Methods and Models in Automation and Robotics*.
14. Schneider, F. E. and Wildermuth, D. (2004). Using an extended kalman filter for relative localisation in a moving robot formation. In *4th International Workshop on Robot Motion and Control (RoMoCo 2004)*.
15. Schulz, D., Burgard, W., Fox, D., and Cremers, A. B. (2001). Tracking multiple moving objects with a mobile robot. In *Proceedings of the 2001 IEEE Computer Society Conference on Computer Vision and Pattern Recognition*.
16. Shumway, R. H. and Stoffer, D. S. (2000). *Time Series Analysis and Its Applications*. Springer.
17. van Keuk, G. (1971). Zielverfolgung nach Kalman–anwendung auf elektronisches Radar. Technical Report 173, Forschungsinstitut für Funk und Mathematik, Wachtberg–Werthhoven, Germany (in German).
18. Viterbi, A. J. (1967). Error bounds for convolutional codes and an asymptotically optimum decoding algorithm. *IEEE Transactions On Information Theory*, IT–13(2).
19. Zhao, H. and Shibasaki, R. (2005). A novel system for tracking pedestrians using multiple single–row laser–range scanners. *IEEE Transactions on Systems, Man and Cybernetics—Part A: Systems and Humans*, 35(2):283–291.

20

Robot Localisation Methods Using the Laser Scanners

Leszek Podsędkowski and Marek Idzikowski

Institute of Machine Tools and Production Engineering
Technical University of Łódź, ul. Stefanowskiego 1/15 90-924 Łódź, Poland
{lpodsedk|idzik}@p.lodz.pl

20.1 Introduction to Robot Localisation Methods

One of the major problems of the navigation of a mobile robot is finding its position and orientation. The process should be fast and reliable so that the control unit is able to steer the robot along a required trajectory. There are several ways to determine the localisation of a mobile robot, namely:

- dead-reckoning – a method that does not use any external reference points. The most popular one is odometry – acquiring information from wheel encoders [5], [7];
- landmark positioning – finding the exact position by measuring the distances to characteristic obstacles in the environment [3];
- sensor-based navigation – method utilising a variety of range scanners to build a map of a visible area. Then it is compared with the global map to find the current position [1], [2], [6];
- other methods – for example, the application of GPS or vision systems [4], [5].

Comparing the map stored in the robot's memory with the one sensed by the robot is the technique of localisation called sensor-based navigation. Different types of sensors used to create the map are described later on. This method is more complicated than finding landmarks but gives more reliable results. Of course, the robot must operate in at least partially known terrain. In most applications it is an acceptable restriction – the walls of buildings do not change every day. Maps can be stored in two methods. In the first one detected edges are converted to vectors. It is easy to compare them, but this procedure has one restriction – it is assumed that the environment consists of some kind of objects [11–13,15]. The shape of those obstacles can be described by geometric primitives like line or arc. In many cases this assumption works well, for example in office-like workspaces. In our research [11], such a system was also developed. In the second method the data received from range sensors are transformed into a bitmap.

Usually, most common sensors are range scanners, which use many different physical phenomena. Most popular are laser, sonar and radar rangefinders. As a result of measurement we get a set of data containing bearing and distance to the obstacle. Each kind has its own key features. Sonar scanners are known to be

K. Kozłowski (Ed.): Robot Motion and Control, LNCIS 335, pp. 315–332, 2006.
© Springer-Verlag London Limited 2006

inaccurate and operate at a low sampling rate. Information passed by sonar is noisy and difficult to process due to many reflections. The returned signal is wide, and a lot of effort is needed to recognise simple objects such as flat walls or edges. Although some results have been introduced, sonar systems are not capable of navigating an indoor mobile robot precisely due to the low sampling rate. Also two independent systems may interfere with each other. Laser rangefinders are well suited for indoor applications. They operate at high sampling rates because of the light speed. The light beam is very thin and returns to the receiver undisturbed. There are two basic types of laser rangefinders. The first ones operate in one plane, usually parallel to the floor. They emit a radiant bundle of beams with high angular resolution. The other ones can scan three-dimensionally. They are more reliable but operate with smaller sampling rate and require more data processing. Another problem of navigation is the localisation of the obstacles in the workspace in which the mobile robot operates. Methods based on active sensors can be used to detect obstacles. Sonars are useful at close range as they are not accurate but can detect obstacles in a wide area. They can be used in an emergency stop procedure. Nowadays, laser devices are applied very often for the purpose of obstacle detection in mobile robots. In 1998, in the Institute of Machine Tools and Product Engineering a car-like robot was built, which is shown in Fig. 20.1. It is a three-wheel robot with dimensions 0.8×1.3 m. In front of the robot there is a laser scanner protected by a solid bumper and an emergency switch stopping the robot on coming into contact with an obstacle.

Fig. 20.1. The NURT Mobile Robot

The robot was built with the basic aim of testing the methods of robot's trajectory planning with nonholonomic limitations [8], hence the name of the robot: the NURT Nonholonomic Universal Robot for Tests. To accomplish the basic aim it was necessary to equip the robot with a system of defining its location as well as detecting obstacles, so two versions of the system were created and installed on the robot. The basis of the first one is a flat SICK scanner and a vector description of the workspace used for localisation. The other version uses a spatial laser scanner of our own construction and a raster description of the obstacles. The theoretical bases and the results of the experiments of both methods are discussed below.

20.2 Localisation Method Using Vector Description of Workspace

20.2.1 Sensor – a Flat Laser Scanner

Laser scanners are currently one of the most precise sensors which can be used in sensor-based robot's navigation methods. A very popular class of these scanners are flat scanners. Figure 20.2 presents two scanners of this type: produced by the SICK Company and by the AccuRange. Both are equipped with the laser rangefinder and a rotating mirror. The first robot navigation system implemented on NURT is based on a SICK LS200 laser scanner presented in Fig. 20.2a. The scanner has an 80 m distance range and a 180° angular range. It measures the distance from the obstacles with the angular resolution of 0.5° and the accuracy of ±15 mm. The measurement is performed on the height of 0.25 m from the floor. The scanning frequency is 25 Hz.

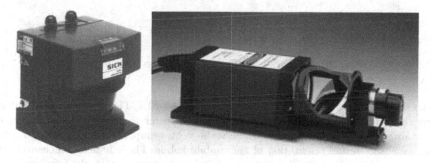

Fig. 20.2. a) LS200 laser rangefinder. b) AccuRange 4000 with a rotating mirror

Another practicable sensor is the AccuRange 4000 laser rangefinder presented in Fig. 20.2b. The measurement of the distance can be done only in the optical axis of the device. A rotating mirror was used to deflect the laser beam, which gives the possibility of a wide range scanning. Due to the fact that a driving motor was installed on a narrow bracket to drive the mirror, the scanner has a comparatively wide angle of sight, about 300°. The device is characterised by high parameters with the range of about 15.5 m (52 ft) and the measurement accuracy of 2.54 mm (0.1 inch). The rangefinder operates with the maximum frequency of 50 kHz, which enables the system to work with high refresh rate maintaining the high angular resolution.

20.2.2 Navigation Module

In the method presented here, defining the localisation of the robot is based on the assumption that there exist flat (straight) fragments of walls. The wall segment is described by a pair of oriented points. The beginning of the wall segment is defined by the point that is observed first. Additionally, few other parameters are calculated, namely: length of the segment, parameters of the line covering the segment, etc.

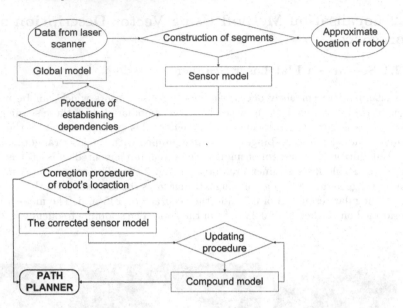

Fig. 20.3. Operation diagram of the navigation module

The module operation diagram is presented in Fig. 20.3. The information about the environment is stored in two maps. One is the raster map used by the trajectory planning module. The other one, which is the vector map, makes it possible to define the localisation and orientation of the mobile robot. The odometric measurements inform us about the relocation of the robot in comparison with the previous cycle. Next, the current location and orientation of the robot can be calculated. We assume that the calculations contain some error which cannot exceed certain values. As there are three coordinates used to describe the location of the robot, all the possible combinations of X, Y and Θ errors create a 3D-space which is called the error space. The information from the environment vector map ("wall-on-map") and the data from the current scanning are used to calculate the correction of the localisation. On the basis of the data collected from the scanner a series of measurement points which form line segments are identified. They are called "the segments observed". Then, on the basis of an approximate localisation of the robot, walls-on-map and the-segments-observed are paired using the following rule: each segment is paired with all the walls which have the same orientation (with the accuracy equal to the maximum angular error) and which is collinear with the accuracy equal to the maximum segment position error. Maximum segment position error is equal to the product of maximum range and maximum angular error increased by maximum robot position error. Each pair has its weight dependent on the wall weight and the length of the segment observed as well as on the number of walls with which a given segment forms pairs. If the segment has more than one pair the weight is reduced.

On the basis of one pair it is possible to define only the correction perpendicularly to the direction of the segment observed. The graphic representation is a straight line in the error space (Fig. 20.4), which is called an error line. To calculate the value of the error precisely, the error space is divided into sub-ranges that create a 3-D

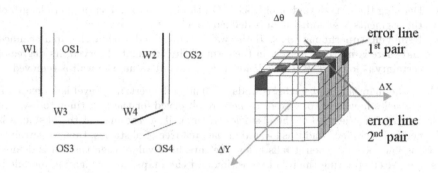

Fig. 20.4. The "segment-observed" (OS) – "wall-on-map" (W) pairs that have been found and the error space

matrix. Then, the weights corresponding to each of the sub-ranges are increased by the weights of the pairs whose error lines cross the given sub-range. The subsection with the highest weight is assumed to constitute the approximate correction of the robot's localisation. To calculate the correction precisely we take into account only the "segment-observed – wall-on-map" pairs whose weights constitute the weight of the chosen sub-range. Differences of angles in the "segment-observed – wall-on-map" pairs are the basis for the calculation of the final robot's orientation correction. Then, the exact value of the relocation correction is calculated. The value of the correction is obtained using the least square method as the point lying closest to all the lines representing the difference of the localisation of the elements in pairs. After calculating the robot's position the system updates the maps. The procedure consists of several phases:

- Modification of wall data, which have been taken into consideration in the "segment-observed" – "wall-on-map" pairs. The parameters of the segments to which only one wall corresponded are used for the correction of the walls' parameters with the exception of the ones defined as invariable. New wall is a linear weighted approximation of the old wall and the segment.

$$\sum_{i=1}^{4} (ax_i + by_i + c)^2 \cdot w_i = \min, \qquad (20.1)$$

where a, b and c are the new line parameters assuming $a^2 + b^2 = 0$, x_i, y_i are the segmnets' endpoints coordinates and w_i are the wall and segment weights. The smaller the factor of the wall's certainty and the greater the length of the segment, the greater the influence of the segment observed. The length of new wall covers old wall and the segment. All four end points are projected onto the new line and the two outer are selected for the begin and the end of the new segment. After modification of the wall its certainty is increased.

- Creation of new walls. If the segment observed did not have its equivalent on the walls' map, then a new wall is created. Its parameters (location and length) are identical with the parameters of the segment observed. The degree of the certainty of such a wall is established on a low level.

- Defining the area devoid of obstacles. On the basis of the location and length of the segments a scanning area is defined which is free of obstacles.
- Removing superfluous walls. If the wall is contained within the area scanned and there is no confirmation in the form of the segment observed, its degree of certainty is lowered. If it falls below the threshold value, the wall is removed.

Then, the map of the obstacles is updated. The cells' certainty level is increased if the obstacles have been observed. If they are observed for the first time, the system places them on the map of the obstacles. If the cell on the map of the obstacles is defined as "occupied" and it is located in the area free of obstacles, then its certainty level is lowered and when it falls below the threshold value, then the cell is defined as free. After updating the robot's position and the maps, the navigation module is ready to process the data from the next cycle.

20.2.3 Identification of the Workspace

As in all kinds of sensor navigation it is necessary to define a model map (the global model) to make it possible to define the robot's localisation. It can be done in several ways. The data concerning the most important obstacles, such as – for example – the walls of rooms, can be introduced from the documentation. The walls on the walls' map that are defined in this way should have the maximum and invariable certainty level. Then, the system creates a bit map of obstacles on this basis. This method gives the best accuracy of the map that is being created, with the assumption that the building was constructed according to plan. Another method involves the robot creating the map of walls and obstacles on its own. In the initial position which is selected so that the biggest number of flat fragments of walls is visible, the robot is making a series of measurements. Then, it changes its position by a short distance, stops and takes another series of measurements. Such procedure enables the creation of a very precise map of walls, and on the basis of that map it can create the map of obstacles. The last method is similar to the previous one. The robot moves from its initial position and takes measurements without stopping. The map which is created in such a way is less precise than in the method with the robot stopping. In our laboratory we have used and tested the second method. During the navigation with the map of the walls formed in such a way, the measurements of the robot's position were stable, and the accuracy of the localisation did not exceed $10mm$.

20.2.4 Laboratory Tests

The system described above has been tested on a mobile robot constructed in our Institute. Figure 20.5 shows the general view of the laboratory where you can see the protective shields which make it possible for the robot to move undisturbed, with the identification system of obstacles being flat. Figure 20.6 presents the workspace in the form of the obstacle map and the wall map. In this complex environment the robot moved smoothly, which confirms the stability of localisation measurements. The accuracy of defining the position did not exceed $10mm$. Taking into account the fact that it is the result of the operation of the whole steering system and the $100 \text{ mm} \times 100$ mm cell size of the raster map, it should be considered high.

Fig. 20.5. The general view of the laboratory workspace

Fig. 20.6. The vector map of walls and the raster map of obstacles

20.3 Localisation Method Using the Raster Map of the Workspace

In the research we have conducted, the navigation system using the vector map has one major restriction. For the proper work there have to exist flat walls long enough (min. 300 mm). Moreover, the system uses a double (raster and vector) description of the environment. For the purpose of the mobile robot localisation we have developed a new method called PLIM. This section presents theoretical bases and experiments of this method.

20.3.1 The 3D Laser Rangefinder

During the experiments with the SICK flat laser scanner, two inconveniences of such a solution have been identified, namely the workspace is being scanned

- only on one level of 0.25 m,
- within the range of only 180°.

For the procedures involving the definition of the robot's position it is entirely sufficient but if the information concerning the workspace in the procedures of defining the localisation is not complete, it may lead to collisions. The most common errors involve the robot trying to move forward between the legs of the tables (with the table top being above the scanning level) and knocking against new obstacles while the robot is moving back. These were the reasons why we have decided to construct a spatial (3-dimensional) laser scanner without such faulty features. Most of the spatial scanners used worldwide have a rotating mirror with the changing angle of inclination. Such constructions usually have a laser rangefinder placed at the bottom and the mirror driving motor on the top mounted on the bracket. This kind of construction results in part of the workspace being obstructed by the bracket. The construction that has been designed, built and patented in IOiTBM (Fig. 20.7) is free of this drawback.

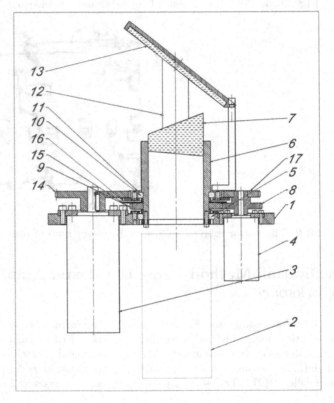

Fig. 20.7. The 3D laser rangefinder

It contains a vertically placed linear laser rangefinder (2) and a mechanism for deflection of the laser beam in two perpendicular planes. The mechanism consists of a rotating mirror (13) and a prism (7) placed in a sleeve, which can revolve. The mirror is responsible for horizontal scanning, and the prism for vertical deflection of the laser beam. Both parts can be driven by independent motors (3) and (4) or by a single one (3) as shown in the picture, using toothed wheel (17). The transmission ratios are matched in such a way that the prism revolves along the mirror. Drives are mounted to the base (1). The 3D laser rangefinder has a 360° horizontal field of view with no dead zones, and a 30° vertical one.

20.3.2 Description of the PLIM Localisation Method

The detailed explanation of the localisation method can be found in [9] and [10]. Here, only the key assumptions and a brief description of the method will be presented. In the localisation method presented here it is assumed that:

- A mobile robot moves on the planar surface and the main axes of the laser rangefinder are perpendicular to the groundwork.
- The 3D laser rangefinder is placed on the constant height, thus a mobile robot position is described by 3 coordinates: x, y and Θ.
- A group of measurements, called series, taken into consideration should contain measurements of the full angle in the horizontal range.
- The laser beam moves along a spiral line with a relatively small angle of inclination. There is no relationship between the z coordinate of the obstacle and the position of the mobile robot, thus a 3D problem can be divided into n 2D problems.
- The initial position (which can be the approximation of the real one) of a mobile robot is known.
- The scanning frequency is high enough to assume that the robot's position is constant during measurement.

The mobile robot localisation method is based on matching the current scan to the map of the environment principle. Measurements R_i are taken in the position which is unknown to the navigation system, see Fig. 20.8. "Obstacle" in Fig. 20.8 is part of a real object that has the same discrete z coordinate. Only the approximate position, for example one calculated in the previous time step, is known. First, a temporary map from the most recent scan is built. Next, a pair of points, which represent the same location in the workspace, is taken into consideration. The first X point is the point where the laser beam hits the obstacle. The second X' point is the image of the first one, created from the estimated position of the mobile robot.

It can be noticed that one point can be transformed into the second one, as shown in Eq. (20.2),

$$X = X' + \overline{B} + \overline{K} \times \overline{R_i}, \tag{20.2}$$

where B is the translation vector of the mobile robot, and $K \times R_i$ is the vector product of the orientation difference of the mobile robot and the given distance to the obstacle. Then, vector P_i, which is perpendicular to the line tangent to the detected obstacle's edge can be found. Vector P_i leads from the X' point to the nearest X'' point in the environment map. Due to the fact that the direction in which the obstacles on the map are being sought is known, the time needed for the search is short. As the sought obstacle on the bitmap usually has edges whose thickness is of a few cells, thus it is the distance to the middle of the edge's section which is sought. An alternative solution involves looking for the distance to the nearest obstacle; however with registering the obstacles in the form of a bitmap, it would result in underestimated values. The maximum length of the P vector

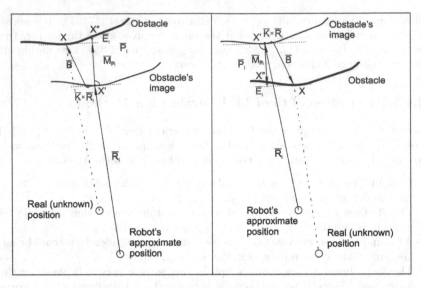

Fig. 20.8. The principle of the localisation method

is limited. If an obstacle cannot be found in a certain range, then this particular range measurement is dropped. This can happen near the obstacles' edges. Next, the displacement vector $K \times R_i + B$ is projected onto the direction of the P_i vector. The M_{P_i} projection vector is equal to the P_i vector with the accuracy of E_i. In this way a relationship between the map, current data and the required parameters of the mobile robot's localisation error B and K can be found. There are basic factors which cause E_i error:

1. The distance measurements contain a certain error, which results in a faulty determination of the direction of the P_i vector.
2. Another factor which influences the value of the error vector is angular relocation of the mobile robot. It is the reason why the line tangent to the obstacle's image in the point being considered is not parallel to the line tangent to the obstacle in point.
3. Next factor is the non-linear shape of the obstacle.

It has been assumed that the distribution of the E_i error vectors is random. It follows from this assumption that for the corrections determined for the robot's position different from the real ones the average value of the E_i error vectors will be increasing. Thus, it is necessary to look for such a position of the mobile robot in which the sum of squares of the E_i error vector lengths is the smallest and it is assumed that this position is as close to the real position of the robot that both can be considered identical. To determine the value of relocation corrections there has been applied the least squares method. The sum of squares of the E_i vector lengths is totalled for the distance measurements taken into account in a given measurement series (n signifies the number of these measurements):

$$S = \sum_{i=1}^{n} \left(|\overline{E_i}|\right)^2 = \min,$$ (20.3)

so we obtain

$$S = \sum_{i=1}^{n} \left(|\overline{P_i} - \overline{M_{P_i}}| \right)^2 . \tag{20.4}$$

Applying the general dependence for the measure of the vector projection on any direction, we obtain:

$$S = \sum_{i=1}^{n} \left(\left| \overline{P_i} - \frac{\left(\overline{B} + \overline{K} \times R_i \right) \circ \overline{P_i}}{\overline{P_i}} \right| \right)^2 . \tag{20.5}$$

Developing Eq. (20.5) further:

$$S = \sum_{i=1}^{n} \left(P_i^2 - 2 \left[P_{ix} \left(B_x + R_{iy} K \right) + P_{iy} \left(B_y - R_{ix} K \right) \right] \right) + $$
$$+ \sum_{i=1}^{n} \left(\frac{\left[P_{ix} \left(B_x + R_{iy} K \right) + P_{iy} \left(B_y - R_{ix} K \right) \right]^2}{P_{ix}^2 + P_{iy}^2} \right) . \tag{20.6}$$

The sum S will reach minimum value when its partial derivatives with respect to B_x, B_y and K which are being searched are equal to zero:

$$\frac{\partial S}{\partial K} = -\sum_{i=1}^{n} P_{ix} R_{iy} + \sum_{i=1}^{n} P_{iy} R_{ix} + $$
$$+ K \left(\sum_{i=i}^{n} \frac{P_{iy}^2 R_{ix}^2}{P_i^2} + \sum_{i=1}^{n} \frac{P_{ix}^2 R_{iy}^2}{P_i^2} - 2 \sum_{i=1}^{n} \frac{P_{ix} P_{iy} R_{iy} R_{ix}}{P_i^2} \right) + $$
$$+ B_x \left(\sum_{i=1}^{n} \frac{P_{ix}^2 R_{iy}}{P_i^2} - \sum_{i=1}^{n} \frac{P_{ix} P_{iy} R_{ix}}{P_i^2} \right) + $$
$$+ B_y \left(\sum_{i=1}^{n} \frac{P_{ix} P_{iy} R_{iy}}{P_i^2} - \sum_{i=1}^{n} \frac{P_{iy}^2 R_{ix}}{P_i^2} \right) , \tag{20.7}$$

$$\frac{\partial S}{\partial B_x} = -\sum_{i=1}^{n} P_{ix} + K \left(\sum_{i=i}^{n} \frac{P_{ix}^2 R_{iy}}{P_i^2} - \sum_{i=1}^{n} \frac{P_{ix} P_{iy} R_{ix}}{P_i^2} \right) + $$
$$+ B_x \sum_{i=1}^{n} \frac{P_{ix}^2}{P_i^2} + B_y \sum_{i=1}^{n} \frac{P_{ix} P_{iy}}{P_i^2} , \tag{20.8}$$

$$\frac{\partial S}{\partial B_y} = -\sum_{i=1}^{n} P_{iy} + K \left(\sum_{i=1}^{n} \frac{P_{ix} P_{iy} R_{iy}}{P_i^2} - \sum_{i=i}^{n} \frac{P_{iy}^2 R_{ix}}{P_i^2} \right) + $$
$$+ B_x \sum_{i=1}^{n} \frac{P_{ix} P_{iy}}{P_i^2} + B_y \sum_{i=1}^{n} \frac{P_{iy}^2}{P_i^2} , \tag{20.9}$$

$$\begin{cases} \dfrac{\partial S}{\partial K} = 0 \\ \dfrac{\partial S}{\partial B_x} = 0 \\ \dfrac{\partial S}{\partial B_y} = 0 \end{cases} \Rightarrow \quad K, B_x, B_y . \tag{20.10}$$

By solving the system of equations (20.10) the correction B_x, B_y of the linear position and the correction K of the angular position can be determined.

20.3.3 Simulations

Before the new localisation method was implemented in the real system, the concept had been tested in a simulator. In the simulations, two simplifications were made. First, a two dimensional laser rangefinder was simulated. Second, the map of the environment was symbolized by vectors. Simulations were run to test the influence of some factors on the localisation accuracy. The accuracy of the laser measurement, the relocation of the mobile robot and the rotation of the mobile robot were considered. The simulations were set up so that all the factors were tested independently. Simulations were run in different types of environment. The sample map is presented in Fig. 20.9. A large number of experiments were conducted. Typically, there were twenty runs with one set of parameters fixed in each test point. The variations of parameters were as follows: the accuracy of range measurement was $0 - \pm 12.5$ cm with step of 0.5 cm, the robot's rotation was $0 - \pm 5°$ with step of 0.2° and the robot's translation was $0 - 50$ cm with step of 1 cm. The results obtained were fitted with polynomial curves. The exemplary dependence is shown in Fig. 20.10. The experiments have shown that:

- The E_{LRF} laser rangefinder accuracy has linear influence on error of orientation E_A calculation and square influence on error of position E_L calculation.
- The length of mobile robot translation M_L has linear impact on error of orientation E_A and position E_L calculations.
- The size of mobile robot rotation M_A has square influence on error of orientation E_A and position E_L calculations.

According to the superposition rule, the accumulated error of position calculation E_L in the simulated environment can be presented by Eq. (20.11):

$$E_L = 0,1756 \cdot E_{LRF} + 0,01 \cdot E_{LRF}^2 + 0,0317 \cdot M_A + 0,0082 \cdot M_A^2 + \\ + 0,0076 \cdot M_L \,[\text{cm}]. \tag{20.11}$$

Similarly, the error of orientation calculation E_A is shown in Eq. (20.12):

$$E_A = 0,0246 \cdot E_{LRF} + 0,0133 \cdot M_A + 0,0031 \cdot M_A^2 + 0,0047 \cdot M_L \,[°]. \tag{20.12}$$

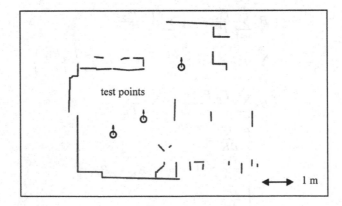

Fig. 20.9. The sample map

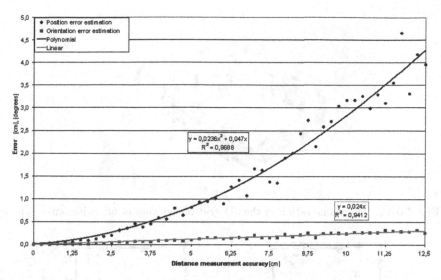

Fig. 20.10. The error of the determination of the robot's positioning and orientation depending on laser accuracy

20.3.4 Stationary Experiments

As the first step of evaluating the new navigation system several stationary experiments were carried out. In the navigation module of the mobile robot's control system, a 3D fine grained grid based map of the environment was used. The cell size was set to $(1, 1, 10)$ cm $[x, y, z]$. For purposes of the stationary tests, the 3D laser rangefinder was placed in the usual workroom. During the experiment it remained rooted in one place. The visualization of this workspace is presented in Fig. 20.11. The stationary tests allowed us to eliminate "almost" all the errors and bugs in the code, as well as to determine the characteristics of the localisation method in the real environment. During this phase of research several short-term and long-term experiments were carried out. The standard deviation of position estimation was 1.10 and 1.32 cm, respectively. The standard deviation of orientation calculation was $0.39°$ and $0.42°$. The short-term experiments (a few minutes long) allow to determine the accuracy of the robot's position and orientation estimation. The map was not updated during the experiments. The long-term stationary experiments, each of which lasted over 90 min, have been conducted with map updating procedure. It proved that the localisation method is very stable and is free from the drift of the map of environment.

20.3.5 Experiments with Mobile Robot

The final verification of the new mobile robot localisation method was completed due to the experiments in industry like environment with the NURT robot. The first thing that was done was the creation of the map of the environment. The process of map creation was as follows:

Fig. 20.11. The visualization of the 3D grid based map of office-like environment

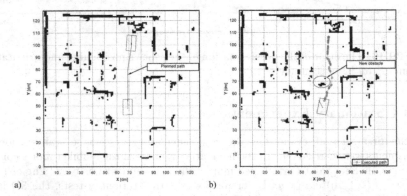

a)   b)

Fig. 20.12. Dynamic environment test: a) planned path b) executed path

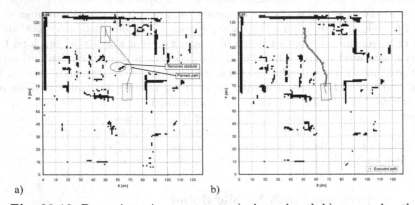

a)   b)

Fig. 20.13. Dynamic environment test: a) planned path b) executed path

1. The initial robot position was set up.
2. The workspace was scanned for a period of time.
3. Collected information was put into the map.
4. The mobile robot was moved to the next location relatively close to the current one.

Steps 2–4 are repeated until the whole area is mapped. Next, the experiments in a dynamic environment were carried out. After successful localisation, the same data are used to update the 3-D occupancy grid map. The update process uses the ray tracing technique to update the state of all the cells of the map the laser beam "travelled" through. The sample runs are presented in Fig. 20.12 and Fig. 20.13. The presented maps were taken from the path planning module of the navigation system, which uses a two-dimensional grid based map with cell size of 10 × 10 cm. The idea of those tests was to evaluate the behaviour of the navigation system in the cases of detecting new obstacles in the path, or detecting new accessible passages. One can see (Fig. 20.12a) that the planned path ran forward in a straight line. After the robot started moving, a carton was thrown into the planned trajectory. The new obstacle was detected and information was stored in the map. Next, the updated two-dimensional map was sent to the path planner module. The Path Planner has planned a new path to avoid the obstacle. The executed path as well as the updated map are shown in Fig. 20.12b. Rounded markers along the robot's path show the successive points in which the navigation module determined the robot's position. Similarly in a situation when the obstacle (a cardboard box) was removed from the workspace (Fig. 20.13). The path along which the robot moved considers the change in the environment. During the experiments with removing the obstacle the robot's speed was reduced. During the robot's movement the localisation and the map updating were done in real time. In comparison to [14] the localisation is performed much faster than the complete 3D scan is done. In the experiments the localisation frequency was 10.5 Hz. It is possible mainly because of scanning technique different from [14]. The laser beam sweeps along the helical path with a relatively small angle of inclination. Due to the considerably increased size of the map, the process of refreshing the information about the obstacles runs much more slowly. The time it takes the robot to react to a new obstacle is relatively short (about 1 s), but the time needed to remove the obstacles from the map is much longer (about 5 s). It results from the fact that at each rotation of the scanner the laser beam has a different angle of inclination, so the system requires a much longer time to check if all the cells that had previously been marked as occupied are now free. The mobile robot stayed localised in the whole previously mapped area. However, the localisation accuracy have not been determined because of the lack of the external reference system.

20.4 Other Localisation Methods

In paper [16] there is presented the method of mobile robots localisation in the office or office-industry environment. The workspace of this type can be easily presented in the form of a collection of lines. Thus, the authors have chosen the vector map. A flat laser scanner of 360 degrees work range was used to collect data about the environment. Generally, localisation involves comparing the current measurement (image) with the environment map. The authors assume, as many others, that the robot's relocation as well as its rotation between subsequent computation cycles are not big. The robot's initial position, which can be approximate, is also known. The robot's localisation procedure begins with finding the nearest linear segment on the map for each point of the image. Then, the congruence is being sought which minimises the sum of squared distances between the image points and their corresponding linear segments. It can be presented in the form of the equation:

$$S = \sum_i \left([R\left(\Theta\right)\left(v_i - c\right) + \left(c + t\right)]' \, u_i - r_i \right)^2 = \min, \qquad (20.13)$$

where Θ means the robot's rotation, t – robot's relocation, v_i – image points, u_i – corresponding linear segments on the map, c – centre of gravity of the image points and $'$ – matrix transposition. The Θ and t values found constitute rotation and relocation of the robot with respect to the last known location. Due to computation power which was then at the authors' disposal, various simplifications and limitations were used, among others – more than 80% of the image points were rejected. The basic differences between the localisation method presented in [16] and the PLIM method discussed in Section 20.3 are the result of the assumption of an unstructured environment described by the raster map as the main type of workspace. In the PLIM method there is minimized the distance of the image points to obstacles without a definite direction as in the case of the vector map. So there has been developed an algorithm seeking the distance to the "blurred" edge of the obstacle in the direction perpendicular to the temporary direction of the obstacle's image. Moreover, in comparison to [16] the approximations of the rotation matrix are not used as well as all the measurements are included in the computations. The authors of [15] present the method of the localisation of mobile robots and the method of creating the map using the information from a flat SICK LMS200 laser scanner. Linear segments have been used to build the map. However, modification of a typical solution has been introduced. They have created a map called a Closed Line Segment, consisting exclusively of linear segments which form a closed region. The lines singled out from the laser scanner measurement are joined with the help of the so-called virtual lines. The method has been tested on the NOMADIC XR4000 mobile platform in the office-like environment of 45m^2. The robot was equipped with the computer based on a 300 MHz Pentium II. In these conditions the times of local and global localisations were 1.6 and 17 ms, respectively. The time needed for updating the map was 170 ms. The accuracy of the localisation was 10 mm and 0.5°, according to the authors. The authors of [14] present the method of mobile robot localisation with the spatial laser scanner and flat maps consisting of linear segments. The spatial scanner that was used is the authors' construction. It consists of a SICK flat scanner mounted on a rotating head. The time of scanning depends on the work parameters and is contained between 1.2 s to 4.8 s. The workspace map (inside buildings) is created by removing the information about the z height from the cloud of n points of three coordinates, x, y, and z. In this way there is formed the so-called 2-D virtual image of the workspace which undergoes further processing. Then, linear segments are isolated which describe mainly the walls of the building used for the localisation of the robot. The method has been tested with the use of the Robot ATRV mobile platform equipped with the on-board computer and scalable processing box for processing the signals from the spatial scanner. The vehicle was radio-controlled by an additional computer. The tests were carried out in an industrial environment of about 1500 m^2. The travelling speed was 0.1 m/s.

20.5 Summary and Conclusions

In the present article two techniques of mobile robots localisation have been presented:

1. using a flat scanner and a vector map of obstacles
2. using a 3-D laser scanner and a raster map

The table below presents the basic similarities and differences of both methods in comparison with the methods taken from literature [14], [15]. Because of the considerable differences in the computer speeds, the table does not include any comparison with [16].

Table 20.1. Comparison of the mobile robot localisation methods

	2D Scanner and vector workspace map	2D Sick scanner [15]	3D scanner and grid map	3D Sick based scanner – virtual 2D [14]
information about obstacles in workspace	partial	partial	full	full (3D point-cloud)
accuracy	10 mm	10 mm, 0.5°	11 mm	?
frequency of localisation	25 Hz	?	5 − 10 Hz	1.2 − 4.8 s
structure of the workspace	with flat walls	with flat walls	unstructured	unstructured + flat walls
time of adding new obstacle	0.1 s	0.17 s	1 s	?
time of removing obstacle	0.2 s	0.17 s	3 − 10 s	?
memory for 100 m²	10 kB	?	10 MB	?
map creation time	short	short	long	long
computational effort	low (P 133 MHz)	medium (PII 300 MHz)	high (2× PIII 1.25 GHz)	high (2 onboard units + 1 remote unit)

Both methods work well in typical office or industry workspaces. However, the second method is more universal but requires a much greater computation effort and is slower. Analysing the behaviour of both scanners and localisation methods, it can be expected that applying a hybrid system will be a good solution: using a 2-D scanner for localisation but with the localisation method based on the raster map (described in Section 20.3). The 3-D scanner could be used to collect the information about the localisation of obstacles. So the Path Planner Module would receive the combined information coming from both scanners. This solution will be tested during the further phase of research.

References

1. Politis Z, Probert P (1998) Perception of an Indoor Robot Workspace by Using CTFM Sonar Imaging. In: Proceedings of IEEE International Conference on Robotics and Automation. Leuven, Belgium
2. Boehmke S K, Bares J, Mutschler E, Lay N K (1998) A High Speed 3D Radar Scanner for Automation. In: Proceedings of IEEE International Conference on Robotics and Automation. Leuven, Belgium
3. Armingol J M, Moreno L, de la Escalera A, Salichs M A (1998) Landmark Perception Planning for Mobile Robot Localisation. In: Proceedings of IEEE International Conference on Robotics and Automation. Leuven, Belgium
4. Sukkarieh S, Nebot E M, Durrant-Whyte H F (1998) Achieving Integrity in an INS/GPS Navigation Loop for Autonomous Land Vehicle Applications. In: Proceedings of IEEE International Conference on Robotics and Automation. Leuven, Belgium
5. Aono T, Fujii K, Hatsumoto S, Kamiya T (1998) Positioning of Vehicle on Undulating Ground Using GPS and Dead Reckoning. In: Proceedings of IEEE International Conference on Robotics and Automation. Leuven, Belgium
6. Crowley J L, Wallner F, Schiele B (1998) Position Estimation Using Principal Components of Range Data. In: Proceedings of IEEE International Conference on Robotics and Automation. Leuven, Belgium
7. Borenstein J (1998) Experimental Evaluation of a Fibre Optics Gyroscope for Improving Dead-Reckoning Accuracy in Mobile Robots. In: Proceedings of IEEE International Conference on Robotics and Automation. Leuven, Belgium
8. Podsędkowski L (1998) Path Planner for Nonholonomic Mobile Robot with Fast Replaning Procedure. In: Proceedings of IEEE International Conference on Robotics and Automation. Leuven, Belgium
9. Idzikowski M (2005) Universal navigation system for mobile robots based on a spatial laser scaner. PhD thesis, Technical University of Łódź. Łódź, Poland (in Polish)
10. Idzikowski M, Podsędkowski L (2003) Theoretical basis of the PLIM method of the mobile robot localisation. In: Proceedings of SYROCO'03. Wrocław, Poland
11. Idzikowski M, Nowakowski J, Podsędkowski L, Visvary I (1999) On-line navigation of mobile robots using laser scanner. In: Proceedings of RoMoCo'99. Kiekrz, Poland
12. Kleeman L, Kuc R (1995) Mobile Robot Sonar for Target Localisation and Classification. International Journal of Robotic Research, Volume 14
13. Skrzypczyński P, Drapikowski P (1999) Environment modelling in a multi-agent mobile system. In: Proceedings of EUROBOT'99. Zurich, Switzerland
14. Wulf O, Arras K, Christensen H, Wagner B (2004) 2D Mapping of Cluttered Indoor Environments by Means of 3D Perception. In: Proceedings of IEEE International Conference on Robotics and Automation. New Orleans, USA
15. Zhang L, Ghosh B K (2000) Line segment based map building and localisation using 2D laser rangefinder. In: Proceedings of IEEE International Conference on Robotics and Automation. San Francisco, USA
16. Cox I, (1990) Blanche: Position Estimation for an Autonomous Robot Vehicle. In: Cox I, Wilfong G (eds) Autonomous Robot Vehicles. Springer-Verlag, Berlin

Part V

Applications of Robotic Systems

21

Complex Control Systems: Example Applications

Piotr Dutkiewicz

Chair of Control and Systems Engineering, Poznań University of Technology
ul. Piotrowo 3a, 60-965 Poznań, Poland `piotr.dutkiewicz@put.poznan.pl`

21.1 Introduction

Recently, there develop intensively applications of control and robotics systems. Special control systems, i.e. robots, are designed for use in environments unfriendly or harmful to human beings. Possible areas of application of robots are continuously repeated tasks or jobs requiring large physical power. A robot may also be used as an assistant of a human (e.g. a robot assistant for a surgeon) or in so-called telerobotic systems, which "extend" human hands. Most dynamical development takes place in mobile robotics. In some tasks, robots should move in a manner resembling humans or animals – they are called walking robots.

In the paper, example applications of special robots are described. Each such robot has a complex control system. The robot consists of three main systems: mechanical, control, and sensor. The mechanical part has to provide locomotion and manipulation functions. The control system with its measurement elements carries out desired tasks. Moreover, in many applications a general purpose measurement system for inspection of the environment is additionally required. The paper discusses problems of development of the special robots.

An example robot SAFARI allows to describe problems of design of mechanical and control structures. Climbing robots create a specific subclass of walking robots. Organization of the software and hardware system of a climbing robot is outlined. The kinematics of such robots, similarly to other mobile robots, belongs to the class of nonholonomic systems, which results from velocity constraints. This implies design of specific control of mechanical systems. As a consequence, communication and computational requirements concerning particular elements of the control system are formulated. Existing distributed software solutions are briefly presented with special attention to the industrial standard 'Corba'. Its properties and specific features of the climbing robots also affect architecture of the control system.

In Section 21.2, mechanical systems of such construction are described. Design of control and sensor systems is discussed in Section 21.3. Specific features of measurement and control system of the SAFARI climbing robot are described. Requirements concerning software of the measurement/diagnosis system for such a robot are discussed. Then, architecture of the control/diagnosis system of the SAFARI robot is described.

K. Kozłowski (Ed.): Robot Motion and Control, LNCIS 335, pp. 335–350, 2006.
© Springer-Verlag London Limited 2006

Separately, vision feedback/measurement in complex control systems is described in Section 21.4. A concept of vision feedback in control of position of surgical tools for Minimal Invasive Surgery (MIS) is presented. Remarks concerning vision measurement of position and orientation of a mobile soccer robot group are quoted. Concluding remarks are presented in Section 21.5.

21.2 Mechanical Systems

The most dynamically developing area in robotics seems to be mobile robotics. There may be various types of locomotion of mobile robots. Most often mobile robots move on wheels or caterpillars. There are also walking, flying or swimming types. The manner of locomotion simply depends on the specific environment of the mobile robot. The kinematics of walking and other mobile robots causes them to be nonholonomic systems, which results from velocity constraints. This implies the necessity of design of specific control systems. A mobile machine gives the possibility to observe and examine the state of the environment in which the machine moves. Because of this, the inspection machine should be autonomous.

Review of Climbing Robots Types

Climbing robots represent a specific kind of walking robots. They are specially designed for moving on vertical surfaces. There are two types of climbing robots depending on the method of attachment to the surface: robots with mechanical connection between its grippers and the environment and robots making use of adhesive forces.

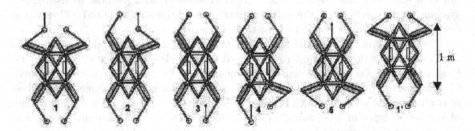

Fig. 21.1. Movement of the Klettermax robot

In literature, many constructions [5, 10, 17, 20] of climbing robots are described. The examples illustrate the ways of adapting them to desired tasks. In most cases a climbing robot moves similarly to a human climber. It should have a support at three points on the surface. This can be seen in Fig. 21.1 presenting the four-leg robot Klettermax. Its movement results from a sequence of movements of its legs. The Klettermax [5, 20] has four legs (of two joints each) and is very light when pneumatic actuators are employed.

Another climbing robot Ninja I [14, 18] has been designed for such tasks as climbing buildings, bridges, and other constructions. It is presented in Fig. 21.2. The

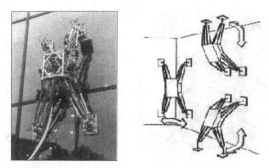

Fig. 21.2. Climbing robot Ninja I

legs of the robot are driven with prismatic actuators, working in parallel (so-called coupled drive), which are vertically oriented. The mass of the leg can be minimized and the mass of the robot is distributed evenly between all elements of its mechanical construction. An extra passive degree of freedom causes the feet of the robot to be always oriented parallel to the ground. The Ninja I robots are equipped with special grippers – pneumatic VM-type (valve-regulated multiple) suction cups. Many small cells of suction cups are controlled independently. This reduces underpressure losses on porous or cracked ground.

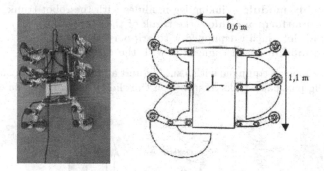

Fig. 21.3. The Rest six-legged robot

The Rest robot [19] shown in Fig. 21.3 is an example of a six-legged robot, which is destined for moving on ferromagnetic surfaces. The legs move forward independently. The body is moved forward when all the legs are located in a final position. In many walking robot applications sliding frame constructions are used. They need special control strategies, but such constructions are especially well suited for climbing applications. An example is the Wally robot [17], presented in Fig. 21.4. The Wally uses two pneumatic actuators. Suction cups are mounted at ends of each actuator. Its movement is ensured by synchronous control of suction cups and actuator movement.

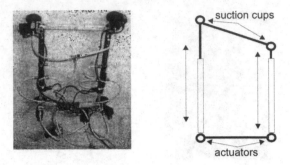

Fig. 21.4. The Wally robot with its kinematic structure

SAFARI Robot

The project 'SAFARI' [10] has been supported by the Polish State Committee for Scientific Research under grant no. 8 T11A 010 17 'An inspection climbing robot for high-wall buildings'. The main aim of the SAFARI robot is inspection of concrete of high-wall buildings walls. The SAFARI belongs to a group of climbing robots with so-called sliding frame drive. Such mechanical construction allows implementing a **modular structure**, consisting of special modular blocks of defined movement functions. The SAFARI consists of the following modules:

- **an assembly module** – linking leg modules with the robot trunk,
- **a sliding platform module** – the trunk of the robot,
- **a leg module**, which is equipped with grippers (four legs),
- **adhesive modules** – for connection with the working surface.

Moreover, the robot is equipped with a supply unit and measurement/control devices for performing particular tasks mounted on the sliding platform module.

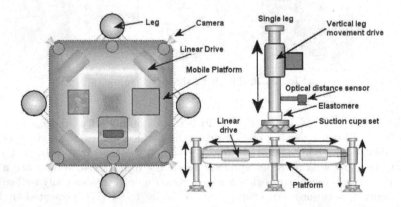

Fig. 21.5. Mechanical structure of the SAFARI

The mechanical structure of the SAFARI is presented in Fig. 21.5: the top view on the left side and one leg on the right. The sliding frame with joints placed at its vertices is of quadrilateral shape [3, 9, 10] and includes extra devices: vacuum pumps, elements of the measurement system and an on-board manipulator. The robot as a whole consists of eight identical cylinders forming the sliding frame and four legs. During its movement the cylinders (prismatic joints) change their lengths. The cylinders are made of steel pipes. A pulling screw placed inside each piston is rotated and changes the length of the cylinder.

Each of the cylinders of the robot is independently driven by a DC motor. The torque is transmitted from the motor via a cog-belt gear to a screw. The linear drives of the sliding frame are connected with four R-type passive degrees of freedom. Such solution eliminates the necessity of taking nonholonomic constraints into consideration in the control algorithm. The kinematic structure for correct work of this robot requires adhesive fastening of three feet to the surface. That means that only one suction cup may be moved at a time.

A leg is connected to two sides of the sliding frame and it can be lifted by moving one of those two linear drives (sides of frame). During feet repositioning, only one leg is lifted up and down. The other three legs must hold the construction on the vertical working surface. This sequence consists of following steps [9]:

1. Release the fixing force by cutting out an airflow,
2. Move up this leg with suction cups,
3. Move the leg to the target position (using two linear drives of sliding frame attached to this leg),
4. Move down the leg,
5. Try to seal the suction cups on the new position by applying underpressure to them,
6. If it is not possible to obtain significant adhering force in the new location, try to place the leg in another position (return to 1),
7. If this is successfully completed, the robot lifts up another leg and moves it to a new position.

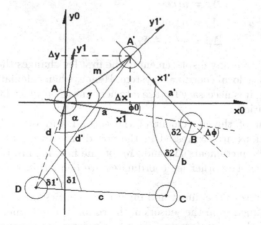

Fig. 21.6. Displacement of the first leg A

Climbing of the robot results from sequential movements of the legs (each separately) [5, 8, 10]. The kinematic model is described in detail in [10]. Here we consider only kinematics of movement of one leg (from start point A). Let A' be the target point of moving this leg. We have to know solutions of two kinematic problems: direct and inverse [15]. Solving the direct kinematics problem consists in finding the position/orientation of the robot given its joint coordinates. On the other side, the inverse kinematics problem requires calculating joint coordinates necessary for the robot to reach the desired position/orientation.

The displacement of the first leg from A to A' is shown in Fig. 21.6. It is assumed that $\varphi_0 < 0$. The inverse kinematics equations have the following form [8]:

$$m = \sqrt{\Delta x^2 + \Delta y^2}, \tag{21.1}$$

$$\gamma = \operatorname{atan2}(\Delta y, \Delta x), \tag{21.2}$$

$$a' = \sqrt{m^2 + a^2 - 2ma\cos(\gamma - \varphi_0)}, \tag{21.3}$$

$$d' = \sqrt{d^2 + m^2 - 2md\cos(\gamma - \varphi_0 + \alpha)}, \tag{21.4}$$

where m is the length of the leg displacement from A to A', Δx and Δy are the coordinates of m measured in the $x_0 y_0$ coordinate frame, γ is the angle between the AA' and the x_0 axis (calculated with a 2-argument version of the 'arctan' trigonometric function). Moreover, a, d, a' and d' are the lengths of sections AB, AD, A'B and A'D, respectively. Based on Fig. 21.6, formulae (21.1)-(21.4) can be easily processed to obtain the direct kinematics equations, which are as follows:

$$m = \sqrt{a'^2 + a^2 - 2a'a\cos(\delta_2' - \delta_2)}, \tag{21.5}$$

$$\sin(\gamma - \varphi_0) = \frac{a'\sin(\delta_2' - \delta_2)}{m}, \tag{21.6}$$

$$\cos(\gamma - \varphi_0) = \frac{m^2 + a^2 - a'^2}{2ma}, \tag{21.7}$$

$$\gamma = \operatorname{atan2}\left(\sin(\gamma - \varphi_0), \cos(\gamma - \varphi_0)\right) + \varphi_0, \tag{21.8}$$

$$\Delta x = m\cos\gamma, \tag{21.9}$$

$$\Delta y = m\sin\gamma, \tag{21.10}$$

$$\Delta\varphi = -(\delta_2' - \delta_2). \tag{21.11}$$

It is worth noting that any displacement of the first leg changes the robot orientation and the offset of the local coordinate system. Hence, when calculating the kinematics for the second leg, it is necessary to take into account the updated value of the desired change of orientation.

A description of the movement of the second and other legs can be found in [10]. Its positions are expressed in the coordinate frame attached to the first leg. Therefore, the increments Δx and Δy for the first leg are used, as in formulae (21.9) and (21.10). The other leg coordinates are defined in the reference frame of the first leg.

The calculations of the direct and inverse kinematics make use of the measurement results obtained from the sensors of the robot: the incremental encoders of the motor shafts and the potentiometers for measurement of configuration angles of the sides of the sliding frame. The control system receives also information about the robot orientation in global coordinates obtained from the acceleration sensor. The following quantities are measured in the mechanical system:

- the lengths of the sides of the sliding frame,
- the angles between these sides,
- the lengths of the legs,
- the force fastening the suction cup to the ground.

The positions of the suction cups relative to the reference frame, attached to the first foot (A), are calculated taking into account the current values of the measured quantities. The position of this frame is specified relative to the base coordinate frame attached to the wall of the building.

In case of the inverse kinematics problem, the new position of the first foot is defined with use of the increments Δx and Δy. For the other legs, the target position in the reference frame attached to the first foot is known. The sequence of leg movements is defined by a special algorithm [8]. The movement path for the robot is divided into a set of repositioning displacements. Single repositioning increments for leg A constitute the desired values of a single step of the whole robot. This algorithm checks all the possible sequences and chooses the first which allows correct robot movement. When the robot cannot obtain significant adhering force in the new location of the leg, it tries to place the leg in another position obtained from this algorithm.

21.3 Design of Sensor and Control System

The control system of the climbing robot is a real-time system. Its observation system should be flexible and easy to extend due to a large spectrum of measurement systems available. This concerns an additional measurement system dedicated for expert tasks as well. As a result, the system should be distributed. In our specific case, the hardware implementation of the control/sensor system [9] of the climbing robot, presented in Fig. 21.7, is as follows:

- a host computer (an industrial PC computer),
- a control system for actuating units (microcontrollers),
- a measurement system of the robot (position, pressure, temperature, distance sensors, etc.),
- an inspection system (cameras, ultrasonic devices, special devices for particular tasks),
- a main control/acquisition console (a PC computer) for an operator supervising the system.

Communication between the console and the climbing robot is implemented via a Wi-Fi network. All devices working in the local network of the robot are connected to a node, which is connected to the access point. The local network of the robot consists of an industrial miniature control computer (PC class), a mirroring computer, and a vision server. The industrial computer is connected to intelligent actuating/mesurement devices via a serial bus (Universal Serial Bus or RS-485). The microcontrollers are connected to the network with implementation of local industrial networks. To make the system secure and reliable, in case of failure of the industrial computer, it is automatically replaced by a microprocessor unit which gives the console access to actuating devices. The control console, placed on the ground, is also equipped with the access point. Moreover, it enables the console–robot communication and is connected to the wide-area network (WAN).

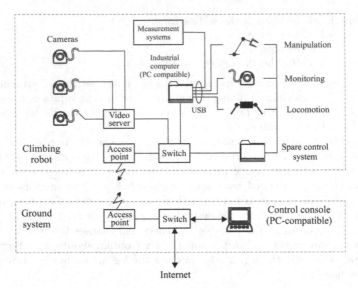

Fig. 21.7. Hardware of control/sensor system of the robot

The robot is constructed for moving on vertical surfaces. To calculate the instantaneous value of the force fixing the robot to the surface, it is necessary to know the value of underpressure in each of the suction cups. Here we propose to use relative pressure sensors (the output voltage is proportional to the pressure difference). Each of the pressure sensors consists of four piezoresistant elements, forming a bridge. The full-scale range of the pressure sensors is ±1 bar, with a transient overload of ±2 bar allowed. The robot has to know the actual distance from the end of the leg to the surface. To perform such measurement, a low cost optical sensor is attached to each leg. In order to verify whether the leg has stable contact with it, up to eight microswitches (or other open-close sensors) are installed around the set of suction cups of each leg.

The SAFARI robot requires a great deal of sensor data to be collected. Sensing elements are mostly of analog type and, since control algorithms are implemented on digital machines, it is necessary to digitize analog values. Besides that, some sensors require non-linearity correction and temperature drift compensation, which are usually easier to implement in a digital system. These are the reasons why all sensors (except the vision system) are connected to digital microcontrollers, equipped with appropriate analog front-ends. Next, the microcontrollers are connected together, forming a higher level of the sensor system (the lowest level being composed of the sensors themselves). Each microcontroller is equipped with a UART (Universal Asynchronous Receiver/Transmitter), so the easiest connection between all the microcontrollers is established by the serial bus. Each of the four robot legs is equipped with the following set of sensors:

- three piezoresistant underpressure sensors,
- up to eight microswitches,
- an infrared optical distance sensor,
- an additional temperature sensor for thermal compensation of sensors.

Another important feature is to have the opportunity to control the orientation of the mobile platform relative to the vertical surface. For this purpose, static accelerometers may be used to determine the angle between the robot's platform and the movement surface. Additionally, a digital inclinometer sensor is used in the SAFARI robot system. Here we use two ADXL202 double-axes digital accelerometers. Those devices, available in the form of small integrated circuits, are built as electronic micromachines, containing small silicon inertial elements, deflecting under applied acceleration. This deflection is represented by a voltage difference, developed inside the circuit and compared with an internal voltage reference source. The voltage difference is then converted into a PWM square wave, appearing on the chip's terminals and measured by a microcontroller, supervising the accelerometer's work. A single ADXL202 chip gives information about accelerations acting along two perpendicular axes X and Y. The values of the acceleration are proportional to the duty cycle of the output signals (PWM). By placing two ADXL202 circuits such that their X or Y axes are parallel while the remaining two axes are perpendicular to each other, we have the opportunity to measure X, Y and Z components of acceleration acting in 3-D space.

To control the position of the robot it is necessary to know how far each leg is pulled out along the axis perpendicular to the plane on which the robot moves. Information about the lengths of the linear drives, used for positioning the legs, is also essential. In our robot, linear movement is achieved by usage of linear drives with ball screws. The screws are driven by DC motors and timing belts, so the easiest option for determining the linear drive pull out is to use incremental encoders, installed on the motor shafts.

The control software of this robot is supervised by the RTLinux ver. 3.1 real-time operating system based on Mandrake 2.4.4. Its inspection, measurement and supervisor/acquisition systems [4, 16] are distributed, flexible and easy to extend. The software system implementation of the control/acquisition system of the climbing SAFARI robot is presented in Fig. 21.8 and consists of the following elements:

- software of the host industrial computer, placed at the robot's body,
- a control system for the actuating units (microcontrollers) – manipulation and locomotion of the robot,
- software of the measurement systems of the robot (measuring position, pressure, temperature, distance, etc.),
- tasks of system inspection (cameras connected via the video server, ultrasonic devices, special devices for particular tasks),
- software of the spare control system of the climbing robot mirroring the host computer to improve its reliability,
- software of the main control/acquisition console.

The software organization assures using multiple channels of data interchange between objects responsible for particular tasks. Channels are elements of data streams, which are generally of two types: the stream of control information and the stream of measurement data. Each element of the software system is implemented as an *agent* – an *object* equipped with *plug-ins*, which can pass signals from outside causing execution of some operations inside the object. The object gives access to a predefined set of plug-ins. The signal is described by a *protocol* consisting of the name of the action and parameters. All objects can pass messages to other

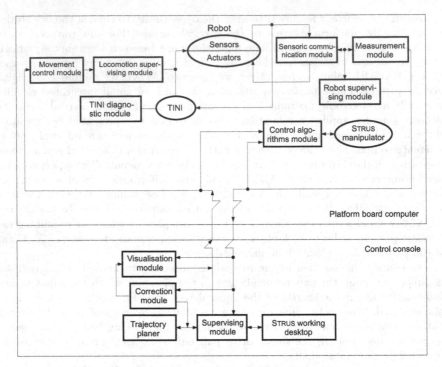

Fig. 21.8. Software organization

objects and receive a response. Objects know only the *interfaces*, i.e. plug-in sets with protocols [2,4]. Particular object implementations can be done in any programming language and any software/hardware platform, which enables using the CORBA standard. The software system has open architecture made of several modules, with well-defined communication between them. It can be easily extended by adding extra modules which will be necessary to incorporate some specific measurement for the purpose of diagnosis.

Fig. 21.9 presents the scheme of sensor and control system organization. The left part of that figure shows sensor arrangements. The right part represents information flow between the elements controlling the actuators. The hierarchical system is a bit more complicated since it is a multiple masters-multiple slaves infrastructure. For example, actuator masters may be responsible for decomposition of complex directives, received from the on-board computer (like "move leg 10cm ahead") and assigning tasks for each of the underlying slaves, directly controlling the actuators' behavior. In such a combination, complexity of tasks, performed by each element of the system, may be easily matched with processing ability of that element (e.g. with computation speed of a microcontroller). Tasks requiring complex computations and large amount of data to be processed, such as general control algorithms for the robot as a whole, are done at a higher level of the structure presented in Fig. 21.11, while simpler tasks are executed at a lower level.

There are specific requirements of the measurement system, resulting from inspection functions of the special robot. Additionally, the inspection robot should

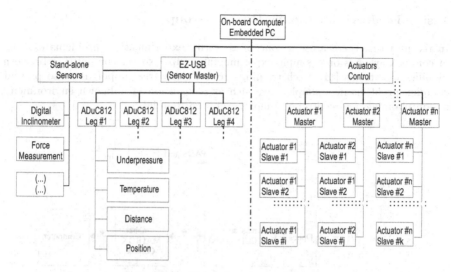

Fig. 21.9. Control and sensor system

be equipped with a system enabling observation of the environment. The gathered signals are used, generally, for orientation and navigation of the robot. Moreover, particular information on the state of walls (concrete corrosion), construction corrosion, reinforcement, etc., should be obtained from measurements. All mentioned tasks should be performed in parallel on-line and, additionally, with data acquisition. This implies special hardware and software organization.

Considering the tasks for the SAFARI robot, a vision subsystem is an important element of the measurement system. It is used for observation of the environment of the robot. The main element of the vision system is a Convision V600 videoserver. This videoserver enables acquisition of image frames, coming from the cameras mounted on the robot's platform. Then it converts the analog PAL video signal into digital frames, compresses the frames using the JPEG format and sends digitized frames via an Ethernet LAN to visualization stations. Data exchange between the videoserver and the remaining parts of the Ethernet network is carried out via TCP/IP with UTP (Unshielded Twisted Pair) cable used as a transmission medium. The videoserver allows to connect up to six video sources. Four cameras are mounted on the mobile platform and may be used by the robot itself, e.g. for detection of obstacles [9].

21.4 Vision System in Complex Control Systems

In this chapter we consider an implementation of the vision system as the vision feedback/measurement system. Below we present two examples: vision feedback for control of a mobile robot group and visual tracking of surgical tools.

Vision Feedback for Mobile Robots Group

In recent years, there arose a concept to verify experimentally problems existing in complex systems on a simple experimental setup – soccer competition between mobile robot teams [11]. Such training ground gives great opportunity to present complex problems in a simple way, such as recognition of unknown environment, action planning, and control of a mobile robot system [6].

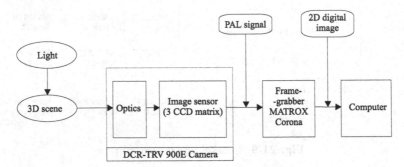

Fig. 21.10. Vision system

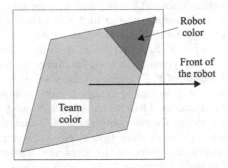

Fig. 21.11. Patch marker of the robot

The important component is a vision system, which may be used for object recognition and measurement of position and orientation of mobile robots on the playground. In this competition, there are three mobile robots of one team, three robots of the opponent team and the ball. Measurement data, coming from vision feedback, are used as input to control algorithms of this robot. The implementation of an extraction algorithm of specific features should satisfy criterion of speed, high reliability of operation and easy characterization of features for objects being recognized. The presented algorithm for object recognition is based on a compromise between widely known methods of filtering and image processing [12], and a necessity of real-time processing [1]. A vision system usually consists of a CCD camera, a framegrabber and a computer as shown in Fig. 21.10.

A robot has a 8 × 8cm marker with two color patches as in Fig. 21.11. One patch indicates team color (common for one team), while the other patch is an individual robot color, which identifies each robot inside the team. The team patch is a uniform spot square with minimum dimensions of 3.5 × 3.5cm, of yellow or blue color. The goal of the vision system is the recognition of robots and then measurement of their orientation and position on playground.

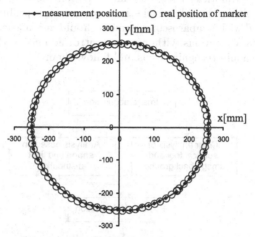

Fig. 21.12. Path of the mobile robot

The algorithm of recognition of robots works as follows. Detection of pixels representing objects is based on colors. Due to an image noise and to make easier specification and allocation of colors to objects, segmentation of available color space is performed at the beginning. Color segmentation from 24-bit color palette (R:G:B components have respectively 8:8:8 bits) is done by an indirect addressing algorithm. As a result of that, a three-dimensional version of a look-up-table technique is obtained. During image searching for a pixel of color that is assigned to objects, two image preprocessing methods are used: a contextless method and logical context filtering. After that, the whole edge of the robot's patch is followed in order to find two sharp corners. For edge detecting and following, a detector based on the Laplace operator is employed. Sharp corners are used to determine the position and orientation of the robot.

The methods of edge following and finding sharp corners of rhombus are selected, to enable real-time image processing. This condition is necessary to control correct objects, which are mobile robots. The simple edge detection compares the luminance of pixels with an appropriate threshold. Use of generally known edge detectors for edge following, based on masks of e.g. 3 × 3 pixels, like initially used Laplace filter, will increase the number of checked pixels almost eightfold.

The dynamic measurements were carried out on the trajectory of the circular path type with constant angular velocity. A path of the robot with real and measured position of the marker is shown in Fig. 21.12. During experiments a situation was observe when the marker was not found. In such a case, the measurement system gives the last correct result of the measured marker position and additionally informs that the marker is not found.

Visual Tracking of Surgical Tools

Medicine, and especially surgery, seems to be one of the most promising areas for applications of robotics. There are many papers devoted to implementation of robotics technologies in solving the robotic assistant problem, e.g. [13]. This assistant, equipped with a laparoscopic camera should automatically cooperate with the surgeon during operations without any control actions of the surgeon's hand. The laparoscope should recognize tools and follow them.

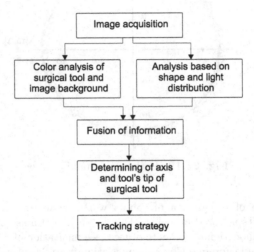

Fig. 21.13. Algorithm of vision tracking of surgical tools

The algorithm of vision tracking of surgical tools described in [7] consists of two parts: color analysis and shape analysis. Its block diagram is shown in Fig. 21.13. Its first part is a preprocessing of images coming from the laparoscopic camera: frequency filtration, space filtration, and color filtration. Additionally, taking into account properties of the living tissue, a redundant component of color is removed – in our case the R (red) coordinate is removed. Then in the feature extraction process, the places where the tool may be potentially placed are determined. This process runs simultaneously for color and shape analysis. Moreover, information on specific light distribution in surgical tools is taken into account. Very useful is also the fact that textures of human tissues and of the tool are different. At the end of processing, as the result of data fusion, the position and orientation of the surgical tool are calculated.

Figure 21.14 presents the laboratory setup, which consists of an industrial manipulator Stäubli RX60 equipped with a laparoscope, a human body model, a laparoscopic camera, a PC computer with a framegrabber card and a communication link to transmit data from the vision system to the robot control unit. The Stäubli

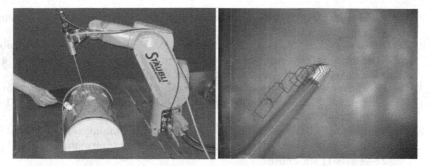

Fig. 21.14. Experimental setup and results of the tracking algorithm

RX60 manipulator is used as the robot assistant. It is a 6dof manipulator driven by permanent magnet synchronous motors.

The same figure presents results of our tracking algorithm. The vision system detects an incorrect position of the tool, which has to be in the center of the image. Then, it determines the current position of the tool and the corrections of orientation of the laparoscope. After transmission of these data to the robot control system, the manipulator moves the laparoscope. The movement of the manipulator causes the view of the (immobile) tool to be moved towards the center of the image. This movement lasts until the view of the tool is located in the middle region of the image defined as a square of size 10 × 10 pixels. The right side of Fig. 21.14 presents the final frame. The intermediate positions of the surgical tool are depicted as a set of black objects.

21.5 Concluding Remarks

In the paper, examples of complex systems, which consist of many parts (mechanics, control, and sensor hardware) and many levels of data processing have been presented. Contributions of mechanical, control, and sensor design in such systems are discussed mainly using the example of the inspection robot SAFARI. The mechanical and control structures have been outlined. Description of the sensor and measurement systems has also been included. A very important source of information about the robot environment seems to be the vision system. This has been discussed using two examples – vision feedback for a mobile robots group and visual tracking of surgical tools. Analyzing all examples of complex systems it can be noticed that desirable manner of their design leads to system modularity. Moreover, modern prototyping tools and advanced knowledge from the areas of electronics and programming technology should be used in design. In future, we plan to design similar constructions for other applications.

References

1. S. Hutchinson, G.D. Hager, P.I. Corke, A tutorial on visual servo control, *IEEE Trans. on Robotics and Automation*, vol. 12, no. 5, pp. 651-670, 1996.

2. Common object request broker: architecture and specification. Object Management Group, Revision 2.3. http://www.omg.org

3. D. Dobroczyński, P. Dutkiewicz, P. Herman, W. Wróblewski, A climbing robot SAFARI for buildings inspection, *Proc. of the 4th Int. Conf. on Climbing and Walking Robots 'CLAWAR'*, Karlsruhe, Germany, 937-944 (2001).

4. D. Dobroczyński, P. Dutkiewicz, W. Wróblewski, Software system of the climbing robot. *Studies in Automation and Information Technology*, 25, 71-82 (2000) (in Polish).

5. D. Dobroczyński, P. Dutkiewicz, W. Wróblewski, Climbing robots in context of walking robots control, *Studies in Automation and Information Technology*. 26, 27-44 (2001) (in Polish).

6. D. Dobroczyński, P. Dutkiewicz, T. Jedwabny, K. Kozłowski, J. Majchrzak, G. Niwczyk: *Mobile robot soccer, Proc. of the 6th Int. Symp. on Methods and Models in Automation and Robotics*, pp. 599-604, (2000).

7. P. Dutkiewicz, M. Kieczewski, M. Kowalski, Visual Tracking of Surgical Tools for Laparoscopic Surgery, *Proc. of the 4th Int. Workshop on Robot Motion and Control*, 23-28, (2004).

8. P. Dutkiewicz, M. Kowalski, M. Ławniczak, M. Michalski, SAFARI inspection robot motion strategy, *Proc. of the 3rd Int. Workshop on Robot Motion and Control*, Bukowy Dworek, Poland, 93-100 (2002).

9. P. Dutkiewicz, M. Ławniczak, Sensor system for SAFARI wall-climbing robot, *Archives of Control Sciences*, 12(1-2), 103-123 (2002).

10. P. Dutkiewicz, K. Kozowski, Wróblewski, Inspection robot SAFARI – construction and control, *Bulletin of the Polish Academy of Sciences Technical Sciences*, vol. 52, no. 2, 129-139 (2004).

11. HTML page: www.fira.net;   fira.cqu.edu.au/rwc2000/mrules.htm

12. R. Gonzales, R. Woods: *Digital Image Processing*, Addison-Wesley Publication Company, MA, (1993).

13. R. Hurteau, S. DeSantis, E. Begin, M. Gagner, Laparoscopic Surgery Assisted by a Robotic Cameraman: Concept and Experimental Results, *Proc. of the IEEE Int. Conf. on Robotics and Automation*, 2286-2289, (1994).

14. S. Hirose, K. Kawabe, Ceiling walk of quadruped wall climbing robot Ninja, Proc. of the 1st Int. Symposium 'CLAWAR', 143-147 (1998).

15. K. Kozłowski, P. Dutkiewicz, W. Wróblewski, *Robot modelling and control*, Polish Scientific Publishers PWN, Warsaw 2003, (in Polish).

16. K. Kozłowski, P. Dutkiewicz, M. Ławniczak, M. Michalski, M. Michałek, Measurement and control system of the climbing robot SAFARI, *Proc. of the 5th Int. Conf. on Climbing and Walking Robots and the Support Technologies for Mobile Machines*, Paris, 1003-1012 (2002).

17. G. Muscato, G. Trovato, Motion control of a pneumatic climbing robot by means of a fuzzy processor, *Proc. of the 1st Int. Symposium 'CLAWAR'*, 113-118 (1998).

18. A. Nagakubo, S. Hirose, Walking and running of the quadruped wall-climbing robot, *Proc. of the IEEE Int. Conf. on Robotics and Automation*, 1005-1012 (1994).

19. M. Prieto, M. Armada, J.C. Grieco, Experimental issues on wall climbing gait generation for a six-legged robot, *Proc. of the 2nd Int. Symposium 'CLAWAR'*, 125-130 (1999).

20. T. Szypiło, *Control algorithms of walking robots*, M.Sc. Thesis, Poznań University of Technology, Poznań 2000 (in Polish).

Examples of Transillumination Techniques Used in Medical Measurements and Imaging

Anna R. Cysewska-Sobusiak and Grzegorz Wiczyński

Institute of Electronics and Telecommunications, Division of Metrology
Poznań University of Technology, ul. Piotrowo 3a, 60-965 Poznań, Poland
Anna.Cysewska@put.poznan.pl, gwicz@et.put.poznan.pl

22.1 Introduction

Transillumination is understood as the phenomenon of transmitting optical radiation with defined parameters by an object, which becomes the carrier of information on the characteristic size of the object [1–7]. In case of biological objects the optical properties of systemic liquids and other tissues are utilized. Transillumination as a method of examination by passage of light through tissues or a body cavity is a diagnostic technique in the course of intensive development at the moment.

As far back as 1876 Karl van Vierordt observed changes in the solar spectrum transmitted by the finger tissues of his own hand. He discovered that after pressure causing ischemia, a change occurred in the spectrum composition obtained, which he related to the changing participation of oxygenated and reduced hemoglobin in the tissues. Information on the first attempts of transillumination with optical radiation appeared in 1929 [1]. Fast development of such techniques, however, occurred in 1980s. This was related to the development of optoelectronic elements and the arrival of new possibilities of numerical transformation. Images of transilluminated objects, however, were not sufficiently clear. Simultaneously some tests were made in relation to transillumination for purposes other than imaging. The most common example of such transillumination is the observation of arterial blood pulsation allowing to measure the blood pulse and blood oxygenation [8–10]. In 1977, a hundred years after van Vierodt's observations, Minolta built the first oximeter based on the transillumination of the ear lobe. In 1985 Nellcor-100 was created in the USA – a model that became a synonym of the term pulse oximeter introduced at that time. Pulse oximetry has been recommended to be used in clinical critical situations, when there is a risk of hypoxia.

Numerous transillumination attempts have been made since then leading to object imaging. At the beginning a technique called diaphanography was used, which consisted in illuminating the object on one side and observation of the same with a camera on the other. The use of a camera instead of a naked eye made illumination with infrared radiation possible. This technique was most frequently used in devices called mammoscopes [11, 12]. The purpose of such processing is the conversion of optic figures into information on the anatomic and functional properties of the tested

K. Kozłowski (Ed.): Robot Motion and Control, LNCIS 335, pp. 351–364, 2006.
© Springer-Verlag London Limited 2006

object in possibly the shortest time. In devices called tomographs (*tomos* – dividing, *graphos* – record) the 3D space is represented by means of a sequence of 2D images.

In 1895 Wilhelm Conrad Roentgen made the first radiogram of a hand, starting the development of noninvasive image diagnostics methods. Other imaging methods appeared only after several dozen years. Thanks to the selective optical properties of tissue cells, optoelectronic measuring methods can be used for determination of essential features of tissue sets, particularly useful in noninvasive diagnostics [12–17]. With the use of optical radiation, harmless to humans, (within near infrared and visible light range), it is possible to determine the parameters of tissues allowing for the reproduction of anatomic and functional properties. The interactions occurring between light and tissues result in scattering, absorption and fluorescence, providing information on the structure, physiology, biochemistry and molecular functions. Optical imaging is used for description of surface and volume structures.

Among the applied methods of tissue parameter measurement, a tendency to develop methods based on detection and analysis of natural and forced biooptical phenomena is significant. Under the noninvasive transillumination and illumination from underneath, it is possible to diagnose and monitor the parameters of tissues and organs examined. This paper includes discussion of selected issues related to the biophysical and optical phenomena used and examples of modern medical applications of tissue transillumination. The transillumination image resolution is lower than that of X-ray images. However, it enables to disclose information on the functional condition, unavailable in the X-ray technique. The scanning times of modern optical tomographs are a few minutes. During that time the test object (i.e. a selected human body part) should remain immobilized in relation to the scanning system. During the scanning process the examined object is illuminated with optical radiation of several wavelengths. The radiation is modulated with a harmonic signal of 70-100 MHz frequencies or creates a sequence of impulses. Modulation with very short impulses lasting 0.4ns or ca. 1 μs can be distinguished. Very short impulses are used in measuring photon flight times.

The techniques developed at present are those of efficient transillumination of thick layers of tissues and effective processing of data obtained that way. Due to strong scattering of the light, the practical implementation of optical transillumination is an immensely difficult task. The major goal of the paper is to present selected examples of current modern application of transillumination technique in medical diagnostics. Advantages and limitations of this technique are discussed.

22.2 Principle of Tissue Layers Transillumination

The systems for transillumination of layers of tissues described in literature, built according to the general block diagram from Fig. 22.1, differ in the method of optical input function generation and their detection. Two basic groups of systems can be differentiated:

- measuring the amplitude and phase of the transilluminated optical radiation;
- using the photon time-of-flight analysis.

The first group includes systems with processing the frequency area allowing to measure of the amplitude and optical output signal phase values related to the input

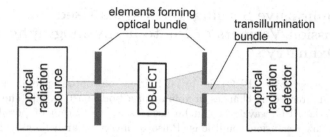

Fig. 22.1. Basic configuration of a system for tissue transillumination

signal. The systems with impulse modulation with the amplitude value of the optical output signal measured constitute a subgroup. A separate and specific group includes systems in which the photon times of flight through the transilluminated layer of tissues are measured and analyzed. Their specific characteristics is the necessity to measure time below 1ns.

The basic issues related to the transillumination of optically thick layers of tissues are strong scattering and absorption [18–21]. During transillumination the radiation source illuminates the examined object from one side. The detector placed on the opposite side (in relation to the source) collects part of the radiation coming out of the objects. The basic application difficulty in effective transillumination of thick tissue layers is the low power of radiation subject to detection. Therefore, it is necessary to force the optical power of the source and to apply sensitive photodetectors. The power increase is restricted by the permitted energy density of 329 mJ/cm^2. The photodetectors are most frequently photomultipliers or – seldom avalanche photodiodes.

During transillumination of soft tissues or when the transillumination process is long enough, it is appropriate to position the object in relation to the measuring system. The immobilization may not influence the object's functional condition maintaining the examined person's comfort at the same time. The methods used in positioning objects in relation to the sources and optical radiation sources are presented in Fig. 22.2.

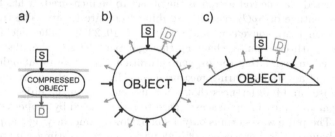

Fig. 22.2. Methods of positioning an object in relation to optical radiation sources S and detectors D [13]

22.3 Noninvasive Sensing Techniques Used in Transmission Variants of Photoplethysmography and Pulse Oximetry

Photoplethysmography (PPG) is by definition an optoelectronic method for measuring and recording changes in volume of a body part [9, 22]. The shape and stability of the PPG waveform can be used as an indication of possible motion artifacts or low perfusion conditions. Pulse oximetry smartly joins rules of both in vivo spectrophotometry and transmission or reflection photoplethysmography to monitor the arterial blood oxygen saturation. Biophysical and optical principles of noninvasive sensing and measurement procedures are illustrated in Figs 22.3 and 22.4.

The noninvasive optoelectronic sensor can give an electrical signal that is proportional to the transmitted light intensity, representing pulsatile changes in the arterial blood volume. Cyclical changes in light intensity appear according to selective changes in the object transmittance. Peak and bottom values of pulse components, at two wavelengths 660 nm and 940 nm that are needed to detect two absorbers Hb and HbO_2, are used to estimate the oxygen saturation SaO_2. Transmittance sensors of pulse oximeters employ a light emitter containing two alternately powered LEDs, and the photodiode as the photodetector which is placed in line with the emitter to detect the maximum amount of the transmitted light. What is unique in pulse oximetry is the possibility to sense the global oxygen saturation of human body arterial blood by noninvasive transillumination of only a peripheral tissue set, which allows us to see a "representative" arterial blood in other tissue components. The principle is simple as based on spectrophotometry rules, referring to idealized blood consisting of only two absorbers to be detected. In practice, the measurements are accomplished with optoelectronic sensors placed on "living cuvettes", which can strongly affect the reliability of final results.

Several locations on the body such as the finger-tip, toe, ear-lobe, tongue, lip, cheek wall, nasal-bridge and wing of the nostril in adults, and the foot, palm, or even arm in infants, are the useful sites to place the optoelectronic sensors which are applied directly and very often in prolonged duration. When the reflection variant of light-tissue interaction is used, the forehead, chest, back, or other places, more centrally located, can be considered as the object to be measured. It has been noted that the percentage of SpO_2 reading can differ from site to site. The transmission variant is often more convenient and sensitive [9, 10, 23, 24]. The thickness of the object varies with each pulse, changing the light path length, the effects of which are eliminated in estimating the oxygen saturation. Also, the input light intensity is negligible as a variable during measurements. However, there are always some artifacts, noises and interferences that can cause changes in the object parameters. The raw signals acquired from a sensor need to be enhanced by proper conditioning [10, 25, 26]. The pulse wave-form is only a small, slowly time-varying component of a given signal and requires processing of high performance to estimate its parameters.

The majority of the manufactured commercial pulse oximeters, which are portable models or pocket-size units, display values of oxygen saturation and pulse rate, while others can display pulse strength, present the trend from a past period of time, and display specific waveforms. The models which are only slightly larger

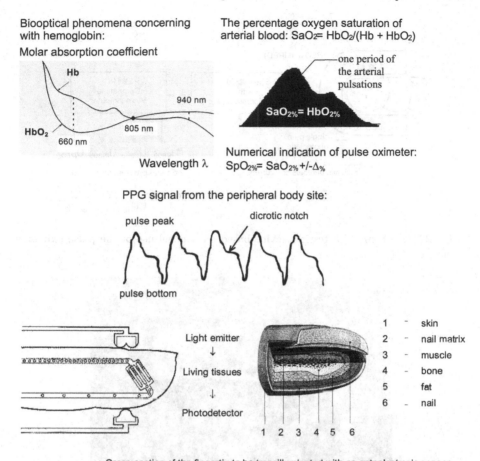

Biooptical phenomena concerning with hemoglobin:

Molar absorption coefficient

Hb

HbO₂

660 nm

805 nm

940 nm

Wavelength λ

The percentage oxygen saturation of arterial blood: SaO₂= HbO₂/(Hb + HbO₂)

one period of the arterial pulsations

SaO₂%= HbO₂%

Numerical indication of pulse oximeter: SpO₂%= SaO₂% +/-Δ%

PPG signal from the peripheral body site:

pulse peak   dicrotic notch

pulse bottom

Light emitter
↓
Living tissues
↓
Photodetector

1 - skin
2 - nail matrix
3 - muscle
4 - bone
5 - fat
6 - nail

1 2 3 4 5 6

Cross-section of the fingertip to be transilluminated with an optoelectronic sensor

Fig. 22.3. Illustration of PPG and pulse oximetry measurement signals and arrangement of a transmission noninvasive optoelectronic sensor placed on a fingertip – a pulse oximeter indication $SpO_{2\%}$ estimates the true value $SaO_{2\%}$ with the uncertainty $\Delta_\%$

than most reusable finger sensors have been designed especially for evacuation situations. The alarms mean the critical functions in a pulse oximeter, alerting of a potentially dangerous situation that usually include high oxygen saturation, low oxygen saturation, high pulse rate, and low pulse rate. Figure 22.5 shows an example of the most modern devices.

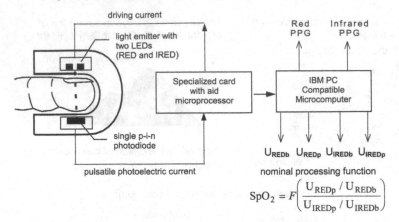

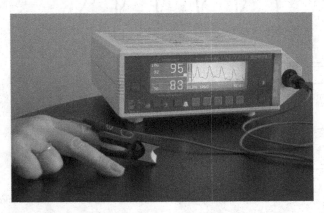

$$SpO_2 = F\left(\frac{U_{REDp} \, / \, U_{REDb}}{U_{IREDp} \, / \, U_{IREDb}}\right)$$

Fig. 22.4. Scheme of a testing device based on a novel use of the pulse oximetry concept [24]

Fig. 22.5. Modern pulse oximeter used for continuous noninvasive monitoring of changes in PPG waveform and the arterial blood oxygen saturation

22.4 Practical Usefulness of Tissue Illumination and Transillumination

A biological object examined is a homogeneous or complex mixture of solids, liquids and gases. Being an integral part of the measuring link, it plays a dual role during optic wave propagation: the input side of the object receives the signal emitted from the source and the output side is a secondary transmitting surface of the transmitted radiation. The radiation is subject to absorption and scattering of intensity depending on the object thickness, extent of non-homogeneity and individual optical properties. In the particular elements of the measuring track some determined fractions of the photon jet are absorbed. Depending on the type of interactions, the energy of the photons absorbed may be transformed into another type of energy, with or without ionization of atoms.

Different imaging methods allow the detection of different properties of tissues through a variety of utilized phenomena. For example, videoendoscopy makes possible evaluating attributes of a pathological object only when a detected tissue change grows up into the bore of the investigated digestive tract. However, it is impossible to get information about the degree to which this abnormal process has spread beyond the tract wall. On the contrary, an ultrasound probe incorporated at the videoendoscope tip or independent ultrasound probe inserted through the endoscopic operating channel allow evaluating this process – the depth of the tumor invasion may be estimated [27].

Transillumination allows supporting detection of pathological formations in some tissue collections, particularly those located on body perimeter. A frequently used source of light is the halogen lamp of set radiation intensity value. Useful applications of transillumination and illumination from underneath of tissue collections include, for example:

- hydrocephalus diagnostics in infants – translucence of heads of infants with secondary megalocephaly due to excessive cumulating of cerebrospinal fluid in the cerebral ventricles is observed;
- supporting dental procedures such as e.g. dental root canal treatment control or early caries imaging (the caries area is characterized with double refraction [2]); the polarization originating from double refraction combined with polarization filters allows reducing the influence of the detector's overloads on the teeth edges. The signal from the object is processed in the focal plane array (FPA) matrix InGaAs of 318×252 pixel size. A 1310 nm band transfer filter of 50 nm bandwidth filters the radiation processed in the FPA matrix. In order to obtain the best possible contrast between the area of pathological lesions and surrounding enamel without saturating the FPA matrix, the light intensity, the distance between the source and the object and the shutter diameter are selected each time;
- detection of small veins in hands and monitoring obliterating therapy of varicose veins [5];
- illumination from underneath of pathological lesions such as lipomas, cysts, neoplasms, haematomas, calcifications, exostoses, gouty nodules, in order to distinguish them among surrounding healthy tissues, particularly in hand [5]. In case of a lipoma no conspicuous changes in transparency were observed, cysts filled with fluid are more translucent than healthy tissues, the neoplasm area shows not too conspicuous shadowing, in other cases the less transparent areas were conspicuous compared to their surrounding.

New advanced illumination, illumination from underneath and transillumination techniques may be useful not only in obliterating therapy of varicose veins located not too deeply, but also in surgery of veins in outpatient clinics, facilitation of access to veins in pediatrics, neonatology, dermatology, and medical rescue procedures. Some selected transillumination results performed with an optoelectronic-electronic measuring system used in human fingers imaging are presented below.

22.5 Examples of Transillumination Used in Finger Tissue Imaging

Due to the ease of setting the location in relation to the measuring system and due to the variability of the optical properties, a convenient object in the transillumination tests is a finger. The selection of fingers made it possible to carry out tests with the use of a simple as well as efficient optoelectronic-electronic measuring system. The mechanical structure of the measuring system is constructed in the form of the letter C fixed to the robot's arm (Figs 22.6 and 22.7) that consists of segments interconnected by joints [28]. The robot arm assembly is flexible and contains the motorization, brakes, motion transmission mechanisms, cable bundle, pneumatic and electrical circuits for the user. A scanning system is joined with the robot arm.

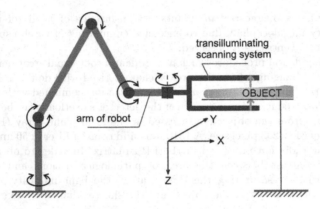

Fig. 22.6. Illustration of kinematics of a scanner consisting of an arm robot and an appropriate scanning system

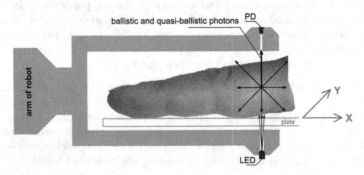

Fig. 22.7. Structure of a transilluminating scanning system made for finger tissue imaging

During the scanning time the test object (i.e. selected human hand fingers) should remain immobilized in relation to the scanning system. The immobilization may not disturb the object's function, maintaining simultaneously the examined person's comfort. The hand examined is laid on a transparent plastic panel stabilizing its position. The scanning system is moved in accordance with the trajectory presented in Fig. 22.8. According to the robot software, scanning procedure of the object is accomplished in the Cartesian coordinate system (z = const, x = var, y = var). A light emitting diode (LED) and photodiode (PD) – without the additional systems focusing the optical bundle – are incorporated at the structure ends in optical channels of 3 mm diameter and ca. 20 mm length.

Various high power LED diodes are used in the tests. The results presented have been obtained for an ELJ-880-228B transmitter with $\lambda = 880$ nm fed with current impulses $I_m = 7$ A, $T_i = 1$ μs, $T_{rep} = 1$ ms. The receiver is the PIN BPW24R photodiode. The transverse motion of the scanning system in relation to the fingers has been input.

Figure 22.9 presents the standardized S_{OUT} values of the converse of the converted output signal from the photodiode for several x values. The results obtained as transillumination images are shown in Fig. 22.10. The grayness intensity is represented by the output signal S_{OUT} values of the fingers examined (F1, F2, F3, F4).

The transillumination image resolution is lower than that of X-ray images. However, it enables to disclose information on the functional condition, unavailable in the X-ray technique. With the use of optical radiation (within near infrared and visible light range harmless to humans), it is possible to determine the parameters of tissues allowing the reproduction of anatomic and functional properties. The experimental transillumination of a body peripheral is possible in a simple enough sending-receiving system presented in Fig. 22.12.

The optical part of the transmitting-receiving system is an LED placed opposite the PIN photodiode PD. Several radiation lengths have been used for the tests, but the best effects have been found for the 870 nm and 660 nm radiation. The LED diodes have been fed with current impulses of ca. 1 μs duration and repetition frequency 1 kHz. As the optical transmission measure, the amplitude of voltage impulses has been assumed. Despite the simplicity of the measuring system the imaging obtained was as anticipated. Differences between the $U_I(t)$ amplitude variability for fingers without and with joint degenerations were observed.

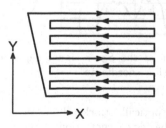

Fig. 22.8. Trajectory of scanning the object (z = const, x = var, y = var)

Fig. 22.9. View of a performed transillumination system joined with the robot arm

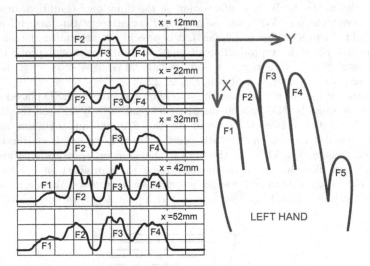

Fig. 22.10. Specification of the S_{OUT} signal dependency on the y location for selected cross-sections with x coordinate

To transilluminate some optically thicker body parts it is necessary to increase the system's sensitivity, better concentration of the optical bundle and possibility to select the detectable photons. Nevertheless, the optical radiation modulation method used is an alternative for solutions with modulation by means of a harmonic waveform.

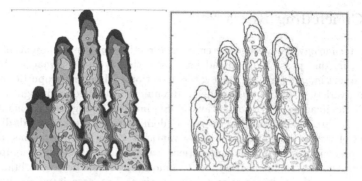

Fig. 22.11. Transillumination images obtained for hand fingers

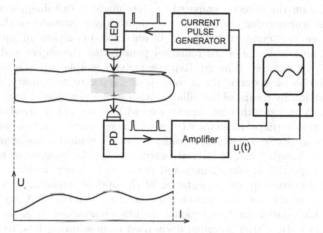

Fig. 22.12. Impulse transillumination system of a finger

The practical performance of transillumination is possible thanks to transmission supported by forward scattering. However, the backward scattering supports illumination of the object from underneath, useful in the diagnostic-therapeutic purposes mentioned above. In cases of peripheral body areas, it is possible to apply transillumination as a simple low-cost method to assist in distinguishing pathological lesions that appear e.g. in fingers, hands, feet and teeth.

Transillumination allows supporting detection of pathological formations in some tissue collections, particularly those located on body peripheral sites. The finger transillumination test results show that effective transillumination is possible even in a simple system and indicate that further development of the measuring system is reasonable. Upon the test result analysis the necessity arises to configure the optical part so that a better resolution can be obtained with capacity of effective transillumination of optically thicker objects. The basic application difficulty in an effective transillumination of thick tissue layers is the low power of radiation subject to detection. Therefore, it is necessary to force the optical power of the source and to apply sensitive photodetectors.

22.6 Concluding Remarks

Modern technology knowledge determines directions in development of current medicine, making possible to spread into areas that have been inaccessible up to now. Different imaging methods allow the detection of different properties of tissues through a variety of utilized phenomena. In clinical tests medical imaging based on X-ray (X-ray imaging), ultrasonic waves (USG imaging), magnetic resonance (MRI imaging), infrared radiation (thermography) dominate. All these methods differ from the point of view of advantages and disadvantages, range of applications, degree of invasive or noninvasive interaction, patient ballast, and procedural complication. The interest in development of optical techniques in biomedicine to obtain images of tissues and organs is great and still growing. The associated application of various methods is presently developed, allowing to obtain a more complete set of information on the object, compared to tests made in one diagnostic technique only. In the transmission variant of the light-object interaction the objective information on the quantity measured is obtained by means of optoelectronic sensor containing a source of the radiation penetrating the object and receiver of the radiation transmitted through. Representative examples of such examinations are e.g. detection of systemic fluids components (spectrophotometry, oximetry in vivo), localization by means of transillumination or illumination of veins, cysts, and neoplasms from underneath, transillumination with white light (instead of X-raying), imaging and monitoring of pulse wave (photoplethysmography), transmission variant of pulse oximetry. Pulse oximetry is considered one of the most eminent achievements made in medical engineering since electrocardiography. An imaging technique using the optical properties of transilluminated tissues is also optical tomography. Due to strong light scattering, implementation of the optical tomographs in practice is difficult. The resolution of the image obtained is lower than that of X-ray images, but may provide information on the functional condition unavailable in X-ray techniques.

At present optical transillumination is used in monitoring blood oxygenation, hemorrhage detection, brain imaging, Alzheimer disease diagnostics, mammography, and rheumatism and joint inflammable condition monitoring. This diagnostic technique is under intensive research. It concentrates on development of effective transillumination of thick layers of tissues and on building efficient and stable algorithms representing anatomic and functional properties. Numerous issues related to the measurement result interpretation still remain unsolved. New advanced illumination, illumination from underneath and transillumination technique may be useful not only in obliterating therapy of varicose veins located not too deeply, but also in surgery of veins in outpatient clinics, facilitation of access to veins in pediatrics, neonatology, dermatology, and medical rescue procedures.

References

1. Cutler, M.: *Transillumination of the breast*, Surg. Gynecol. Obstet. 48, 1929, pp. 721-727.
2. Jones R.S., Huynh G.D., Jones G.C., Fried D.: *Near-infrared transillumination at 1310nm for the imaging of early dental decay*, Optics Express 11, 2003.

3. Cysewska-Sobusiak A.: *One-dimensional representation of light-tissue interaction for application in noninvasive oximetry*, Optical Engineering 36, 1997, pp.1225-1233.
4. Hanspaul S.M., Frieden I.J.: *Transillumination of a cystic lymphatic malformation*, The New England Journal of Medicine, vol. 349, 2003.
5. Eaton Ch.: *Clinical example flashlight transillumination for tumor diagnosis.*, In: The electronic textbook of hand surgery, www.eaton.hand.com.
6. Cysewska-Sobusiak A., Wiczyński G.: *Medical applications of human tissue transillumination.*, Proc. of the Fifth International Workshop on Robot Motion and Control RoMoCo'05, Poznań, Poland, 2005, pp. 69-72.
7. Bauer J., Boerner E. Podbielska H., Suchwalko A.: *Pattern recognition methods of transillumination image for diagnosis of rheumatoid arthritis*, Proc. SPIE vol. 5959, 2005 (in press).
8. Cysewska-Sobusiak A.: *Modeling and simulation of a tissue response to light transmission*, Med. Biol. Eng. Comput. 37, suppl. 1, 1999, pp. 216-217.
9. Enderle J., Blanchard S., Bronzino J.: *Introduction to Biomedical Engineering*, Academic Press, San Diego, 2000.
10. Cysewska-Sobusiak A.: *Powers and limitations of noninvasive measurements implemented in pulse oximetry*, Biocybernetics and Biomedical Engineering vol. 22, 2002, pp. 79-96.
11. Klingenbeck K., Schuty O.A., Oppelt A.: *Mammography with light*, Aktuelle Radiol. vol. 5, 1995, pp. 115-119.
12. Swartling J., Andersson-Engels S.: *Optical mammography - a new method for breast cancer detection using ultra-short laser pulses*, DPOS-NYT 4-2001, 2001, pp. 19-21.
13. Pogue B.W., McBride T.O., Osterberg U.L., Paulsen K.D.: *Comparison of imaging geometries for diffuse optical tomography of tissue*, Optics Express 4/8, 1999, pp. 270-286.
14. Tu Khanh Trinh M.: *Pushing the limits of optical tomography using a silicon micromachined collimator array*, Simon Fraser University, 2002, pp. 1.1-2.6.
15. Becker W., Wabnitz H.: *Optical tomography: TCSPC imaging of female breast*, Becker & Hickl GmbH, 1999.
16. Yhou J., Bai J.: *Spatial location weighted optimization scheme for DC optical tomography*, Optics Express vol. 11, 2003, pp. 141-150.
17. Arridge S.R.: *Optical tomography in medical imaging*, Inverse Problems 15, 1999, pp. R41-R93.
18. Beuthan J. et. al.: *Light scattering study of rheumatoid arthritis*, Quantum Electronics vol. 32, 2002, pp. 945-952.
19. Arridge S.R., Hiraoka M., Schweiger M.: *Statistical basis for the determination of optical pathlengths in tissue*, Physics in Medicine and Biology vol. 40, 1995, pp. 1539-1558.
20. Cheong W.F., Prahl S.A., Welch A.J.: *A review of the optical properties of biological tissues*, IEEE J. Q. Electron. vol. 26, 1990, pp. 2166-2185.
21. Tuchin V.V.: *Handbook of optical biomedical diagnosis*, SPIE Press, Bellingham, 2002.
22. Bołtrukiewicz M., Cysewska-Sobusiak A.: *A novel approach to processing of the finger photoplethysmographic signals*, Med. Biol. Eng. Comput. vo. 37, suppl. 2, 1999, pp. 522-523.

23. Cysewska-Sobusiak A.: *Noninvasive optoelectronic monitoring of the living tissues vitality*, Proc. of the Third International Workshop on Robot Motion and Control RoMoCo'02, Poznań, Poland, 2002, pp. 13-19.
24. Cysewska-Sobusiak A., Wiczyński G., Jedwabny T.: *Specificity of software co-operating with an optoelectronic sensor in a pulse oximeter system*, Proc. SPIE vol. 2634, 1995, pp. 172-178.
25. Kingston R.H.: *Optical sources, detectors, and systems. Fundamentals and applications*, Academic Press London, 1995.
26. Challis R.E., Kitney R.I.: *Biomedical signal processing (in four parts)*, Med. Biol. Eng. Comput. vol. 28-29, 1990-1991.
27. Cysewska-Sobusiak A., Skrzywanek P., Sowier A.: *Utilization of mini-probes in modern endoscopic ultrasonography*, IEEE Sensors Journal, 2005 (in press).
28. *Arm RX60B Family Characteristics*, Stäubli Faverges, 2001.

Telerobotic Simulator in Minimally Invasive Surgery

Piotr Sauer[1], Krzysztof Kozłowski[1], and Wojciech Waliszewski[2]

[1] Chair of Control and Systems Engineering, Poznań University of Technology
ul. Piotrowo 3a, 60-965 Poznań, Poland
piotr.sauer@put.poznan.pl, krzysztof.kozlowski@put.poznan.pl
[2] Department of General and Laparoscopic Surgery, Hospital J. Strusia, Poznań, Poland

23.1 Introduction

Minimally Invasive Surgery (MIS) is a technique which was established in the 1980s. The advantages for patient treated by the MIS compared to open (classic) surgery are the following: small incisions reduce pain and trauma, shorter residence at hospital and shorter rehabilitation time, cosmetical advantage due to small incisions. At the same time the Minimally Invasive Surgery has several disadvantages for the surgeons: reduced sight, reverse motion (chop-stick effect), restricted motion because of pivot point (trocar kinematic), reduce tactile and force–feedback because of long instruments, amplification of the tremor due to long instruments. The above mentioned disadvantages are the main reasons why MIS is restricted to a small number of applications. In Minimally Invasive Surgery, the surgeon views the anatomy from inside. The camera is controlled by a surgical assistant. The laparoscope itself consists of a chain of lens optics to transmit the image of the operation site to the CCD camera connected to its outer end, and optical fibers to carry light to illuminate inside. An image of the operation site is displayed on a high resolution CRT screen. The laparoscope and instruments used for the operation are inserted through trocars placed at the incisions of the abdomen. Consequently, the surgeon has no direct control over his/her viewing direction, and the laparoscopic image often is unstable because of tremor and sudden movements of the surgical assistant. To overcome those drawbacks robot surgery plays an important part, because Minimally Invasive Surgery is dedicated to telepresence. The surgical telepresence system helps the surgeon overcome barriers, such as the patient's chest or distances, if the surgeon and the patient are located in different rooms or even hospitals. Minimally Invasive Surgery is one of the fields where telerobotics enlarges human possibilities. Surgical robots allow more precise movements of surgical instruments. In robotic telesurgery, the robotic tools are not automated robots but teleoperated systems under direct control of surgeon. Previous research on medical robotics covered the development of an endoscopic manipulator [11]. The telerobotic assistant for laparoscopic surgery was developed by Taylor et al. [9]. For example, the telesurgery experiments were performed at Jet Propulsion

K. Kozłowski (Ed.): Robot Motion and Control, LNCIS 335, pp. 365–373, 2006.
© Springer-Verlag London Limited 2006

Laboratory, California Institute of Technology and Polytechnic University of Milan, Italy [6]. In September 2001, Marescaux et al. successfully carried out a remote laparoscopic cholecystectomy on a 68-year old female. They used telerobotic system named ZEUS. The surgeon was in New York while the patient was in Strasbourg [5].

In the paper, we present the first results of our project "Multi-level Control System of a Manipulator for Laparoscopy in Surgery". This project was financed by the Ministry of Science and Informatics. The multi-level control system is a result of cooperation between the Chair of Control and Systems Engineering, Poznań University of Technology, and the Department of General and Laparoscopic Surgery, Hospital J. Strusia in Poznań. In general, the proposed control system consists of three parts:

- slave system (teleoperator side),
- master system (operator side),
- communication system.

The goal of this project is to develop robotic tools to replace manual control of the laparoscope in surgery. In this project we make use of a Stäubli robot. The presented system replaces the surgical assistant during laparoscope cholecystectomies. The manipulator performs some task done by the surgical assistant, returns camera control to the surgeon, and stabilizes the laparoscopic image. The multi-level control system has been named ASYSTANT. This subject is in general not new in telesurgery, but is new in Poland.

This work consists of five sections. Section 23.2 describes a mathematical model of the laparoscope motion planning. Section 23.3 presents details of the structure of the proposed multi-level control system for the manipulator which has been used as a robotic assistant during laparoscopic cholecystectomy. Section 23.4 describes simulation programs of a laparoscope motion controller. Finally, Section 23.5 is devoted to conclusions.

23.2 The Proposed Model of the Task

The proposed system operates under the surgeon's direct teleoperation of the precision movements needed to manoeuvre the laparoscope within the patient's body. The input device system translates the joystick movements to robot Cartesian coordinates. For this action the telerobotic system has to know the model of the task. The manipulator is placed nearby the stretcher with the patient and has its end effector equipped with a laparoscope. The model of the teleoperated task is used to calculate the precise relationship between the base of the manipulator and the insertion point where the surgery tool (laparoscope) goes into the patient's body. This relationship depends on the following problems:

- the inverse and forward kinematics of the robot assistant,
- the measurement of distance along the optic axis between the camera and the insertion point.

The teleoperated model is determined by the homogeneous transformation $^{B}T_O$ which is defined as:

$$^{B}T_O = {}^{B}T_E \times {}^{E}T_H \times {}^{H}T_O \tag{23.1}$$

where

- $^{B}T_E$: robot forward kinematic model which defines the relationship between the base frame B and the end effector frame E affixed to the robot last link,
- $^{E}T_H$: the relationship between the end effector frame E and the holder frame H,
- $^{H}T_O$: the relative location of the insertion point frame O with respect to the holder frame H.

The last homogeneous transformation is unknown since the distance L from the holder frame H to the insertion point frame O along the optic axis must be calculated. The frames H and O have the same orientation. The scheme of the model of the task is presented in Fig. 23.2. The model of the task (23.1) is realized under the assumption that the laparoscope is already in the patient's body. To insert the laparoscope into the patient's body we use force control algorithm. In the first step, the robot is placed at an initial point above the patient. Next, the surgeon moves the end of the manipulator and puts pressure on it. When the force exceeds the force threshold, the laparoscope moves in the direction of the force action [8].

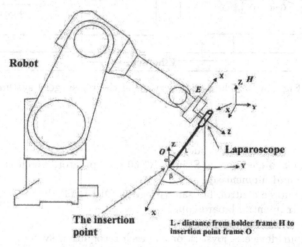

Fig. 23.1. Scheme of the teleoperation task

The forward kinematics problem calculates the position and orientation of the end effector using a video camera. This kinematics problem is computed using the Denavit-Hartenberg notation. The forward kinematics of the manipulator is also necessary for computing position errors based on force feedback. The inverse kinematics of the robot assistant has to determine the joint angles of the actuated joints of the manipulator given the desired movement of the end effector. Its desired movement is obtained by appropriate movement of the joystick. In order to control the system, the inverse and forward kinematics of the Stäubli robot have to be solved [4].

23.3 The Concept of the Multi-level Control System

The current design is a telerobotic system (Fig. 23.3), which consists of the following parts:

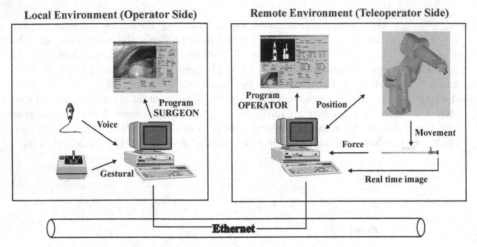

Fig. 23.2. Schematic overview of the telesurgery system

- slave system (teleoperator side)
  - the robot: a commercial Stäubli RX60 manipulator, which is equipped with a force and moment sensor,
  - control: force control, fail safe (program OPERATOR),
  - the instrument: a laparoscope,
- master system (operator side)
  - the input device: a joystick or a speech recognition system,
  - the vision system,
  - the trajectory generator (program SURGEON)
- communication system
  - flexible: different master/slave stations,
  - an independent network.

The overall structure of the proposed controller for telesurgery is shown in Fig. 23.3.

The teleoperation unit executes all the processes allowing communication with the surgeon as well as data transmission with the robot system module. The trajectory generator allows the surgeon to apply two command modes for controlling the robotic assistant movements on the base of the task model described in Section 23.2. This module has to define complex laparoscope trajectories. The graphics processing modules are used for sending the system status to the visual display unit. The motion control module translates the command transmitted to it by the trajectory generator into mechanical displacements of the arm to which laparoscope

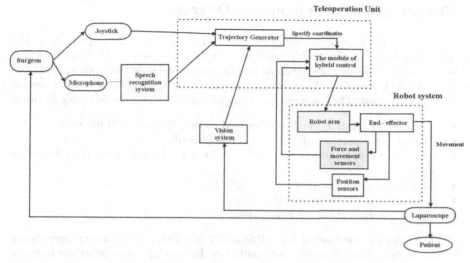

Fig. 23.3. Structure of the proposed controller for telesurgery

is attached. The module of hybrid control is used as force feedback control law with estimation of the environment stiffness. Force feedback is realized based on data from a force and torque sensor JR3, which is mounted between the Stäubli RX60 robot and the laparoscope. The control algorithm consists of two algorithms:

- a force control algorithm,
- a control algorithm with force protection [8].

We have build module of hybrid control on the basis of experimental results on liver's pig. During this research we measured properties of soft tissues [7]. The control algorithm with force protection plays the role of protection. It controls values of force and torques during robot movements. The system measures forces, which act at the end of the laparoscope. When the values of force exceed their threshold the system is stopped.

The speech recognition system is used to control the robot movement. This unit is an important interface between the surgeon and the robot because it allows the surgeon to communicate with the telerobotic system using natural commands. These commands are quite similar to those used during normal laparoscopic surgery without a robotic system. We propose the following set of commands: "robot left", "robot right", "robot up", "robot down", "robot closer", and "robot further". In order to recognize the commands we use a method based on pattern matching known as Dynamic Time Warping (DTW) [8].

The vision system recognizes the surgical tool and sends back information about the position of the tool's tip to the main unit of the system. It tries to manipulate the robot with the laparoscope to keep the tool in the center part of the camera image [3].

23.4 Programs: Surgeon and Operator

The multi-level control system for telesurgery has been developed on a PC compatible computer under Microsoft Windows 95, 98, 2000 or XP. We have made two simulation applications: Surgeon and Operator. We may test both applications in a virtual environment. These programs could be used in real operation to control the robot assistant. The programs Surgeon and Operator have the following features:

- control via a joystick which is connected to the PC by a USB interface,
- possibility to control with the help of a keyboard,
- program Surgeon can be connected with program Operator via the Internet in a client-server system,
- the kinematics of Stäubli robot is applied,
- monitoring of the trajectory generator,
- control parameters can be monitored.

The software is written in C++ Builder 5.0 and Delphi 5.0. These programs can be run in two modes: training and simulation. In training mode programs Surgeon and Operator run separately. Program Surgeon is designed for surgeon training. He/she can practice control of the laparoscope. The program makes use of picture from inside body in a JPG file. The surgeon can move the picture (laparoscope) with the help of a joystick. The state of the joystick is written in structure *joyinfo_tag*. This structure is declared in the following form:

typedef struct joyinfo_tag {
UINT wXpos; / x position */*
UINT xYpos; / y position */*
UINT wZpos; / z position */*
UINT wButtons; / buttom states */*
*} JOYINFO, *PJOYINFO, NEAR *NPJOYINFO, FAR *LPJOYINFO.*

We use function *JoyGetPos(UINT uJoyID, LPJOYINFO pji)* to read the state of the joystick. Variable *uJoyID* makes possible the choice of the joystick (first or second). *LPJOYINFO* is the pointer of structure *joyinfo_tag*, where joystick parameters are written. The program Surgeon is presented in Fig. 23.4. The program Operator is designed for verification of robot movement by the operator. The operator can control robot by the joystick. This program uses the same structure as program Surgeon. The program Operator is presented in Fig. 23.4. In this program the operator can carry out calibration of the robot. This application presents coordinates of start and current positions. These data are used to verification of the system.

In simulation mode, these programs run on two computers at the same time. Programs Surgeon and Operator operate via the Internet. These programs use TCP/IP protocol and network components of Delphi.

23.5 Conclusions

A robotic assistant has been proposed as a help aid to surgeons in laparoscopic cholecystectomy. Our system does not require any modification of a standard

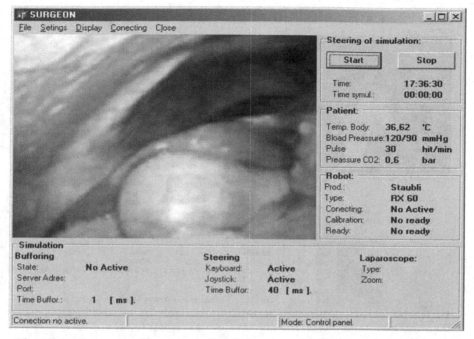

Fig. 23.4. Surgeon – simulation program for control of the Stäubli RX60 robot

operation mode: furniture or surgery tools. The robotic assistant is characterized by the precision for positioning the laparoscope within the abdomen.

The surgeon can control the system with the help of a joystick, voice commands and commands from the vision system. We have built simulation programs in order to test the control system. In future we will build a surgical simulator to simulate the reaction of the human body during surgical procedures. One of the main problems of the surgical simulator is the development of realistic physical models of organs and soft tissue. Our research in the area of surgical simulation will be mainly focused on developing 3D geometrical models of the body from 2D medical images, visualization of internal structures, and graphical display of soft tissue behaviour in real time. There are essentially three different approaches presented in literature to carry deformable tissue modeling:

- lumped element models [10],
- linear finite element models [1],
- nonlinear continuum models [12].

We plan to use linear elasticity as a way to obtain a good approximation of the behaviour of a deformable body. The stress-strain relations are important in this project because they give a physical link between the deformation of the body and the force induced in the force feedback system. We will test haptic rendering algorithms for simulating laparoscopic instruments – soft tissue interactions. In surgery simulation we will use a fast algorithm to display tissue cutting in virtual environments with applications to laparoscopic surgery, which is described in work

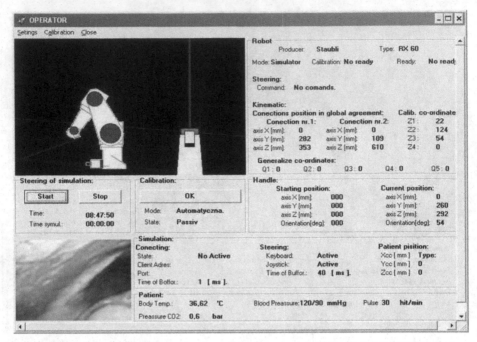

Fig. 23.5. Operator – simulation program for control of the Stäubli RX60 robot

[2]. We want to network independent communication between robots and input devices (or two computers – programs) realized with CORBA.

23.6 Acknowledgements

This research was supported by the Ministry of Science and Informatics grant, No 4T11A 026 22 titled "Multi-level Control System of a Manipulator for Laparoscopy in Surgery".

References

1. M.Bro-Nielsen, Finite Element Modeling in Surgery Simulation, *Proceedings of the IEEE*, 86(3), pp. 490-503, March 1998
2. Cagatay Basdogan, Chih-Hao Ho, Mandayam A. Srinivasan, Simulation of Tissue Cutting and Bleeding for Laparoscopic Surgery Using Auxiliary Surfaces, *Medicine Meets Virtual Reality*, pp. 38-44, 1999.
3. P.Dutkiewicz, M.Kiełczewski, M.Kowalski, W.Wróblewski, Experimental Verification of Visual Tracking of Surgical Tools, *Proceedings of the 5th International Workshop on robot motion and control RoMoCo'05*, Dymaczewo, Poland, pp. 237–242, 2005.

4. W.Kowalczyk, M.Ławniczak, Study on opportunity to use industrial manipulator in surgical–laparoscopic application, *Proceedings of 5th International Workshop on Robot Motion and Control RoMoCo'05*, Dymaczewo, Poland, pp. 63–68, 2005.

5. J.Marescaux, et al., Transatlantic Robot-Assisted Telesurgery. *Nature*, Vol. 413, pp. 379-380, 27 Sept. 2001.

6. A.Rovetta, R.Sala, X.Wen A.Togno, Remote Control in Telerobotic Surgery, *IEEE Transactions on Systems, Man and Cybernetics - Part A: Systems and Humans*, 26(4): pp. 438-443, July 1996.

7. P.Sauer, K.Kozłowski, J.Majchrzak, W.Waliszewski, Measurement System of Force Response for Minimal Invasive Surgery, *Computer–Aided Medical Intervantions: tools and applications SURGETICA'2005*, Chambery, France, pp. 139–147, January 2005.

8. P.Sauer, K.Kozłowski, D.Pazderski, W.Waliszewski, P. Jeziorek, The Robot assistant system for Surgeon in Laparoscopic Interventions, *Proceedings of the 5th International Workshop on robot motion and control RoMoCo'05*, Dymaczewo, Poland, pp. 55–62, 2005.

9. R.H.Taylor, J.Funda, B.Eldrige, S.Gomory, K.Gruben, D.Larose, M.Talamini, L.Kavoussi, J.Anderson, A Telerobotics Assistant for Laparoscopic Surgery. *IEEE Engineering in Medicine and Biology Magazine*, 14(3): pp. 279-288, May-June 1995.

10. D.Terzopoulos, J.Platt, A.Barr, K.Fleischer, Elastically Deformable Models, *Proccedings of SIGGRAPH 87: 14th Annual Conference on Computer Graphics*, pp. 205-214, ACM, July 1987.

11. J.Wendlandt, S.S.Sastry, Design and Control of a Simplified Steward Platform for Endoscopy, *Procceedings of the IEEE Conference on Decision and Control*, volume 1, pp. 357 362, 1994.

12. Y.Zhuang, J.Canny, Haptic Interaction with Global Deformations, *Proceedings of the IEEE International Conference on Robotics and Automation (ICRA 2000)*, pp. 2428-2433, 2000.

Human-robot Interaction and Robot Control

João Sequeira and Maria Isabel Ribeiro

Institute for Systems and Robotics / Instituto Superior Técnico
Av. Rovisco Pais 1, 1049-001 Lisbon, Portugal {jseq|mir}@isr.ist.utl.pt

24.1 Introduction

The ever increasing desire for fully autonomous robotics triggered in recent years the interest in the study of the interactions between robots and between humans and robots. The long term goal of this research field is the operation of heterogeneous teams of robots and humans using common interaction principles, such as a common form of natural language.

The paper explores the fundamentals of a robot control architecture tailored to simplify the interactions between a robot and the external environment, containing humans and other robots. The approach followed defines (i) a set of objects that capture key features in human-robot and robot-robot interactions and (ii) an algebraic structure with operators to work on this space of the objects.

In a broad class of robotics problems, such as surveillance in wide open areas and rescue missions in catastrophe scenarios, the interactions among robots and humans are a key issue. In such real missions contingency situations often arise which may force robots to request help from an external agent, in most cases a human operator. The use of HRI that mimics human-human interactions is likely to improve the performance of the robots in such scenarios.

Interactions among humans and robots in semi-autonomous systems are often characterized by the loose specification of objectives. This is also a common feature in natural languages used in human-human interactions and accounts both for ambiguity and semantics[1]. The framework described in the paper handles robot-robot and human-robot interactions in a unified way thus avoiding the need for different skills for each of them.

The paper is organised as follows. Section 24.2 briefly describes how human-robot interaction has been accounted by relevant paradigms in the literature. Section 24.3 presents key concepts from semiotics to model human-human interactions and motivates their use to model human-robot interactions. These concepts are then used in Section 24.4 to propose an architecture first in terms of free mathematical objects

---

[1] Ambiguity is related to errors in the meaning associated with an object, e.g., the precise meaning of the object is not accurately known. Semantics expresses the ability of a concept to have different correct meanings.

K. Kozłowski (Ed.): Robot Motion and Control, LNCIS 335, pp. 375–390, 2006.
© Springer-Verlag London Limited 2006

and next in terms of concrete objects. Section 24.5 presents a set of experiments that illustrate the main ideas developed along the paper. Section 24.6 presents the conclusions and some research directions.

24.2 A Brief State of the Art on HRI

HRI has always been present in robotics either explicitly, through interfaces to handle external commands, as in [13], or implicitly, through task decomposition schemes that map high level mission goals into motion commands, as in [22].

Behavioral techniques have been used in [27] to input mission specifications for a robot, in [23] to have a robot capable of expressing emotions and in [2] to make a robot join a group of persons behaving as one of them. In [20] robots are equipped with behaviors that convey information on their intentions to the outside environment. The CAMPOUT architecture for groups of heterogeneous robots, [11], is also supported on a hierarchy of behaviors. These are constructed using behavior composition, coordination and interfacing subsystems. The interaction among the robots is achieved with the exchange of both explicit data, such as state, and implicit, by having behaviors in charge of observing the environment for changes caused by the teammates. The MACTA architecture, [3], is also behavior based, with the HRI handled by a reflective agent that interfaces the robot and the environment. This agent decomposes a high level goal into planning primitives and the corresponding execution schedule. The MAUV architecture, [1], uses a sense-process-act loop, based on artificial intelligence techniques, to perform task decomposition from high level goals to low level actions. A hybrid, deliberative/reactive architecture is presented in [15], based on a functional hierarchy with planning, sequencing and skill managing layers. The HRI is implemented through standard viewing and teleoperation interfaces.

Often, humans are required to have specific skills to properly interact with robots, e.g., be aware of any motion constraints imposed by the kinematics. High level primitives can be used to encapsulate such aspects and reduce the required skills. A robot for disabled persons, working with a reduced set of high level motion primitives, has been presented in [22].

High-level commands have been used as a crude natural language for HRI. Human factors, such as the anthropomorphic characteristics of a robot, are a key subject in HRI as humans tend to interact better with robots with human characteristics, [14]. These behaviors allow a form of implicit communication between agents such as the robot following a human without having been told explicitly to do so. Natural language capability is undoubtedly one of such characteristics, this being an issue currently being tackled by multiple researchers. In [4] a spatial referencing language is used by a human to issue commands to navigate a robot in an environment with obstacles. The basic form of language developed in [12] converts sensor data gathered by multiple robots into a textual representation of the situation that can be understood by a human.

If the humans are assumed to have enough knowledge on the robots and the environment, standard (imperative) computer languages can be used for HRI. This easily leads to complex communication schemes relying on protocols with multiple abstraction layers. As an alternative, declarative context-dependent languages, like Haskell [21] and FROB [9,10], have been proposed to simulate robot systems and also as a means to interact with them. BOBJ was used in [8] to illustrate examples on human-computer interfacing. RoboML, [16], supported on XML, is an example of a language explicitly designed for HRI, accounting for low complexity programming, communications and knowledge representation.

24.3 A Semiotic Perspective for HRI

In general, robots and humans work at very different levels of abstraction. Humans work primarily at high levels of abstraction whereas robots are usually programmed to follow trajectories, hence operating at a low level of abstraction. Common architectures implement the mapping between abstraction levels using a functional approach by which a set of interconnected building blocks exchange information. The composition of these blocks maps the sensorial data into the actuators. Using category theory (CT) terminology,[2] an architecture lifts the information from the environment to the robot, as in the following diagram,

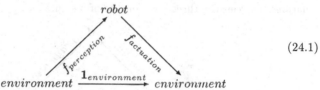

$$(24.1)$$

and simultaneously retracts the information from the environment to the robot,

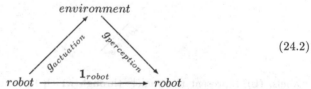

$$(24.2)$$

The above category diagrams show two different perspectives of the well-known sense-process-act loop. Diagram (24.1) represents the way the environment[3] sees the robot whereas diagram (24.2) represents the same for the robot. The maps $f_{actuation}$ and $g_{perception}$ represent the maps implemented by the architecture. The effect of the environment on the robot, represented by the maps $f_{perception}$ and $g_{actuation}$, has to be known for the above diagrams to commute. Thus, the architecture design corresponds to the classical CT problem of, given an architecture proposal defined through $f_{actuation}$ and $g_{perception}$, solving the corresponding determination and choice problems.

In a sense, diagrams (24.1) and (24.2) establish a sort of bounds on the design of robot control architectures. Namely, $f_{perception}$ represents the limits, set by the environment, on the perception of the environment by the robot. Similarly, $g_{actuation}$ represents the constraints, imposed by the robot, on the perception of the robot by the environment.

---

[2] Throughout the paper CT is used as the underlying tool supporting the proposed architecture, and clarifying the relations among the objects therein.

[3] The environment contains any relevant entity external to the robot, e.g., other robots and humans.

Semiotics is a branch of general philosophy that studies the interactions among humans, such as the linguistic ones, (see for instance [7] for an introduction to semiotics). Along the last decades semiotics has been brought to intelligent control and naturally spread to robotics (see for instance [18]). Different paradigms motivated by semiotics have been presented to model such interactions. See, for instance [17] on algebraic semiotics and its use in interface design, [19] on the application of hypertext theory to World Wide Web, or [6] on machine-machine and human-human interactions over electronic media (such as the Web).

The idea underlying semiotics is that humans communicate with one another (and with the environment) through signs. Slightly different definitions of sign have been presented in the literature on semiotics. Roughly, a sign encapsulates a meaning, an object, a label and the relations between them. Sign systems are formed by signs and the morphisms defined among them (see for instance [17] for a definition of the sign system) and hence, under reasonable assumptions on the existence of identity maps, map composition, and composition association, can also be modeled by CT. The following diagram, suggested by the "semiotic triangle" diagram common in the literature on semiotics (see for instance [5]), expresses the relations between the three components of a sign.

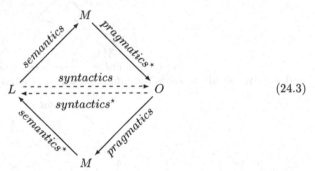

$$(24.3)$$

Labels, (L), represent the vehicle through which the sign is used, for instance an algorithm. Meanings, (M), stand for what the users understand when referring to the sign. The Objects, (O), stand for the real objects signs refer to.

The morphisms in diagram (24.3) represent the different perspectives used in the study of signs. Semiotics currently considers three different perspectives: semantics, pragmatics and syntactics, [19]. Semantics deals with the general relations among the signs. For instance, it defines whether or not a sign can have multiple meanings. Pragmatics handles the hidden meanings that require the agents to perform some inference on the signs before extracting their real meaning. Syntactics defines the structural rules that turn the label into the object the sign stands for. The starred morphisms are provided only to close the diagrams (the "semiotic triangle" is usually represented as an undirected graph).

Following C.S. Pierce, signs can be of three classes [6,17]: (i) symbols, expressing arbitrary relationships, such as conventions, (ii) icons, such as images, (iii) indices, such as indicators of facts or conditions. Symbols are probably the most common form of signs in robotics. For instance, state information exchanged among robots in a team is composed of symbol signs. Icons are often used by humans in their interactions, (e.g., an artistic painting can be used to convey an idea) and are also often found in robotics. For example, topological features can be extracted from an image and used for self-localisation. Indices are also often used among humans, e.g., in literary texts and when inferring a fact from a sentence. As a typical example, "the robot has no batteries" can be inferred from the observation "the robot is not moving". Similarly, "the robot is moving away from the predefined path" can lead to the deduction that "there must be an obstacle in the path".

The HRI model considered in this paper uses primarily symbols as basic data units for communication between a robot and its environment. Iconic information (images) will also be used, though morphed into symbols.

24.4 An Architecture for HRI

This section introduces the architecture by first defining a set of context free objects and operators on this set. Next, the corresponding realizations for the free objects are described.

Diagram (24.3) provides a roadmap for the design of a control architecture and a tool for the verification that the sign system developed in the paper is coherent with the semiotic model of human-human interactions.

The proposed architecture includes three classes of objects: motion primitives, operators on the set of motion primitives and decision making systems (on the set of motion primitives). The sign model (24.3) is used to design the motion primitives objects.

The ambiguities common in human-human interactions amount to say that different language constructs can be interpreted equivalently, that is as synonyms. Semantics often performs a sort of smoothing of the commands, by removing features that may be not relevant, before they are sent to the robot motion controller. The ability to cope with semantics is thus a key feature of an HRI language and the main focus of the framework described in this section. Standard computer languages tackle this issue using several constructs to define general equivalence classes among symbols.

The first free object, named *action*, defines primitive motions using simple concepts that can be easily used in an HRI language. The actions represent motion trends, i.e., an action represents simultaneously a set of paths that span the same bounded region of the workspace. These paths are equivalent in the sense that they drive the robot through the same region of the workspace.

Definition 1 (Free action). *Let k be a time index, q_0 the configuration of a robot where the action starts to be applied and $a(q_0)|_k$ the configuration at time k of a path generated by action a.*

A free action is defined by a triple $A \equiv (q_0, a, B_a)$, where B_a is a compact set and the initial condition of the action, q_0, verifies,

$$a(q_0)|_0 = q_0, \tag{24.4}$$

$$\exists_{\epsilon > \epsilon_{\min}} : \mathcal{B}(q_0, \epsilon) \subseteq B_a, \tag{24.4b}$$

with $\mathcal{B}(q_0, \epsilon)$ a ball of radius ϵ centered at q_0, and

$$\forall_{k \geqslant 0} \ a(q_0)|_k \in B_a. \tag{24.4c}$$

$\square$

Definition 1 creates an object — the action — able to enclose different paths with similar (in a wide sense) objectives. Paths that can be considered semantically equivalent, for instance because they lead to a successful execution of a mission, may be enclosed within a single action.

Representing the objects in Definition 1 in the form of a diagram it is possible to establish a correspondence between free actions and the sign model (24.3),

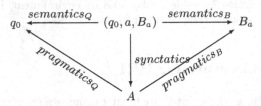

The a label represents the algorihm that generates the paths for the robot to follow. The projection maps $semantics_B$ and $semantics_Q$ express the fact that multiple paths starting in a neighborhood of q_0 and lying inside B_a may lead to identical results. The $pragmatics_B$ map expresses the fact that given the action being executed it may be possible to infer the corresponding bounding region. Similarly, $pragmatics_Q$ represents the maps that infer the initial condition q_0 given the action being executed. The $synctatics$ map simply expresses the construction of an action through Definition 1.

Following model (24.3), different actions (with different a labels) can yield the same meaning, that is, two actions can produce the same net effect in a mission. This amounts to require that the following diagram commutes,

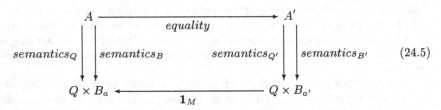

$$(24.5)$$

where $\mathbf{1}_M$ stands for the identity map in the space of meanings, M.

Diagram (24.5) provides a roadmap to define action equality as a key concept to evaluate sign semantics.

Definition 2 (Free action equality).

Two actions (a_1, B_{a_1}, q_{0_1}) and (a_2, B_{a_2}, q_{0_2}) are equal, the relation being represented by $a_1(q_{0_1}) = a_2(q_{0_2})$, if and only if the following conditions hold

$$a_1(q_{0_1}), a_2(q_{0_2}) \subset B_{a_1} \cap B_{a_2}, \qquad (24.6)$$

$$\forall_{k_2 \geqslant 0}, \exists_{k_1 \geqslant 0}, \exists_\epsilon : a_1(q_{0_1})|_{k_1} \in \mathcal{B}(a_2(q_{0_2})|_{k_2}, \epsilon) \subset B_{a_1} \cap B_{a_2}. \qquad (24.6b)$$

□

The realization for the free action of Definition 1 is given by the following proposition.

Proposition 1 (Action).

Let $a(q_0)$ be a free action. The paths generated by $a(q_0)$ are solutions of a system in the following form,

$$\dot{q} \in F_a(q), \tag{24.7}$$

where F_a is a Lipschitzian set-valued map (see Appendix A) with closed convex values verifying,

$$F_a(q) \subseteq T_{B_a}(q), \tag{24.7b}$$

where $T_{B_a}(q)$ is the contingent cone to B_a at q (see Appendix B for the definition of this cone).

□

The demonstration of this proposition is just a re-statement, in the context of this paper, of Theorem 5.6 in [26] on the existence of invariant sets for the inclusion (24.7).

■

The convexity of the values of the F_a map must be accounted for when specifying an action. The Lipschitz condition imposes bounds on the growing of the values of the F_a map. In practical applications this assumption can always be verified by proper choice of the map. This condition is related to the existence of solutions to (24.7), namely it implies upper semi-continuity (see [26], Proposition 2.4).

Proposition 2 (Action identity). *Two actions a_1 and a_2, implemented as in Proposition 1, are equal if,*

$$B_{a_1} = B_{a_2}, \tag{24.8}$$

$$\exists k_0 \; : \; \forall_{k>k_0}, \; F_{a_1}(q(k)) = F_{a_2}(q(k)). \tag{24.8b}$$

□

The demonstration follows by direct verification of the properties in Definition 2.

By assumption, both actions verify the conditions in Proposition 1 and hence their generated paths are contained inside $B_{a_1} \cap B_{a_2}$, which implies that (24.6) is verified.

Condition (24.8b) states that there are always motion directions that are common to both actions. For example, if any of the actions a_1, a_2 generates paths restricted to $F_{a_1} \cap F_{a_2}$ then condition (24.6b) is verified. When any of the actions generates paths using motion directions outside $F_{a_1} \cap F_{a_2}$ then condition (24.8b) indicates that after time k_0 they will be generated after the same set of motion directions. Both actions generate paths contained inside their common bounding region and hence the generated paths verify (24.6b).

■

A sign system is defined by the signs and the morphisms among them. The action equality induces an equality morphism. Two additional morphisms complete the algebraic structure: action composition and action expansion.

Definition 3 (Free action composition). *Let $a_i(q_{0_i})$ and $a_j(q_{0_j})$ be two free actions. Given a compact set M, the composition $a_{joi}(q_{0_i}) = a_j(q_{0_j}) \circ a_i(q_{0_i})$ verifies,*

if $B_{a_i} \cap B_{a_j} \neq \emptyset$

$$a_{joi}(q_{0_i}) \subset B_{a_i} \cup B_{a_j} \tag{24.9}$$

$$B_{a_i} \cap B_{a_j} \supseteq M \tag{24.9b}$$

otherwise,   the composition is undefined.

□

Action $a_{joi}(q_{0_i})$ resembles action $a_i(q_{0_i})$ up to the event marking the entrance of the paths into the region $M \subseteq B_{a_i} \cap B_{a_j}$. When the paths leave the common region M the composed action resembles $a_j(q_{0_j})$. While in M the composed action generates a link path that connects the two parts.

Whenever the composition is undefined the following operator can be used to provide additional space to one of the actions such that the overlapping region is non empty.

Definition 4 (Free action expansion). *Let $a_i(q_{0_i})$ and $a_j(q_{0_j})$ be two actions with initial conditions at q_{0_i} and q_{0_j} respectively. The expansion of action a_i by action a_j, denoted by $a_j(q_{0_j}) \boxtimes a_i(q_{0_i})$, verifies the following properties,*

$$B_{j \boxtimes i} = B_j \cup M \cup B_i, \quad with \ M \supseteq B_i \cap B_j, \tag{24.10}$$

where M is a compact set representing the expansion area and such that the following property holds

$$\exists_{q_{0_k} \in B_j} : a_i(q_{0_i}) = a_j(q_{0_k}) \tag{24.10b}$$

meaning that after having reached a neighborhood of q_{0_k}, $a_i(q_i)$ behaves like $a_j(q_j)$.

□

The use of M emphasizes the nature of the expansion region (alternatively, it can be included directly in B_{a_j}). M can be defined as the minimum amount of space that is required for the robot to perform any maneuver.

Proposition 3 (Action composition). *Let a_i and a_j be two actions defined by the inclusions*

$$\dot{q}_i \in F_i(q_i) \quad and \quad \dot{q}_j \in F_j(q_j)$$

with initial conditions q_{0_i} and q_{0_j}, respectively. The action $a_{joi}(q_{0_i})$ is generated by $\dot{q} \in F_{joi}(q)$, with the map F_{joi} given by

$$
F_{joi} = \begin{cases} F_i(q_i) & \text{if } q \ni B_i \backslash M, & (3) \\ F_i(q_i) \cap F_j(q_j) & \text{if } q \in M, & (3b) \\ F_j(q_j) & \text{if } q \in B_j \backslash M, & (3c) \\ \emptyset & \text{if } B_i \cap B_j = \emptyset & (3d) \end{cases}
$$

for some $M \subset B_j \cap B_i$.

Outside M the values of F_i and F_j verify the conditions in Proposition 1. Whenever $q \in M$ then $F_i(q_i) \cap F_j(q_j) \subset T_{B_j}(q)$.

□

The first trunk of the resulting path, given by (3), corresponds to the path generated by action $a_i(q_{0_i})$ prior to the event that determines the composition. The second trunk, given by (3b), links the paths generated by each of the actions. Note that by imposing that $F_i(q_i) \cap F_j(q_j) \subset T_{B_j}(q_j)$ the link paths can move out of the region M. The third trunk, given by (3c), corresponds to the path generated by action $a_j(q_{0_j})$.

By Proposition 1, each of the trunks is guaranteed to generate a path inside the respective bounding region and hence the overall path verifies (24.9).

■

The action composition in Proposition 3 generates actions that resemble each individual action outside the overlapping region. Inside the overlapping area the link path is built from motion directions common to both actions being composed. The crossing of the boundary of M defines the events marking the transition between the trunks.

Whenever $F_i(q_i) \cap F_j(q_j) = \emptyset$ it is still possible to generate a link path, provided that M has enough space for maneuvering. The basic idea, presented in the following proposition, is to locally enlarge either $F_i(q_i)$ or $F_j(q_j)$. Iterative procedures can be used for this purpose (see [25] for details).

Proposition 4 (Action expansion).

Let a_i and a_j be two actions defined after the inclusions

$$
\dot{q}_i \in F_i(q_i) \quad \text{and} \quad \dot{q}_j \in F_j(q_j).
$$

The expansion $a_{j\boxtimes i}(q_{0_i})$ verifies the following properties

$$
F_{i\boxtimes j} = \begin{cases} F_i & \text{if } q \ni B_i \backslash M, & (24.11) \\ F_j \cup F_i & \text{if } q \in B_i \cap B_j \cup M, & (24.11b) \end{cases}
$$

where $M \supseteq B_i \cap B_j$ is the expansion set chosen large enough such that $F_j \cup F_i$ verifies (24.7b).

□

Condition (24.11) generates paths corresponding to the action $a_i(q_{0_i})$. These paths last until an event, triggered by the crossing of the boundary of M, is detected. This crossing determines an event that expands the overall bounding region by M and the set of paths, by F_j, as expressed by (24.11b).

Assuming that $F_j \cup F_i \subset T_{B_i \cap B_j \cup M}$, that is, it verifies (24.7b), the complete path is entirely contained inside the expanded bounding region. After moving outside M paths behave as if generated by action a_i, as required by (24.10b).

■

Instead of computing a priori M, the expansion operator can be defined as a process by which action a_i converges to action a_j in the sense that $F_i(q_i) \rightarrow F_j(q_j)$ and M is the space spanned by this process.

Additional operators may be defined in the space of actions. For the purpose of the paper, i.e., defining the properties of a set of actions sufficient to design successful missions, action composition and expansion are the necessary and sufficient operators.

The overall architecture can be represented as shown in Figure 24.1. This architecture naturally yields a system of signs the robots can use to interact with each other much like a basic natural language. The components of this crude language are the elements forming the actions. The role of the supervisor is (i) to trigger the application of the composition and expansion operators whenever specific events are detected and, (ii) to manage the interaction with the external environment using this language, namely selecting bounding regions and initial conditions for the actions.

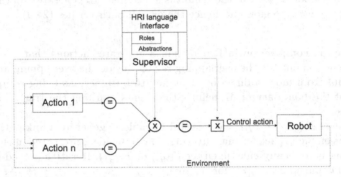

Fig. 24.1. The architecture for HRI

For the case of the composition operator the triggering events can be the crossing of some a priori defined region. The expansion operator is triggered when an event indicating that it is not possible to link the current action with the next chosen one is detected.

The supervisor block can be implemented as a finite state automaton. The states shape the actions bounding regions. The transitions can be identified with sequences of composition and expansion operations. For single robot missions each state basically sets a goal region for the robot to reach and shapes it to account for the external environment and mission. For team missions each state shapes the bounding regions using also the information from the other robots. Different automata yield different roles for the robot, each emphasizing specific skills, e.g., tracking moving people using sensor data, wandering, and systematically exploring the environment.

24.5 Experimental Results

This section presents two basic simulation experiments on HRI. Both experiments illustrate the behavior of the system when the robots are moving towards a goal region. The experiments are extremely simple as the emphasis of the paper is not on high level behavior design, which is related to the supervisor design. The experiments emphasize the visual quality (absence of complex maneuvering) of the trajectories obtained and the ability of the robots to move according motion trends instead of specific paths. The basic idea behind the experiments is that some external agent, human or robot, issues linguistic commands that lead the robot towards a goal region.

The setup considered can be compared to others described in the literature for HRI experiments. For instance, in [22] a robot is simply made to navigate around a standard environment, with static obstacles. In [2] a robot simply approaches a group of people wandering in the neighborhood as if joining them to take part in a conversation. In [12] robots are made to wander around in an exhibition scene, collecting and interpreting the sensor data to a human-comprehensible textual description. In [4] a robot is controlled using spatial references (e.g., go behind the table) to generate the adequate motion trend.

Unicycle robots, moving in a synthetic scenario, are considered in both experiments. The robots use a single action defined as

$$F(q) = (G - q) \cap H(q), \tag{24.11}$$

$$B(q) = \{p | p = q + \alpha G(q), \quad \alpha \in [0, 1]\}, \tag{24.12}$$

where q is the configuration of the robot, G stands for the goal set, and $H(q)$ stands for the set of admissible motion directions at configuration q (easily obtained from the well-known kinematics model of the unicycle). This action simply yields a motion direction pointing straight to the goal set from the current robot configuration. Often, the set of admissible motion directions that lead straight to the goal region is empty, $F(q) = \emptyset$, resulting in a singleton bounding region (the current configuration q) and no admissible control. In such cases the bounding region must be expanded using operator 4. The expansion action is simply given by

$$F_{i \boxtimes j} = \begin{cases} H_i(q) \cup \left\{ \begin{array}{c} \text{set of motion} \\ \text{directions} \end{array} \mid d(H_i, G - q) \to 0 \right\} & \text{if } q \ni B_i \backslash M, \\ F_i(q) & \text{otherwise.} \end{cases}$$

where $d(,)$ stands for a distance between the sets in the arguments. This action corresponds to having $H(q)$ converging to $G - q$. For the presented experiments the algorithm chosen is described in [24]. The same set of actions and the supervisors is used by both robots.

The simulation environment is implemented as a multi-thread system. Each robot simulator thread runs at 10 Hz whereas the architecture thread runs at 1 Hz. Data is recorded by an independent thread at 10 Hz.

24.5.1 Mission 1

The first experiment demonstrates the operation of two robots operating independently, trying to reach the same goal region. No information is exchanged between the robots and no obstacle avoidance behavior is considered.

Figure 24.2 shows the trajectories followed by the robots superimposed on the synthetic image representing the test environment. The irregular shapes in the upper part of the images represent the region to be reached by the robots. The goal region is defined as a circle centered at the centroid of the convex hull of the countour of these shapes (basic image processing techniques were used). This circle is shown in light colour superimposed on the corresponding shape. The symbol o marks the position of each robot along the mission. Marks connected by a dashed line were recorded at the same time.

Both robots start at the lower part of the image, with 0 rad orientation. This immediatelly forces the use of the expansion operator as the admissible motion directions lie outside the cone defined by $G - q_0$. Nevertheless, the trajectories obtained show a fairly acceptable behavior namely in the initial stage where the basic sequence expansion-composition is constantly being triggered.

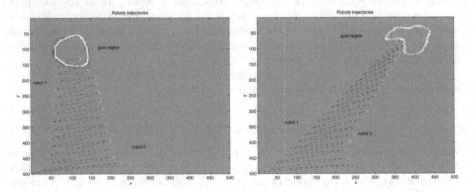

Fig. 24.2. Independent robots – two runs

24.5.2 Mission 2

The proposed mission is classical in the context of robotics experiments. Two robots have to move in a loose formation towards a common goal region while avoiding direct contact which each other. Maintaining such a formation requires the loose form of interaction between the robots that is easily modeled with the framework developed.

Figure 24.3 illustrates two runs of this mission. The supervisor at each robot shapes the action bounding regions to avoid contact between the robots. A basic shaping procedure is considered. The action bounding region of each robot is obtained by removing any points belonging also to that of the teammate. This results in a much smaller bounding region that constrains significantly the trajectories the action generates.

While interacting with the teammate, each robot replaces the original mission goal by intermediate goal regions placed inside the shaped bounding regions. Once an intermediate goal is reached the robot stops whereas the teammate continues towards the original mission goal (as it does not need to shape its own action bounding region).

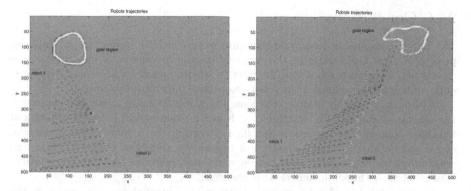

Fig. 24.3. Interacting robots – two runs

Figure 24.4 illustrates an alternative shaping of the action bounding region used when an obstacle is present in the environment. In this case an intermediate goal region is chosen far from the obstacle such that the new action bounding region allows the robots to move around the obstacle. This strategy has close connections to well-known path planning schemes widely used in robotics.

Robot 0 is the first to reach the intermediate goal and proceeds to the final goal. In the leftmost plot the interaction between the robots leads to trajectories passing far from the obstacle. In the righmost plot, the bounding regions of robot 0 are not influenced by the obstacle at the beginning of the mission and hence the robot tends to approach the obstacle (without colliding – robots where not given physical dimensions). Meanwhile, the interaction with robot 1 is clearly visible as the trajectory passes far from the obstacle and robot 0. The slight oscilations in the initial stage, clearly visbible in robot 1, are due to the interaction between the robots through the shaping of the action bounding regions.

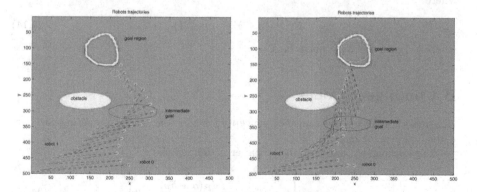

Fig. 24.4. Interacting robots in the presence of a static obstacle

24.6 Conclusions

The paper presented an algebraic structure to model HRI supported on semiotics principles. The key feature of this work is that is handles in a unified way any interaction between a robot and its external environment. Furthermore, the basic data units exchanged among the robots have straighforward meanings.

Although extremely simple, the simulation experiments presented capture a key feature of linguistic interactions, both among robots and between robots and humans. Namely, the motion is specified according to a motion trend, instead of a rigid path. The results illustrate acceptable trajectories both for single and team missions. The initial configurations do not promote straight line motion. Nevertheless, no harsh maneuvering is observed.

Future work includes (i) analytical study of controllability properties in the framework of hybrid systems with the continuous state dynamics given by differential inclusions and, (ii) study of the intrinsic properties for the supervisor building block, currently implemented as a finite state automata, that may simplify design procedures.

Acknowledgments

This work was supported by the FCT project POSI/SRI/40999/2001 – SACOR and Programa Operacional Sociedade de Informação (POSI) in the frame of QCA III.

A Lipschitz Set-valued Maps

A set-valued map F is said to be Lipschitz if it verifies.

$$\exists_{\epsilon \geqslant 0} \; : \; \forall_{x_1, x_2 \in X}, \; F(x_1) \subset F(x_2) + \epsilon |x_1 - x_2|_X \mathcal{B}_Y, \tag{24.13}$$

where

$$\mathcal{B}_Y = \{ y \in Y \; : \; |y| \leqslant 1 \}, \tag{24.14}$$

where $| \cdot |_X$ stands for a norm in X.

B Contingent Cones

Nonsmooth analysis uses tangency concepts for which a variety of contingent cones is defined (see for instance [26]).

The contingent cone used in the paper is defined as

$$T_B(q) = \left\{ v \; : \; \lim_{h \to 0+} \inf \frac{d_B(q + hv)}{h} = 0 \right\}, \tag{24.15}$$

where

$$d_B(q) = \inf_{p \in B} |p - q|_Q. \tag{24.16}$$

References

1. J. Albus. A control system architecture for intelligent systems. In *Procs. of the 1987 IEEE Intl. Conf. on Systems, Man and Cybernetics*, October, 20-23 1987. Alexandria, VA.

2. P. Althaus, H. Ishiguro, T. Kanda, T. Miyashita, and H. Christensen. Navigation for human-robot interaction tasks. In *Procs. of the 2004 IEEE Int. Conf. on Robotics and Automation, ICRA'04*, 2004. New Orleans, USA, April.

3. R. Aylett and D. Barnes. A multi-robot architecture for planetary rovers. In *Procs. of the 5th European Space Agency Workshop on Advanced Space Technologies for Robotics & Automation*, December 1998. ESTEC, The Netherlands.

4. S. Blisard and M. Skubic. Modeling spatial referencing language for human-robot interaction. In *Procs. of the 14th IEEE Int. Workshop on Robot and Human Interactive Communication, RO-MAN 2005*, 2005. Nashville, USA, August 13-15.

5. D. Chandler. *Semiotics, The basics*. Rutledge, 2003.

6. P. Codognet. The semiotics of the web. In *Leonardo*, volume 35(1). The MIT Press, 2002.

7. U. Eco. *Semiotics and The Philosophy of Language*. Indiana University Press, Bloomington, 1984.

8. J. Goguen. Semiotic morphisms, representations and blending for interface design. In *Procs of the AMAST Workshop on Algebraic Methods in Language Processing*, 2003. Verona, Italy, August 25-27.

9. G. D. Hager and J. Peterson. Frob: A transformational approach to the design of robot software. In *Procs. of the 9th Int. Symp. of Robotics Research, ISRR'99*, 1999. Snowbird, Utah, USA, October 9-12.

10. Paul Hudak. Modular domain specific languages and tools. In *Proceedings of Fifth International Conference on Software Reuse*, pages 134–142, June 1998.

11. T. Huntsberger, P. Pirjanian, A. Trebi-Ollennu, H.D. Nayar, H. Aghazarian, A. Ganino, M. Garrett, S.S. Joshi, and P.S. Schenker. Campout: A control architecture for tightly coupled coordination of multi-robot systems for planetary surface exploration. *IEEE Trans. Systems, Man & Cybernetics, Part A: Systems and Humans*, 33(5):550–559, 2003. Special Issue on Collective Intelligence.

12. B. Jensen, R. Philippsen, and R. Siegwart. Narrative situation assessment for human-robot interaction. In *Procs. of the IEEE Int. Conf. on Robotics and Automation, ICRA'03*, 2003. Taipei, Taiwan, Sept. 14-19.

13. K. Kazuhiko, N. Phongchai, M. Kazuhiki, J. Adams, and C. Zhou. An agent-based architecture for and adaptive human-robot interface. In *Procs. of the 36th Hawaii Int. Conf. on System Sciences, HICSS'03*, 2003.

14. S. Kiesler and P. Hinds. Introduction to the special issue on human-robot interaction. *Human-Computer Interaction*, 19(1-2), 2004.

15. D. Kortenkamp, R. Burridge, R.P. Bonasso, D. Schrekenghost, and M.B. Hudson. An intelligent software architecture for semi-autonomous robot control. In *Procs. 3rd Int. Conf. on Autonomous Agents - Workshop on Autonomy Control Software*, May 1999.

16. M. Makatchev and S.K. Tso. Human-robot interface using agents communicating in an xml-based markup language. In *Procs. of the IEEE Int. Workshop on Robot and Human Interactive Communication*, 2000. Osaka, Japan, September 27-29.

17. G. Malcolm and J.A. Goguen. Signs and representations: Semiotics for user interface design. In *Procs. Workshop in Computing*, pages 163–172. Springer, 1998. Liverpool, UK.

18. A. Meystel and J. Albus. *Intelligent Systems: Architecture, Design, and Control.* Wiley Series on Intelligent Systems. J. Wiley and Sons, 2002.

19. M. Neumüller. Applying computer semiotics to hypertext theory and the world wide web. In S. Reich and K.M. Anderson, editors, *Procs. of the 6th Int. Workshop and the 6th Int. Workshop on Open Hypertext Systems and Structural Computing*, Lecture Notes in Computer Science, pages 57–65. Springer-Verlag, 2000.

20. M. Nicolescu and M. Matarić. Learning and interacting in human-robot domains. *IEEE Transactions on Systems, Man, and Cybernetics, Part A: Systems and Humans*, 31(5):419–430, September 2001.

21. John Peterson, Paul Hudak, and Conal Elliott. Lambda in motion: Controlling robots with haskell. In *First International Workshop on Practical Aspects of Declarative Languages (PADL)*, January 1999.

22. N. Phongchai, P. Rani, and N Sarkar. An innovative high-level human-robot interaction for disabled persons. In *Procs of the IEEE Int. COnf. on Robotics and Automation, ICRA'04*, 2004. New Orleans, USA, April.

23. P. Rani and N. Sarkar. Emotion-sensitive robots - a new paradigm for human-robot interaction. In *Procs. of the IEEE-RAS/RSJ Int. Conf. on Humanoid Robots (Humanoids 2004)*, 2004. Nov. 10-12, Santa Monica, Los Angeles, CA, USA.

24. J Sequeira and M.I. Ribeiro. Hybrid control of a car-like robot. In *Procs of the 4th Int. Workshop on Robot Motion and Control*, June 17-20 2004. Puszczykowo, Poland.

25. J. Sequeira and M.I. Ribeiro. Hybrid control of semi-autonomous robots. In *Procs. of the 2004 IEEE/RSJ Int. Conf. on Intelligent Robots and Systems*, September 28 - October 2 2004. Sendai, Japan.

26. G. Smirnov. *Introduction to the Theory of Differential Inclusions*, volume 41 of *Graduate Studies in Mathematics*. American Mathematical Society, 2002.

27. Georgia Tech. Real-time cooperative behavior for tactical mobile robot teams – subsystems specification. Technical Report A002, Georgia Tech College of Computing and Georgia Tech Research Institute, October 1984.

On Electrical Analogues of Mechanical Systems and their Using in Analysis of Robot Dynamics

Edward Jezierski

Institute of Automatic Control, Technical University of Łódź
ul. Stefanowskiego 18/22, 90-246 Łódź, Poland   vrecedu@sir.p.lodz.pl

25.1 Introduction

Walking animals and human beings are characterized by smoothness of all movements as they exploit the natural dynamics of the body, especially the limbs [14, 16]. This natural dynamics is closely related to oscillations that are well known from electrical engineering. There are well worked out theoretical foundations of harmonic oscillations in linear systems, as well as the theory of general non-linear oscillations. A good example of using this theory is generation of gait rhythm of a walking machine [17].

There are a few efficient software packages devoted for analysis, simulation and synthesis of electrical and electronic systems. Thus, the tendency of exploiting all this potential in mechanical, or more general in mechatronic systems, seems to be a natural extension, and could be an efficient tool for testing new control algorithms. For a linear dynamic system the concept of impedance or admittance could be used, which allows to analyze the features of the system both in s-domain and ω-domain. Thus well known methods from electrical engineering could be easily transformed into mechatronic systems. A good example is application of Thevenin or Norton theorems to robotic systems that leads directly to the origin of the term „impedance control" [6, 7]. It gives a new look on the control of robotic manipulators in contact with the environment [1]. Recently the impedance control has been applied with good results to manipulators with elastic joints [9] as well as to biped mobile robots [13]. Another example is using the theory of operational amplifiers to receive a controlled stiffness of a robotic actuator [10, 12].

25.2 Review of Velocity-Current Analogy

25.2.1 Basic Relationships

In electrical engineering there are two types of elements that could store energy: capacitors and inductors. The basic equivalents of these elements in mechanics are inertial elements and springs.

K. Kozłowski (Ed.): Robot Motion and Control, LNCIS 335, pp. 391–404, 2006.
© Springer-Verlag London Limited 2006

Let us start with consideration of two simple linear systems without dissipative elements, as presented in Fig. 25.1. Each of them is characterized by a pair of parameters: mass m and stiffness k, or capacitance C and inductance L. It is easy to show that all variables in both systems like position, velocity or acceleration in the mechanical system, and voltage or current in the electrical system, fulfill the second order linear differential equation of the form

$$\frac{d^2 z}{dt^2} + \omega^2 z(t) = 0, \tag{25.1}$$

where $z(t)$ denotes one of the above variables. The natural frequency of harmonic oscillations observed in these systems depends on the above-mentioned parameters in the following way

$$\omega_m = \sqrt{\frac{k}{m}}, \qquad \omega_e = \frac{1}{\sqrt{LC}}. \tag{25.2}$$

From this observation, well-known for decades, researches have tried to find some analogues between mechanical and electrical systems. Additionally, taking into account the expressions for instantaneous power delivered in mechanical systems $p_m(t) = f(t)v(t)$ and instantaneous power delivered in electrical systems $p_e(t) = v(t)i(t)$ we can see that this analogy is evident.

The first attempt to specify this analogy is based on similarities between the formulae $v(t) = dx/dt$ for mechanical systems and $i(t) = dq/dt$ for electrical systems, and it could be found in the classical text-books in physics, e.g. [5]. Here the consecutive symbols denote velocity, position, current and charge. Additionally, taking into account quite similar expressions for energy stored in a moving mass $E_m = \frac{mv^2}{2}$, and energy stored in an inductor $E_L = \frac{Li^2}{2}$, it is concluded that velocity $v(t)$ and current $i(t)$ are equivalent variables, and moreover, mass m and inductance L are equivalent parameters of both systems.

These equivalences could be extended by introducing the definition of two conjugate variables: *a flow* (linear velocity, angular velocity or current), and *an effort* (force, torque or voltage). Along each degree of freedom, the instantaneous power that is transmitted between physical systems can always be defined as the product of these conjugate variables. Up to date most of researches working in the field of position/force or impedance control of manipulators have used this analogy [8].

To complete the comparison of both systems presented in Fig. 25.1, it is enough to recall the expression for energy stored in a spring $E_s = \frac{kx^2}{2}$, and the expression

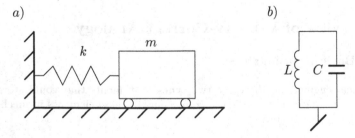

Fig. 25.1. Basic mechanical and electrical oscillation systems

for energy stored in a capacitor $E_C = \frac{q^2}{2C}$. To systemize *the velocity-current analogy* the set of equivalent variables and parameters are compiled in Table 25.1.

Table 25.1. Basic relationships between mechanical and electrical values in velocity-current analogy

Mechanical variable or parameter	Electrical variable or parameter
Velocity $v(t)$	Current $i(t)$
Force $f(t)$	Voltage $u(t)$
Mass m	Inductance L
Stiffness coefficient k	Inverse of capacitance $\frac{1}{C}$

On the basis of equivalents presented in the above table it is possible to compare some more complicated mechanical systems and electrical circuits, as shown in Table 25.2.

Table 25.2. Comparison of parallel/serial connections of elements in velocity-current analogy

Basic mechanical systems		Electrical equivalents	
Structure	Description	Structure	Description
	$f = f_1 + f_2$ $\Delta x = \Delta x_1 = \Delta x_2$ $k = k_1 + k_2$		$u = u_1 + u_2$ $i = i_1 = i_2$ $\frac{1}{C} = \frac{1}{C_1} + \frac{1}{C_2}$
	$f = f_1 = f_2$ $\Delta x = \Delta x_1 + \Delta x_2$ $\frac{1}{k} = \frac{1}{k_1} + \frac{1}{k_2}$		$u = u_1 = u_2$ $i = i_1 + i_2$ $C = C_1 + C_2$
	$f = f_1 + f_2$ $v = v_1 = v_2$ $m = m_1 + m_2$		$u = u_1 + u_2$ $i = i_1 = i_2$ $L = L_1 + L_2$

Figure 25.2 shows how a much more complicated mechanical system is transformed into its electrical equivalent.

25.2.2 Weak Points of Velocity-Current Analogy

The velocity-current analogy is not the only choice of equivalent values in mechanical and electrical systems, because conclusions about the relations between some scalar factors in the base equality of their products $a \cdot b = c \cdot d$ are not unique. Thus another analogy could be introduced, namely *velocity-voltage analogy*.

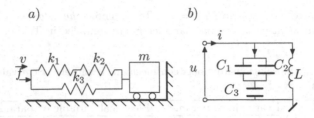

Fig. 25.2. Equivalent systems in velocity-current analogy

Such a comparison is also known from the literature, and probably for the first time was presented in [4]. It is based on the assumption that the basic variables in physical systems could be classified into one of the groups: *the through-variable* like force, torque, current, fluid volumetric flow rate, heat flow rate, and *across-variable* like translational velocity, angular velocity, voltage, difference of pressures, difference of temperatures [3]. However, the velocity-voltage analogy is nearly not noticed in the current literature. To examine the utility of both types of mechanical-electrical analogies let us start with listing two fundamental doubts of using the velocity-current analogy.

The first doubt is connected with transformation of parallel/serial connection of basic mechanical elements into electrical equivalents. From Table 25.2 one can conclude that parallel connection of mechanical dynamic elements is equivalent to serial connection of the convenient electrical elements, and vice versa. This feature makes more difficult the transformation of more complicated systems, as presented in Fig. 25.2.

The second doubt stems from the one-to-one relation between the mass (or inertia in the case of rotational movement) and the electrical inductor. It is worth recalling the following observation. The mass (inertia) is not loosing its accumulated energy when it is not in contact with an environment. In a similar way one can state: an ideal inductor is not loosing its accumulated energy only when it is closed by the ideal switch. Taking into account these two statements, a comparison of transient states in the two simple analogue systems could be done, which is presented in Fig. 25.3. For $t < 0$ the flywheel and the spiral spring are disconnected from each other, and for $t \geqslant 0$ the coupling links both elements. At $t = 0$ a transient state begins that consists of changes of the accumulated kinetic energy of the flywheel into the elastic strain energy of the spring. The electrical equivalent of this system is presented in

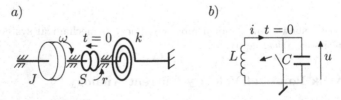

Fig. 25.3. Comparison of equivalent mechanical and electrical systems during transient state

Fig. 25.3b. However, in this case the switch is closed for $t < 0$, and open for $t \geqslant 0$. It is evident that to obtain similar behaviors of both systems, the coupling in Fig. 25.3a and the switch in Fig. 25.3b have to be in opposite states. This fact additionally complicates the analysis.

25.3 Review of Velocity-Voltage Analogy

25.3.1 Basic Assumptions

Table 25.1 is built on the basic assumption of correspondences between the mechanical velocity and the electrical current, and consequently between the force and the voltage. Other analogies, like equivalents of dynamic elements, are simple consequences of the main assumption. However, this assumption cannot be treated as a unique foundation, and in the sequel it will be shown that quite opposite assignment is also possible, and has some advantageous features.

As most actuators used in robotics, or generally in mechatronics, are based on transformation of electrical energy into mechanical energy, it would be worth recalling two important laws of electromagnetism that are fundamentals of all electrical generators and motors.

From the Faraday's law

$$e(t) = \frac{d\Phi}{dt}, \tag{25.3}$$

the first basic formula follows

$$e(t) = Blv(t). \tag{25.4}$$

This formula describes the voltage induced between both ends of a conducting bar of length l, placed in a magnetic field having constant and uniform flux density B, while the bar is moving with velocity $v(t)$. It is assumed that vectors B and $v(t)$, as well as the direction of the moving bar, are perpendicular to each other. This formula, known as „Blv law", is a fundamental relation exploited in all types of electrical machines. As the consequence, the voltage generated in a coil of a rotating machine is proportional to the angular velocity.

On the other hand, from Lorentz force equation

$$f(t) = qv(t) \times B, \tag{25.5}$$

that describes the force acting on charge q moving in a magnetic field, the second basic formula follows in the form

$$f(t) = Bli(t). \tag{25.6}$$

Here f is the force acting on a moving bar in a similar experiment to the previously described one when the current-loop is closed. This formula, known as „Bli law", allows to calculate the force generated by all types of electrical motors.

From the above considerations it follows that the foundation of *velocity-voltage analogy* is the assumption about mutual equivalence of mechanical and electrical variables as below

This analogy could be also called natural analogy.

velocity $v(t) \iff$ voltage $u(t)$;   force $f(t) \iff$ current $i(t)$.

25.3.2 Equivalence of Basic Dynamic Elements

Now we are in a position to easily find the equivalents for basic dynamic elements. Such a comparison is presented in Table 25.3. The new symbols denote: J – inertia in rotational movement, ω – angular velocity, τ – torque, $c = 1/k$ – compliance coefficient of the spring.

Table 25.3. Basic relationships between mechanical and electrical values in velocity-voltage analogy

Mechanical value or formula		Electrical value or formula
Inertia		Capacitance
$f(t) = m\frac{dv}{dt}$ $E = \frac{1}{2}mv^2$	$\tau(t) = J\frac{d\omega}{dt}$ $E = \frac{1}{2}J\omega^2$	$i(t) = C\frac{du}{dt}$ $E = \frac{1}{2}Cu^2$
Elasticity		Inductance
$v(t) = c\frac{df}{dt}$ $E = \frac{1}{2}cf^2$	$\omega(t) = c\frac{d\tau}{dt}$ $E = \frac{1}{2}c\tau^2$	$u(t) = L\frac{di}{dt}$ $E = \frac{1}{2}Li^2$

Most of the formulae shown in the table are evident, but the formula characterizing the effect of elasticity should be explained. The description of a tensional spring and a torsional spring have similar standard forms

$$f = k \cdot \Delta x \quad \text{or} \quad \tau = k \cdot \Delta\Theta, \tag{25.7}$$

where k denotes the stiffness coefficient. This description could be transformed to time dependent forms as

$$f(t) = k \int_0^t v(\bar{t})d\bar{t} + f(0) \quad \text{or} \quad \tau = k \int_0^t \omega(\bar{t})d\bar{t} + \tau(0), \tag{25.8}$$

and after differentiation to the final forms

$$v(t) = \frac{1}{k}\frac{df}{dt} = c\frac{df}{dt} \quad \text{or} \quad \omega = \frac{1}{k}\frac{d\tau}{dt} = c\frac{d\tau}{dt}. \tag{25.9}$$

25.3.3 Parallel and Serial Connection of Elements

Taking into account the results of the previous subsection it is easy to find the convenient description of different connections of basic dynamic systems using the natural analogy. This is summarized in Table 25.4. The structures of mechanical systems presented on the left hand side of the table are the same as those presented in Table 25.2. However, the electrical equivalents are quite different from those presented in Table 25.2.

By comparison of mechanical and electrical systems presented in Table 25.4 one can observe that the biggest advantage of the natural analogy is the preservation of the structure when mechanical systems are transformed into electrical equivalents

Table 25.4. Comparison of parallel/serial connections of elements in velocity-voltage analogy

Basic mechanical systems		Electrical equivalents	
Structure	Description	Structure	Description
	$f = f_1 + f_2$ $\Delta x = \Delta x_1 = \Delta x_2$ $\frac{1}{c} = \frac{1}{c_1} + \frac{1}{c_2}$		$u = u_1 = u_2$ $i = i_1 + i_2$ $\frac{1}{L} = \frac{1}{L_1} + \frac{1}{L_2}$
	$f = f_1 = f_2$ $\Delta x = \Delta x_1 + \Delta x_2$ $c = c_1 + c_2$		$u = u_1 + u_2$ $i = i_1 = i_2$ $L = L_1 + L_2$
	$f = f_1 + f_2$ $v = v_1 = v_2$ $m = m_1 + m_2$		$u = u_1 = u_2$ $i = i_1 + i_2$ $C = C_1 + C_2$

and vice versa. In particular, it means that a parallel connection of springs (or masses) is transformed into a parallel connection of inductors (or capacitors). On the other hand, a serial connection of sprigs is transformed into a serial connection of inductors. This feature allows to easily built equivalent circuits of more complicated systems, as shown for instance in Fig. 25.4. As shown in Fig. 25.3, when using the velocity-current analogy the open mechanical coupling linked in series with other mechanical elements has to be transformed into a closed electrical switch, connected in parallel with the electrical equivalents. Quite different situation occurs when the natural analogy is used. In Fig. 25.5 an electrical analogue circuit of the previous system is presented.

In electronic engineering there are known a few elements that are used as controlled or uncontrolled switches. The simplest one is a semiconductor diode, that is usually used to link two parts of an electric circuit. The diode allows for a natural commutation of the current under condition that the voltage along the

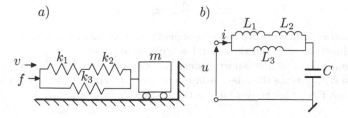

Fig. 25.4. Equivalent systems in velocity-voltage analogy

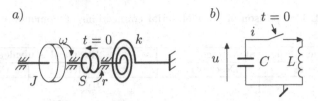

Fig. 25.5. Comparison of equivalent systems during transient state in velocity-voltage analogy

diode is positive, and disconnects both parts of the system when the polarization of the diode is opposite. A similar role plays a ratchet coupling known in mechanical engineering. It transfers the input force while it is moving in one direction, and disconnects both parts of the system while it is moving in the opposite direction.

The basic dissipative electrical element is a resistor described by a simple relationship $u(t) = Ri(t)$ or $i(t) = Gu(t)$. In nonlinear systems a much more complicated situation could arise and in such cases concepts of differential resistance or conductance at the operating point have to be used. In this case the model of the dissipative element consists of a resistor (or a conductor) and a voltage (or current) source. Quite similar situation occurs in mechanics, where dissipation of energy is caused by friction. In regions of work where this effect could be treated as linear, the model of viscous friction is used, usually in the form $f(t) = bv(t)$. We see that the friction coefficient b plays a similar role as the conductance G in electrical circuits.

25.4 Impedance of Kinematic Chain of the Robot

The term *impedance* is widely used in electrical engineering to describe the features of a linear two-port system under sinusoidal input signals. Such a signal (voltage or current) is described as

$$x(t) = X_m \sin(\omega t + \alpha), \tag{25.10}$$

and it could be treated as an imaginary part of the complex function

$$X_m \exp(j(\phi_x + \omega t)) = X_m(\cos(\phi_x + \omega t) + j\sin(\phi_x + \omega t)). \tag{25.11}$$

Thus, the symbolic representation of the sinusoidal signal (25.10) is a following complex value

$$x(\omega) = \frac{X_m}{\sqrt{2}} \exp(j\phi_x). \tag{25.12}$$

This complex value contains two components: one of them is a root-mean-value of periodic function $x(t)$, and the second is a phase angle ϕ_x.

The impedance of a two-port system is a complex number $Z(\omega)$ that describes at a quasi steady-state the relationship between the symbolic representations of the current that flows via the system and the voltage between both ports of the system

$$u(\omega) = Z(\omega)i(\omega). \tag{25.13}$$

The impedance depends in general on angular frequency ω and it could be presented in either an algebraic or an exponential form

$$Z(\omega) = R + jX(\omega) = |Z(\omega)|e^{j\phi(\omega)}, \qquad (25.14)$$

where R is the resistance of the two-port and $X(\omega)$ is the reactance of the two-port.

The impedance could be treated as an extension of the term resistance in cases of sinusoidal input functions applied to the system. However, it also has a wider application. For example, it also allows to discuss such terms as resonance frequencies of the system and damping ratios. In this sense it could be used in robotics to analyse dynamic behaviour of the kinematics chain of a robot.

Unfortunately, the robot is a multi-input multi-output system, and its dynamics is described by a set of nonlinear equations. Thus the impedance of the robot has to be considered only in a neighbourhood of a static point of work.

In a general case, the movements of the end-effector are described by a velocity vector $v(\cdot) \in R^6$, and similarly the forces between the end-effector and the external environment are described by the vector $f(\cdot) \in R^6$. Thus in the neighbourhood of the static point of work the following relationship holds

$$v(\omega) = Z(\omega)f(\omega), \qquad (25.15)$$

where $Z(\omega)$ is a 6×6 matrix that describes the impedance features of the kinematics chain. By comparison of the last equation with (13) one can see that the velocity-voltage analogy has been applied.

The elements of the impedance matrix could be derived from a mathematical model of the robot, or they could be obtained from an experiment on a real robotic stand. It would be worth showing a simple example that illustrates the origins of the impedance features of the robot.

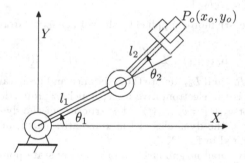

Fig. 25.6. Planar 2DOF manipulator

Let us consider a 2-DOF planar manipulator presented in Fig. 25.6. Under the assumption that both links are ideally rigid bodies, the dynamics of the kinematics chain is described by the equation

$$B(q)\ddot{q} + C(q, \dot{q})\dot{q} + h(q) = \tau, \qquad (25.16)$$

where $q \in R^2$.

To simplify further considerations it is additionally assumed that the inertia of each link is represented by a point mass (m_1 and m_2, respectively) placed at the end of the link. In this case the mathematical model of the manipulator takes the form

$$\begin{bmatrix} m_1l_1^2 + m_2(l_1^2 + 2l_1l_2\cos(\theta_2) + l_2^2) & m_2l_2(l_1\cos(\theta_2) + l_2) \\ m_2l_2(l_1\cos(\theta_2) + l_2) & m_2l_2^2 \end{bmatrix} \begin{bmatrix} \ddot{\theta}_1 \\ \ddot{\theta}_2 \end{bmatrix}$$

$$+ \begin{bmatrix} -m_2l_1l_2(2\dot{\theta}_1\dot{\theta}_2 + \dot{\theta}_2^{\;2})\sin(\theta_2) \\ m_2l_1l_2\dot{\theta}_1^{\;2}\sin(\theta_2) \end{bmatrix}$$

$$+ \begin{bmatrix} m_1l_1\cos(\theta_1) + m_2(l_1\cos(\theta_1) + l_2\cos(\theta_1 + \theta_2)) \\ m_2l_2\cos(\theta_1 + \theta_2) \end{bmatrix} g = \begin{bmatrix} \tau_1 \\ \tau_2 \end{bmatrix}. \quad (25.17)$$

The torque generated by the i-th drive is given by the formula

$$\tau_i = k_i \left(c_{si}i_{ai} - J_{ri}\frac{d\omega_{ri}}{dt} \right) - \tau_{i friction}, \quad i = 1, 2, \quad (25.18)$$

where k_i is the gear ratio, c_{si} is the torque constant of the motor, i_{ai} denotes the armature current, J_{ri} describes the inertia of the rotor, and finally ω_{ri} is the angular velocity of the rotor. Taking additionally the relation $\omega_{ri} = k_i\dot{\theta}_i$, the description of the manipulator dynamics could be presented as follows

$$\begin{bmatrix} m_1l_1^2 + m_2(l_1^2 + 2l_1l_2\cos(\theta_2) + l_2^2) + k_1^2J_{r1} & m_2l_2(l_1\cos(\theta_2) + l_2) \\ m_2l_2(l_1\cos(\theta_2) + l_2) & m_2l_2^2 + k_2^2J_{r2} \end{bmatrix} \begin{bmatrix} \ddot{\theta}_1 \\ \ddot{\theta}_2 \end{bmatrix}$$

$$+ \begin{bmatrix} -m_2l_1l_2(2\dot{\theta}_1\dot{\theta}_2 + \dot{\theta}_2^{\;2})\sin(\theta_2) \\ m_2l_1l_2\dot{\theta}_1^{\;2}\sin(\theta_2) \end{bmatrix}$$

$$+ \begin{bmatrix} m_1l_1\cos(\theta_1) + m_2(l_1\cos(\theta_1) + l_2\cos(\theta_1 + \theta_2)) \\ m_2l_2\cos(\theta_1 + \theta_2) \end{bmatrix} g + \begin{bmatrix} \tau_{1 friction} \\ \tau_{2 friction} \end{bmatrix}$$

$$= \begin{bmatrix} k_1c_{s1} & 0 \\ 0 & k_2c_{s2} \end{bmatrix} \begin{bmatrix} i_{a1} \\ i_{a2} \end{bmatrix}. \quad (25.19)$$

The dynamics of the i-th motor supplied by the voltage $\bar{u}_{ai}(t)$ from a PWM converter is described by

$$R_{ai}i_{ai}(t) + L_{ai}\frac{di_{ai}(t)}{dt} + e_{ai}(t) = \bar{u}_{ai}(t). \quad (25.20)$$

In this formula R_{ai} and L_{ai} denote the resistance and inductance of the armature winding, and $e_a(t)$ is the electromotive force, which is proportional to the angular velocity of the rotor $e_{ai}(t) = c_{si}\omega_{ri}(t)$. The structure of the dynamics of the robot consists of two parts: the dynamics of the kinematics chain, and the dynamics of the drives, as presented in Fig. 25.7.

For a sufficiently small neighbourhood of a steady state point, one can use new variables $\tilde{\theta}_1(t)$ and $\tilde{\theta}_2(t)$, defined as

$$\tilde{\theta}_1(t) = \theta_1(t) - \theta_{1o}, \quad \text{and} \quad \tilde{\theta}_2(t) = \theta_2(t) - \theta_{2o}. \quad (25.21)$$

The next step could be a linearization of the description of the manipulator dynamics. In the case when $\theta_{2o} = 0$, the linear model takes the form

$$\begin{bmatrix} m_1l_1^2 + m_2(l_1 + l_2)^2 + k_1^2J_{r1} & m_2l_2(l_1 + l_2) \\ m_2l_2(l_1 + l_2) & m_2l_2^2 + k_2^2J_{r2} \end{bmatrix} \begin{bmatrix} \ddot{\tilde{\theta}}_1 \\ \ddot{\tilde{\theta}}_2 \end{bmatrix} + \begin{bmatrix} b_1 & 0 \\ 0 & b_2 \end{bmatrix} \begin{bmatrix} \dot{\tilde{\theta}}_1 \\ \dot{\tilde{\theta}}_2 \end{bmatrix}$$

$$+ \begin{bmatrix} -(m_1l_1 + m_2(l_1 + l_2))g\sin(\theta_{1o}) & -m_2l_2\sin(\theta_{1o}) \\ -m_2l_2g\sin(\theta_{1o}) & -m_2l_2\sin(\theta_{1o}) \end{bmatrix} \begin{bmatrix} \tilde{\theta}_1 \\ \tilde{\theta}_2 \end{bmatrix}$$

$$+ \begin{bmatrix} (m_1l_1 + m_2(l_1 + l_2))g\cos(\theta_{1o}) \\ m_2l_2g\cos(\theta_{1o}) \end{bmatrix} = \begin{bmatrix} k_1c_{s1}i_{a1} \\ k_2c_{s2}i_{a2} \end{bmatrix}. \quad (25.22)$$

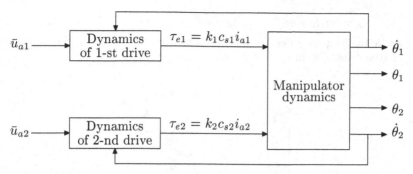

Fig. 25.7. Block diagram of the whole dynamics of the robot

The linearized model of the manipulator could be transformed into an electrical equivalent using the velocity-voltage analogy. Both parts of the robot model could be presented graphically as in Fig. 25.8. The model is described by the set of parameters:

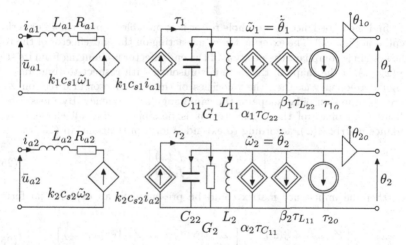

Fig. 25.8. Equivalent model of the robot dynamics in a neighborhood of a steady state

$$C_{11} = m_1 l_1^2 + m_2(l_1 + l_2)^2 + k_1^2 J_{r1}, \qquad C_{22} = m_2 l_2^2 + k_2^2 J_{r2},$$

$$L_{11} = \frac{-1}{(m_1 l_1 + m_2(l_1 + l_2))g\sin(\theta_{1o})}, \qquad L_{22} = \frac{-1}{m_2 l_2 g\sin(\theta_{1o})},$$

$$\alpha_1 = \frac{m_2 l_2(l_1 + l_2)}{m_2 l_2^2 + k_2^2 J_{r2}}, \qquad \alpha_2 = \frac{m_2 l_2(l_1 + l_2)}{m_1 l_1^2 + m_2(l_1 + l_2)^2 + k_1^2 J_{r1}},$$

$$\beta_1 = 1, \qquad \beta_2 = \frac{m_2 l_2}{m_1 l_1 + m_2(l_1 + l_2)},$$

$$G_1 = b_1, \qquad G_2 = b_2,$$

$$\tau_{1o} = (m_1 l_1 + m_2(l_1 + l_2))g\cos(\theta_{1o}), \qquad \tau_{2o} = m_2 l_2 g\cos(\theta_{1o}).$$

The whole model gives a deep insight into the transfer and accumulation of energy in all parts of the system.

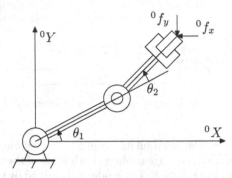

Fig. 25.9. External forces acting on the end-effector of the robot

To find the impedance of the whole robot is it possible to perform an experiment presented in Fig. 25.9. The external forces are acting on the end-effector of the robot. As the system is compliant the position of the end-effector is changing from its steady state (x_o, y_o). If the shape of the force is sinusoidal with respect to time, described by angular frequency ω, then the velocities of the end-effector in direction X and Y have also the sinusoidal shape of the same angular frequency. By measuring the amplitude and phase of these velocities it is possible to find all elements of the impedance matrix $Z(\omega)$, according to extended form of relationship (25.15)

$$\begin{bmatrix} v_x(\omega) \\ v_y(\omega) \end{bmatrix} = Z \begin{bmatrix} f_x(\omega) \\ f_y(\omega) \end{bmatrix}. \tag{25.23}$$

Further the impedance matrix could be presented in an exponential form, as follows

$$Z(\omega) = \begin{bmatrix} Z_{xx}(\omega) & Z_{xy}(\omega) \\ Z_{yx}(\omega) & Z_{yy}(\omega) \end{bmatrix} = \begin{bmatrix} |Z_{xx}|\exp(j\phi_{xx}) & |Z_{xy}|\exp(j\phi_{xy}) \\ |Z_{yx}|\exp(j\phi_{yx}) & |Z_{yy}|\exp(j\phi_{yy}) \end{bmatrix}. \tag{25.24}$$

As an example, a laboratory model of a 2DOF robot, equipped with DC drives, has been tested using Matlab software. The basic parameters of the robot are as follows: $l_1 = 0.8$ m, $l_2 = 0.6$ m, $m_1 = 5.0$ kg, $m_2 = 2.0$ kg, $R_{a1} = R_{a2} = 1.22\,\Omega$, $L_{a1} = L_{a2} = 2.5$ mH, $k_1 = k_2 = 160$, and $J_{r1} = J_{r2} = 2.71 \cdot 10^{-4}$ kgm^2. In Table 25.5 the values of all the elements of the impedance matrix for the neighborhood of a static point of work $\theta_{1o} = -30$ deg and $\theta_{2o} = -45$ deg are compiled.

Finally it would be worth showing how to take into account the features of harmonic gears that are very popular in robots. According to [2, 15] the main difference between this type of gear box in comparison to others (for instance a planetary gear) is relatively low stiffness. Thus the dynamic features of such a gear have to be modelled using an additional inductor (L_{hg}) and capacitor (C_{hg}), as shown in Fig. 25.10.

Table 25.5. Elements of the impedance matrix of a 2DOF robot

	$f_x = 0,5\sin(\omega t)$ [N] $f_y = 0$ [N]				$f_y = 0$ [N] $f_y = 0,5\sin(\omega t)$ [N]			
ω	$|Z_{xx}|[\frac{m}{sN}]$	$\phi_{xx}[deg]$	$|Z_{yx}|[\frac{m}{sN}]$	$\phi_{yx}[deg]$	$|Z_{xy}|[\frac{m}{sN}]$	$\phi_{xy}[deg]$	$|Z_{yy}|[\frac{m}{sN}]$	$\phi_{yy}[deg]$
1	3.993E-02	-7.8	2.192E-02	-8.8	2.155E-02	-13,7	2.332E-02	-14.5
2	5.976E-02	-35.6	4.112E-02	-47.4	3.811E-02	-57.1	4.023E-02	-58.7
5	2.983E-02	-127.2	7.560E-03	-179.7	4.842E-03	-174.7	4.993E-03	-164.4
10	5.149E-03	-167.3	1.006E-03	-187.4	9.187E-04	-176.9	1.093E-03	-172.9

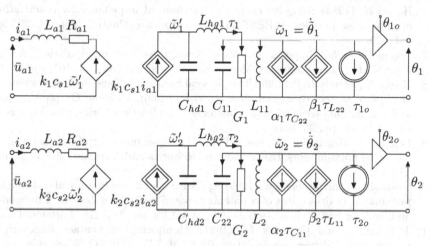

Fig. 25.10. Equivalent model of 2DOF robot equipped with harmonic gears

25.5 Conclusion

In the paper the method of building the electrical analogues of a mechanical system was examined. It allows to easily link such analogue elements with the other parts of the mechatronic system. The main goal of the paper was to show that the method based on natural analogy between mechanical velocity and electrical voltage, and, as a consequence, analogy between force and current is the best choice from the two possibilities. By comparison of the rules and convenient examples the superiority of the natural analogy over the velocity-current analogy in robotics application was proved. Considerations were mainly restricted to linear or linearized systems. However, it was shown how some types of non-linear couplings, so characteristic for manipulators, can be modelled by using the controlled current sources.

Also a potential exists in exploiting the features of the natural analogy in problems of impedance control of manipulators.

References

1. Anderson R.J. and Spong M.W. (1988) Hybrid impedance control of robotic manipulators, IEEE Journal of Robotics and Automation, vol. 4, 1988, pp. 549–556.
2. De Luca A. and Tomei P.: Elastic joints. In Theory of Robot Control, Canudas de Wit C., Siciliano B., Bastin E. (Eds) (1996) Springer Verlag, London, pp. 179–217.
3. Dorf R.C. (1989) Modern Control Systems, Addison Wesley Publ. Comp., Reading, MA.
4. Firestone F.A. (1933) A new analogy between mechanical and electrical system elements. J. of the Acoustic Society of America. vol. 3, pp. 249–267.
5. Halliday D. and Resnick R. (1963) Physics for Students of Science and Engineering, John Wiley and Sons, Inc., New York.
6. Hogan N. (1984) Adaptive control of mechanical impedance by coactivation of antagonist muscles, IEEE Trans. on Automatic Control, vol. AC-29, pp. 681–690.
7. Hogan N. (1985) Impedance control: An approach to manipulation, ASME Journal of Dynamic Systems, Measurements, and Control, vol. 107, pp. 1–23.
8. Hogan N., Breedveld P. (1999) The physical basis of analogies in network models of physical system dynamics. Proc. 1999 Int. Conf. on Bond Graph Modeling and Simulation. Western Multiconference Simululation Series, San Francisco, vol. 31, No 1, pp. 96–104,
9. Ferretti G., Magnani G. A., Rocco P. (2004) Impedance control for elastic joints industrial manipulators. IEEE Trans. on Robotics and Automation, vol. 20, pp. 488–498.
10. Granosik G. and Jezierski E. (1999) Application of a maximum stiffness rule for pneumatically driven legs of a walking robot, Proc. of the 2nd Int. Conference on Climbing and Walking Robots, CLAWAR 99, pp. 213–218, Portsmouth.
11. Jezierski E. and Granosik G. (1997) Monitoring of contact forces in a pneumatically driven manipulator, Proc. of XIV IMEKO World Congress, Tampere, 1997, vol. IXB, pp. 272–277.
12. Jezierski E. (2000) Features and control of mobile robot drive systems, Proceedings of AVT Fall 2000 Symposium on Unmanned Vehicles for Aerial, Ground and Naval Military Operation, Ankara, (in CD).
13. Park J.H. (2001) Impedance control for biped robot locomotion. IEEE Trans. on Robotics and Automation, vol. 17, pp. 870–882.
14. Pratt J.E. and Pratt G.A. (1999) Exploiting natural dynamics in the control of a three-dimensional bipedal walking simulation, Proc. of the 2nd International Conference on Climbing and Walking Robots CLAWAR 99, Portsmouth, pp. 797–807.
15. Tuttle D.T., Seering W.P. (1996) A nonlinear model of a harmonic drive gear transmission. IEEE Trans. on Robotics and Automation, vol. 12, pp. 368–374.
16. Witte H., Hackert R., et al (2001) Transfer of biological principles into the construction of quadruped walking machines. Proc. of the Second Workshop on Robot Motion and Control, Bukowy Dworek, pp. 245–249.
17. Zielińska T. (1996) Coupled oscillators utilized as gait rhythm generators of a two-legged walking machine. Biological Cybernetics, vol. 6, pp. 263–273.

Lecture Notes in Control and Information Sciences

Edited by M. Thoma and M. Morari

Further volumes of this series can be found on our homepage:
springer.com